Scalable
Continuous Media
Streaming Systems

Scalable Continuous Media Streaming Systems

ARCHITECTURE, DESIGN, ANALYSIS AND IMPLEMENTATION

Jack Y. B. Lee

The Chinese University of Hong Kong, Hong Kong SAR, China

John Wiley & Sons, Ltd

Other Wiley Editorial Offices

John Wiley & Sons Inc., 111 River Street, Hoboken, NJ 07030, USA

Jossey-Bass, 989 Market Street, San Francisco, CA 94103-1741, USA

Wiley-VCH Verlag GmbH, Boschstr. 12, D-69469 Weinheim, Germany

John Wiley & Sons Australia Ltd, 33 Park Road, Milton, Queensland 4064, Australia

John Wiley & Sons (Asia) Pte Ltd, 2 Clementi Loop #02-01, Jin Xing Distripark, Singapore 129809

John Wiley & Sons Canada Ltd, 22 Worcester Road, Etobicoke, Ontario, Canada M9W 1L1

Wiley also publishes its books in a variety of electronic formats. Some content that appears
in print may not be available in electronic books.

Library of Congress Cataloging-in-Publication Data

Lee, Jack Y. B.
    Scalable continuous media streaming systems : architecture, design,
analysis and implementation / Jack Y.B. Lee.
        p.   cm.
    Includes bibliographical references and index.
    ISBN 0-470-85754-4
    1. Streaming technology (Telecommunications)   I. Title.
TK5105.386.L44 2005
006.7'876–dc22   2005002759

British Library Cataloguing in Publication Data

A catalogue record for this book is available from the British Library

ISBN-13 978-0-470-85754-0 (HB)
ISBN-10 0-470-85754-4 (HB)

Typeset in 10/12pt Times by TechBooks, New Delhi, India
Printed and bound in Great Britain by Antony Rowe Ltd, Chippenham, Wiltshire
This book is printed on acid-free paper responsibly manufactured from sustainable forestry
in which at least two trees are planted for each one used for paper production.

*This book is dedicated to my wife Man-chu and our daughter Lok-sze
for their love and support*

Contents

Preface

This book addresses the architecture, design, analysis, and implementation of scalable and reliable continuous media streaming systems. This is an intermediate to advanced book aimed at senior undergraduate students, postgraduate students, researchers, and developers of continuous media systems in the industry.

Continuous media refers to media data that have a time specification for correct presentation. Common examples are video data and audio data, of which both require decoding and playing back the data at specific time instants or rates. *Streaming* refers to the way data are delivered from a server to a client for playback. In contrast to the download model, where a data object such as a video file, are completely received before playback, in streaming the client software begins playback before receiving the whole data object, and keeps receiving data from the server during playback. Finally, the term *system* refers to the collection of components in a complete continuous media streaming application, including the server, the network, and the client.

Building continuous media streaming systems are not difficult. The real challenge is to build systems that can scale up to support thousands or even millions of concurrent users. Additionally, given the scale of these systems, individual component failures are inevitable. Thus, it is essential that these systems are fault tolerant, i.e., with the capability to sustain non-stop service when there are one or more component failures. Business issues aside, these are the biggest hurdles to overcome before deploying large-scale commercial continuous-media services such as video-on-demand can become feasible.

Continuous media systems are a relatively new technology area, with only a decade of research. Thus it is no surprise that there is currently no book available that comprehensively covers the area in general, and the two key challenges, namely scalability and reliability, in particular. It is our intention to fill this gap with this book, drawing from cutting-edge research conducted in the past decade by researchers around the world as well as our own research group. In addition to theoretical issues, our research group had implemented several generations of streaming systems, ranging from client-server unicast-based architecture, parallel-server architectures, multicast-based architectures, to the latest peer-to-peer architectures. We have also designed and developed systems for live video multicasting over the Internet, as well as video streaming in 2.5G/3G and WLAN mobile networks. These experiences in building practical systems, of which some have been deployed in commercial use, have brought us insights into the many constraints and design tradeoffs that are otherwise not apparent in theoretical studies. Many of these findings are incorporated into this book in the hope that the presented materials not only will be useful to researchers, but also practitioners in the field.

Structure of the Book

This book is composed of three parts. Part I of this book deals with the concepts, basic principles, and issues in designing and implementing a media streaming system. The eight chapters cover topics in media compression, media data storage and retrieval, I/O scheduling, fault tolerance, as well as streaming protocols and algorithms. These chapters serve as introduction to, or revision of, the fundamentals of media streaming systems for the readers and pave the way for the rest of the book.

Part II of the book focuses on the use of parallel-server architectures to tackle the two key challenges in building large-scale media streaming systems, namely scalability and fault-tolerance. The seven chapters cover various architectural and implementation alternatives and their tradeoffs, as well as issues in system self-healing and expansion.

In addition to the server bottleneck, which can be addressed using the parallel-server architectures, the network itself can also become the bottleneck when the user population grows. Part III of the book investigates the use of network multicast and multicast streaming algorithms to address this challenge. The five chapters first introduce the principles of multicast streaming, and then describe in detail different types of multicast streaming algorithms, their tradeoffs, and issues in the implementation of a multicast streaming server.

Acknowledgements

First and foremost, this book will not be possible without the cutting-edge research done by the many renowned researchers around the world. It is an honor for the author to be part of this exciting field of research. Second, much of our work included in this book is drawn from many fruitful collaborations. In particular, the author would like to thank his colleagues and collaborators L. K. Chen, D. M. Chiu, S. C. Liew, John C. S. Lui, P. C. Wong, W. H. Yeung, O. C. Yu, and his graduate students T. K. Ho, H. L. Lai, L. S. Lam, C. H. Lee, W. C. Liu, W. Wang, and Y. W. Wong. Finally, much of the research work included in this book is made possible by funding received from the Hong Kong Research Grants Council (CUHK4211/03E, CUHK4328/02E, CUHK4209/01E, Direct Grants) and the Area-of-Excellence in Information Technology (AoE/E–01/99) – established under the University Grants Council of the Hong Kong Special Administrative Region, China.

Acknowledgements

The following materials have been adapted and revised from the IEEE papers as shown. These materials appear © 2004 IEEE. Reprinted, with permission, from the following publications.

Chapter 5

Jack Y. B. Lee and John C. S. Lui, "Automatic Recovery from Disk Failure in Continuous-Media Servers," *IEEE Transactions on Parallel and Distributed Systems*, vol. 13, no. 5, May 2002, pp. 499–515.

Chapter 7

H. L. Lai, Jack Y. B. Lee, and L. K. Chen, "A Monotonic-Decreasing Rate Schedulaer for Variable-Bit-Rate Video Streaming," *IEEE Transactions on Circuits and Systems for Video Technology*, vol. 15(2), Feb 2005, pp. 221–231.

Chapter 8

L. S. Lam, Jack Y. B. Lee, S. C. Liew, and W. Wang, "A Transparent Rate Adaptation Algorithm for Streaming Video over the Internet," *Proc. 18th International Conference on Advanced Information Networking and Applications (AINA 2004)*, Fukuoka, Japan, March 29–31, 2004.

Chapter 9

Jack Y. B. Lee, "Parallel Video Servers – A Tutorial," *IEEE Multimedia*, vol. 5, no. 2, June 1998, pp. 20–28.

Chapter 10

Jack Y. B. Lee, "Concurrent Push – A Scheduling Algorithm for Push-Based Parallel Video Servers," *IEEE Transactions on Circuits and Systems for Video Technology*, vol. 9, no. 3, April 1999, pp. 467–477.

Chapter 11

Jack Y. B. Lee, "Supporting Server-Level Fault Tolerance in Concurrent-Push-Based Parallel Video Servers," *IEEE Transactions on Circuits and Systems for Video Technology*, vol. 11, no. 1, January 2001, pp. 25–39.

Chapter 12

Jack Y. B. Lee, "Staggered Push – A Linearly Scalable Architecture for Push-Based Parallel Video Servers," *IEEE Transactions on Multimedia*, vol. 4, no. 4, December 2002, pp. 423–433.

Chapter 14

Jack Y. B. Lee and P. C. Wong, "Storage Rebuild for Automatic Failure Recovery in Video-on-Demand Servers," *Proc. IEEE ISCE'97*, Singapore, Dec. 2–4, 1997, pp. 258–261.

Chapter 15

T. K. Ho and Jack Y. B. Lee, "A Row-Permutated Data Reorganization Algorithm for Serverless Video-on-Demand System," *Proc. 3rd IEEE/ACM International Symposium on Cluster Computing and the Grid (CCGrid 2003)*, Tokyo, Japan, May 12–15, 2003, pp. 44–51.

Chapter 18

W. C. Liu and Jack Y. B. Lee, "Constrained Consonant Broadcasting – A Generalized Periodic Broadcasting Scheme for Large Scale Video Streaming," *Proc. IEEE International Conference on Multimedia and Expo*. Baltimore, USA, July 6–9, 2003.

Chapter 19

Jack Y. B. Lee and C. H. Lee, "Design, Performance Analysis, and Implementation of a Super-Scalar Video-on-Demand System," *IEEE Transactions on Circuits and Systems for Video Technology*, vol. 12, no. 11, November 2002, pp. 983–997.

Chapter 20

P. H. Chan and Jack Y. B. Lee, "An Efficient Disk-Array-Based Server Design for a Multicast Video Streaming System," *Proc. IEEE 14th International Conference on Application-specific System, Architectures and Processors*, The Hague, Netherlands, June 24–26, 2003, pp. 262–272.

Part One

Fundamentals

1

Introduction

Rapid advances in computing and the Internet have spawned many new services and applications. Among them, applications such as the World Wide Web (WWW) have achieved great success and transformed many facets of the society. With the continuous improvements in network bandwidth, computing power, and storage capacity, existing network services are evolving from the delivery of texts and graphics towards sophisticated multimedia contents combing high-quality audio and video.

Delivery of audio and video, however, poses far greater challenges than data applications such as the WWW. Moreover, unlike web pages, multimedia contents often occupy significantly more space for storage, and bandwidth for retrieval and delivery. Coupled with the demand for serving thousands or even tens of thousands of concurrent users, the challenge of designing scalable, reliable, and yet cost-effective multimedia systems has been an area of intense research in the past decade.

In this introductory chapter we will first study the properties of multimedia data and explain the key challenges in building high-performance yet cost-effective systems for multimedia content delivery. The rest of the chapters in Part I will introduce the fundamental concepts in media compression, storage, retrieval, scheduling, fault tolerance, and streaming.

1.1 Elements of a Multimedia System

In this book our focus is on systems for delivering multimedia data over a communication network such as the Internet or broadband residential networks. This system approach is desirable as it takes into account the interaction between various system components to achieve the stringent performance required in multimedia data presentation.

Figure 1.1 shows a generic client-server model for end-to-end multimedia data delivery. At the source we have encoded/compressed media data stored in storage devices such as hard disks. Through media server software these media data are then retrieved according to user requests from the disk to the main memory for transmission over the network. A media application/transport protocol is used to deliver the media data to the client hosts, where the

Scalable Continuous Media Streaming Systems   Jack Y. B. Lee
© 2005 John Wiley & Sons, Ltd.

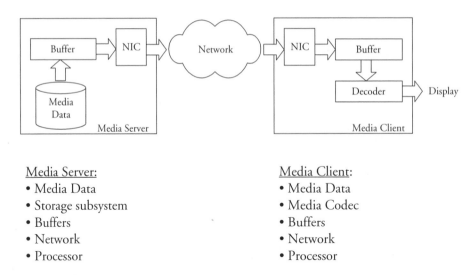

Media Server: Media Client:
- Media Data   - Media Data
- Storage subsystem   - Media Codec
- Buffers   - Buffers
- Network   - Network
- Processor   - Processor

Figure 1.1   Basic building blocks of a multimedia system

media data are first buffered in main memory and/or disk storage, and then eventually decoded and presented to the end user.

A unique characteristic of this system model is that the system components involved work in tandem in the data delivery process. Thus, a problem in any one of the system components can degrade the performance of the whole chain. In the following text we will investigate these system components and their interactions in more detail.

1.2 Media Data

The 'multi' in multimedia refers to multiple media of same or different types that are authored, delivered, and presented together. There are clearly many different types of media, from the simplest plain text, to formatted text, graphics, images, audio, video, or even tactile information. We can broadly classify these diverse types of media into two main categories, especially in the context of multimedia data delivery.

The first type – discrete media – refers to media data that have no explicit requirement for presentation timings. For example, consider retrieving an image from a web server for display in a web browser. Depending on the network bandwidth availability, the browser may take a variable length of time to receive the image data before they can be decoded for display. This may take, for example, from fractions of a second up to tens of seconds or even longer, depending on the size of the image and the network bandwidth available. Obviously, it is desirable to reduce this delay to as short a time as possible but as long as the image data are all correctly received, rendered, and displayed, the request is considered to be successfully completed. In other words, there is no restriction inherent in the media data that requires the media data to be presented at a certain time or within a certain delay limit. This is also why the network traffic resulting from delivering discrete media is also known as *elastic traffic* to reflect the media's ability to tolerate variations in delivery time.

By contrast, the second type of media data – continuous media – does have explicit presenting timing requirements embedded within the media data. The primary examples are audio and video data. For example, video data are usually encoded into video frames to be displayed sequentially at a certain frequency, such as 25 frames per second (fps) for the PAL video standard or 29.9 fps for the NTSC video standard. Thus, to correctly display a video media object, it is necessary not only to receive the video data correctly, but also to decode and present them according to the specified timings. Failure to do so will substantially degrade the perceived quality of the video data (e.g., resulting in jerky motions) even if the video data are all correctly received [1]. Thus, network traffic for continuous media data are also known as *inelastic traffic* because of the need to maintain the timing integrity.

Therefore, the challenge in multimedia data delivery in general and continuous media data delivery, in particular, is to ensure the integrity in both data as well as presentation timing. Moreover, multimedia content often comprises multiple media data streams composed according to a synchronized presentation schedule. In such a synchronized multimedia data stream, we then not only need to ensure the timing integrity in presenting a single media data stream, but also the relative timing integrity between multiple synchronized media data streams as well.

To solve the latter problem the system will need to schedule the data transmission of individual embedded media data objects, taking into account their relative presentation schedule and the network bandwidth available in order to ensure the media data are available at the receiver for synchronized playback. Alternatively, the multiple synchronized media data streams can be multiplexed into a single data stream before delivery. The multiplexer can then take into account the buffer size available at the decoder, as well as the timing relationships between the embedded media data streams to interleave the media streams so that presentation timing integrity is guaranteed (provided that the multiplexed media stream is received and buffered according to the specification). This approach greatly simplifies the media server as the multiple media data streams can be treated as a single media data stream. The downside is less flexibility, as the media stream composition is fixed and thus cannot be dynamically adjusted (e.g., switching to a lower bit-rate stream when bandwidth is insufficient).

1.3 Media Delivery

Of the two types of media data discussed earlier, we will focus on the delivery of continuous media in the rest of the book. We can broadly classify continuous media data delivery into two categories – real-time delivery and soft-real-time delivery.

Real-time delivery refers to applications where the media data must be delivered from the source and presented at the destination within a given *delay budget*. This is most common in applications where there are interactions between users, such as in Internet phone or video conferencing applications (Figure 1.2).

Take Internet phone [2] as an example, the one-way delay, i.e., the delay from capturing the voice data from the speaking user to the time the voice data are played back to the listening user should be no more than 150ms [3]. Longer delays will lead to talking collisions, i.e., both users trying to speak at the same time as commonly experienced in long-distance telephone conversations, and thus this degrades the service quality.

Clearly this real-time delivery requirement often conflicts with the requirements for data integrity and timing integrity. In fact, for applications such as Internet phone, the requirement

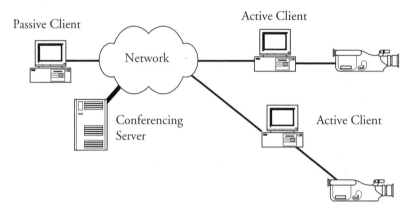

Figure 1.2   Real-time continuous media data delivery in a video conferencing application

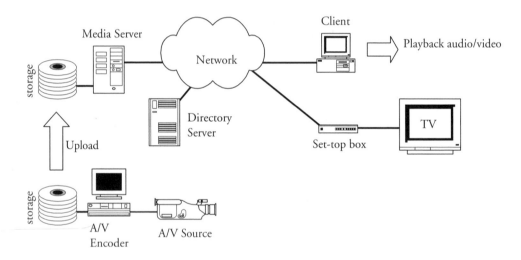

Figure 1.3   Soft-real-time continuous media data delivery in video-on-demand applications

for real-time delivery may even surpass that of data integrity and presentation timing integrity. For example, it may be necessary to allow data loss (or even discard data) and/or playback jitter in order to meet the given delay budget.

On the other hand, for soft-real-time delivery, there is no delay budget given. Instead, the system must deliver the media data so that data integrity and presentation timing integrity are preserved, while reducing the delay as far as possible. Examples of soft-real-time delivery are video-on-demand (VoD) where a user can select and playback a video title from the video collection available at a video server over the network as shown in Figure 1.3. These applications are far more tolerable to longer start-up delays (e.g., in seconds) as long as smooth playback is maintained after playback has started.

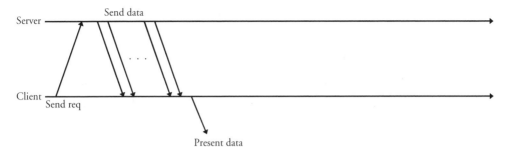

Figure 1.4   Interaction between client and server under the download data delivery model

1.4 Streaming versus Download

Delivering data over a network is not new and there are many different methods already available. Among them, download is the most common method to deliver data from a server to a client. The download model, depicted in Figure 1.4, is relatively straightforward: the client first sends a request to the server indicating the data object to be downloaded; the server then retrieves the data object (e.g., from the local file system) and start sending it over the network to the client using some application/transport protocol. Take the WWW as an example, the web browser first sends a HTTP GET request using TCP to a web server, which then retrieves the required file object and sends it back over the same TCP connection using a HTTP reply message. After completely receiving the data object, the client (e.g., web browser) then decodes and displays the data object to the user.

The key characteristic of the download model is that the data object is first completely received, and possibly cached either in memory buffer or in the local file system, before being decoded and played back. Clearly, as the complete data object is available to the client, the decoding processing and presentation can be done in the same way as local data objects. This download model works well in many applications but, unfortunately, is not very suitable for continuous media data delivery.

Let us reconsider the download process as shown in Figure 1.5. Ignoring processing time, the delay from the instant the user initiates the request to the instant the requested data object can be presented is determined by the size of the data object and the rate at which it is transmitted across the network. For applications such as WWW, the data objects are often text-based HTML web pages or small images/graphics, and thus the delay is relatively small.

Continuous media data objects, however, will likely be significantly larger and thus the delay incurred in downloading, say, a video object will become unacceptably long. Take MPEG2 video as an example. A 2-hour MPEG2 system stream (e.g., a movie) encoded at an average bit-rate of 6Mbps will generate 5.4GB of data. Delivering this amount of data even over broadband access networks, say, at 8Mbps, will take an unacceptably long time (e.g., 5.4GB × 8/8 = 1.5 hours) before playback can start.

The fundamental problem in the download model, as evident in Figure 1.6, is the requirement to wait until the whole video object is downloaded before playback can begin. While this requirement is necessary for many discrete media data types such as image or graphic, continuous media such as video possess the unique characteristics that partial data

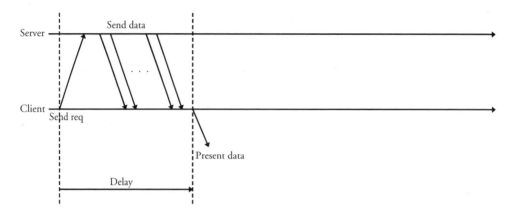

Delay = Data Size/Transmission Rate

Figure 1.5   The start-up delay in the download data delivery model

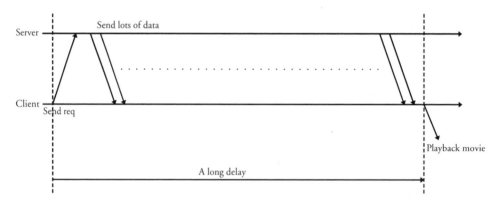

Delay = Data Size/TransmissionRate = 5.4GB × 8/8Mbps = 1.5 hours

Figure 1.6   Calculating the delay for downloading a 5.4GB video for playback

can also be decoded and played back. For example, video data are composed of video frames, which can be played back once all the data of a frame have been received by the client.

Taking advantage of this property of continuous media, we can then modify the download model into a *streaming model* where data are being played back while data reception is in progress, as depicted in Figure 1.7. Specifically, after sending a request to the server to begin the streaming process, the client will wait for the first parcel of data to arrive and then begin playback while receiving the second parcel of data, and so on. Thus, the data transfer and the playback processes are pipelined, therefore significantly shortening the delay to begin media playback.

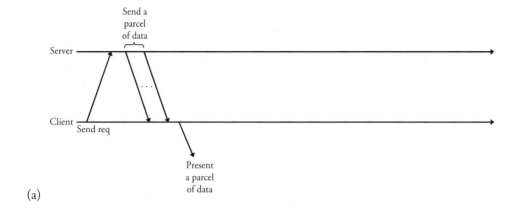

(a)

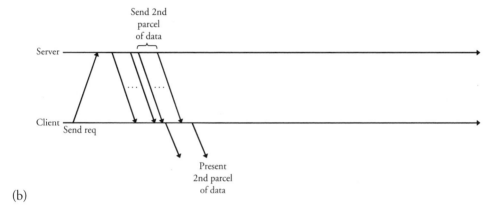

(b)

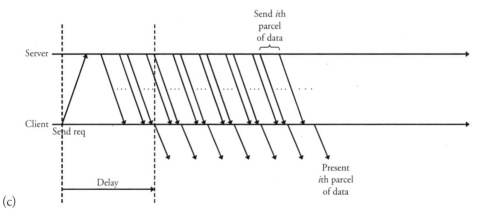

(c)

Figure 1.7   Playback of partial media data in the streaming model

Compared to the download model, there are now two additional requirements for streaming to work. First, the media object must be decomposable into smaller fragments that are independently or progressively (i.e., making use of the current and already received fragments) decodable and presentable. Most continuous media such as audio and video possess this property. Second, to ensure the timing integrity in presenting the media object, we will need to ensure that each and every media fragment can be delivered to the client before their scheduled playback time. This is also known as the *continuity* requirement and is one of the key performance metrics used in designing and evaluating continuous media systems.

With the rapid advances in networking technology one may wonder if in the future our networks will be equipped with so much bandwidth that the transmission time will become insignificant, even using the download model. This is indeed a valid question but in addition to allowing playback to begin earlier, the streaming model also offers another significant advantage – pipelining of multiple concurrent streams.

Specifically, media servers often need to serve many clients concurrently. When multiple clients request service at around the same time, there are two options for the media server if the download model is used – it can either serve the clients one after the other in a sequential manner, or it can serve them simultaneously. In the former, all but the head-of-line client will need to be queued and thus experience additional queueing delay when waiting for service. If the download model is employed, then even in a very high bandwidth network the waiting time will still be significant. In the latter case, serving multiple clients concurrently will reduce the network bandwidth available to each client and thus increases the download time proportionally.

By contrast, the streaming model does not suffer from this problem as playback can begin once an individually decodable fragment of media data is received (Figure 1.8). Moreover, as the media data are often transmitted at the media playback data-rate, the start-up delay is in fact independent of the number of clients requesting service simultaneously, as long as the media server and network capacity are not exceeded. This unique multistream pipelining property can significantly reduce start-up delay especially during high system utilizations.

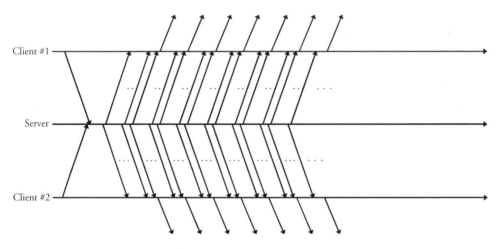

Figure 1.8 Multistream pipelining in the streaming model

1.5 Challenges in Building Continuous Media Streaming Systems

The previous sections outlined the general architecture of continuous media systems and the concept of media streaming. In this section we will discuss the main challenges in the design and implementation of continuous media streaming systems.

1.5.1 Continuity

As briefly mentioned in Section 1.4, once we adopted the streaming model for media data delivery, we will need to ensure that media data fragments are delivered to the client in time to maintain playback continuity. A key decision then is to decide when to begin playback after the streaming process is started. As depicted in Figure 1.9, once playback begins, the playback time for all media data fragments will be fixed (ignoring interactive playback control). Thus, if the client defers the playback start time, the whole playback schedule will be deferred as well, thereby allowing more time for data delivery. We cannot, however, defer playback indefinitely as that will increase the start-up delay experienced by the user – a crucial performance metric. As we will see in the rest of the book, these two conflicting objectives will occur frequently in the design of continuous media streaming systems.

1.5.2 Known and Unknown Variations

A continuous media streaming system comprises many system components. At the highest level we have the media server and a number of media clients (Figure 1.10). For on-demand streaming servers the media data are typically stored in local storage devices such as hard disks, and then retrieved into memory for processing, and finally transmitted through the network interfaces into the network to the clients. The media client will receive the data arriving through a network interface, temporarily store them in memory buffers while waiting for decoding and playback to the user.

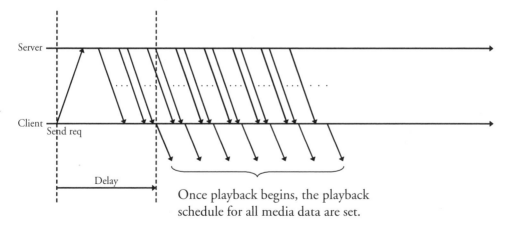

Figure 1.9   Relation between start-up delay and the playback schedule

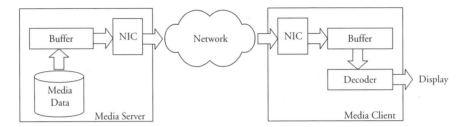

Media data
• playback of media may consume data at varying data-rate.
Media codec
• encoding/compressing media may result in varying data-rate.
Storage subsystem
• the I/O bandwidth may varies w.r.t. the workload.
Network
• the bandwidth available may vary from time to time.
Processor
• the available processor time may vary due to multi-tasking.

Figure 1.10   Known and unknown variations in a continuous media streaming system

If the many system components all operate in a clockwork manner with fixed and known processing time, then there will be little problem in ensuring performance such as delay limit and playback continuity. In reality, these system components seldom, if at all, operate in a clockwork manner. Instead, the performance of the individual system components can and does vary with time. Some of these variations are known while others are unknown and often unpredictable.

Let us begin by considering the preparation of the media data. Data compression is almost always needed to reduce the data rate of the raw media data (e.g., video in hundreds of Mbps) to practical levels (e.g., compressed into a few Mbps). This compression process, which we will cover in more detail in the next chapter, can introduce variations in terms of the amount of data needed for decoding each fragments of the media. Take video, for example, while the raw video frames are of the same resolution and size, the compressed video frames often consume varying amount of bytes for storage and transmission (cf. Section 2.4). Note that for stored video these variations are known and thus can be used in scheduling the media delivery process.

Once stored in the media server's storage, these compressed media data will need to be retrieved in the memory buffer for processing and transmission. The most common storage devices in media servers are hard disks. Unlike memory-based storage, where I/O throughput and access time are relatively constant, disk through and access time can vary substantially depending on the device characteristics, I/O access pattern, as well as the scheduler used in managing the device. These issues will be addressed in Chapter 3.

After successful retrieval from storage, the media server will then process the data (e.g., by adding control information in packet headers) and send them over the network. Depending on the network infrastructure, the time it takes for the media data to traverse the network to arrive at the media client can also vary. The variations are typically caused by temporary congestion

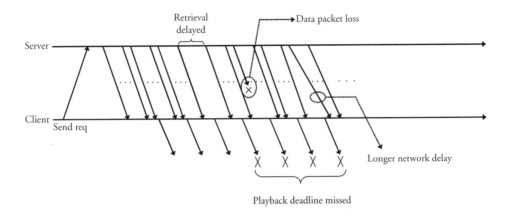

Figure 1.11   Variations in the system can disrupt continuous media playback

in the network or changes in the path the data took to reach the client. For networks that support quality-of-service (QoS) guarantees, the delay and delay variations can be known (and controlled) *a priori*. Otherwise, for public networks such as the Internet, the delay and delay variations are often unknown and unpredictable.

Finally, the media client, upon receiving the media data from the network, will buffer them in local memory to wait for decoding and playback. The decoding process typically takes a variable amount of time, and even the playback speed may sometimes vary. This is especially significant in the PC environment where the decoder/player is implemented in software and has to compete for processor time with other multi-tasking applications.

The above discussions illustrate the fact that variations in a media streaming system are the norm rather than the exception. Any one of these variations can lead to media data arriving too late for playback, i.e., losing playback continuity as depicted in Figure 1.11. In the subsequent chapters we will illustrate some ways to compensate for these variations to ensure continuous media playback.

1.5.3 Real-time Interactivity

In addition to stored media, some applications such as Internet phone and video conferencing send live media streams encoded in real-time to participating users. Unlike stored media, which a few seconds' start-up delay is tolerable, these real-time applications cannot tolerate too long a delay (e.g., 150 ms one way in voice call). This stringent delay requirement consequently puts much tighter constraints on the design and implementation of the media streaming system. It is worth noting that this real-time requirement often conflicts with other design goals, such as continuity or efficiency, and thus careful trade-offs are needed to balance these conflicting requirements.

1.5.4 Efficiency

Continuous media typically generate huge amount of data for storage and delivery. Thus, in designing media streaming systems, efficiency in the various subsystems becomes crucial in determining the service's cost-effectiveness. Many early continuous media streaming services, video-on-demand in particular, have failed to reach wide acceptance partly because the cost of provisioning the service is too high. Along with the recent advances in processor, storage, and network technologies, the cost in storing and delivering vast amount of media data has dropped substantially. Nevertheless, the cost in serving high-quality media contents such as the emerging high-definition video to a large number of users (e.g., in a city) are still very substantial and thus the quest for ever more efficient media streaming system designs continues to be an important research topic.

1.5.5 Scalability

Another challenge related to efficiency is scalability. Specifically, scalability refers to the limit at which one can increase the service capacity of a system, and the rate of increase in the system cost when the system capacity is scaled up. Consider a simple system with one media server as shown in Figure 1.12a. When more and more clients join the system, the media server will eventually become overloaded (Figure 1.12b), thus leading to unacceptable waiting time or service quality. One possible solution is to add a new media server to the system, and replicate all media contents to the new media server as shown in Figure 1.12c. This doubles the system capacity at the expense of doubled system cost.

More generally, Figure 1.13 illustrates three types of cost/capacity relations when we scale up the capacity of a system. In Case #1, the cost per unit capacity increases when one increases the system capacity. For example, if we increase the capacity of a media server by replacing it with a higher-capacity server, then it is quite common that the cost per unit capacity will increase for servers of higher and higher capacity. This is due to the lack of economy of scale in producing the very high capacity servers compared to the mass-produced commodity server platforms.

In Case #2 the cost per unit capacity is constant. Our previous example of replicated media servers falls within this type of scalability. Finally, in Case #3 the cost per unit capacity decreases with increases in the system scale. This is obviously highly desirable as it implies that a service operator can benefit from economy of scale in provisioning media streaming services to a large user population. The multicast streaming architectures to be covered in Part III of this book will cover many streaming architectures that achieve precisely this type of scalability.

1.5.6 Reliability

In addition to scalability, service reliability is another important challenge in provisioning large-scale media streaming services. Starting from the storage subsystem such as a disk array, the failure of a disk will disrupt the operation of the media server unless fault-tolerant mechanisms (e.g., RAID [2]) are employed. In addition to disk failures, the media server itself is also susceptible to many potential failures, including memory failure (some of which can be corrected using error-correcting memory chips), network interface failure, processor failure, power failure, or simply due to hitting a bug in the media server software.

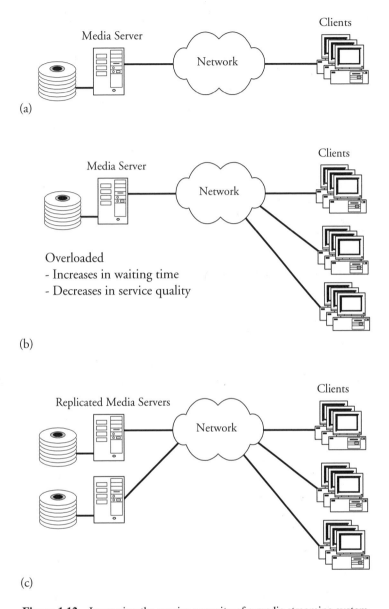

Figure 1.12   Increasing the service capacity of a media streaming system

Clearly, many of the aforementioned failures will render a media server inoperable. To enable the system to survive a server failure, we will need to introduce hardware/capacity redundancy and fault-tolerant mechanisms into the system. The parallel server architectures in Part II of this book provide one approach to addressing the reliability as well as the scalability challenges.

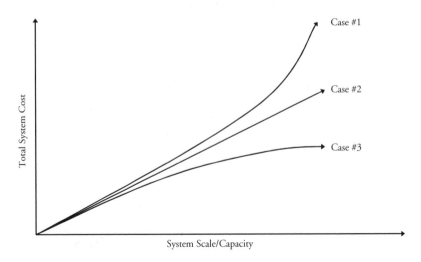

Figure 1.13   Three different types of cost/capacity relations in scaling up a system

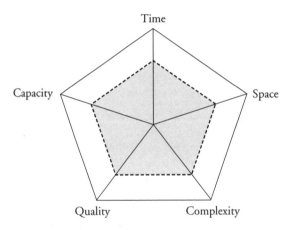

Figure 1.14   Five dimensions of engineering tradeoffs

1.6 Engineering Trade-offs

In devising solutions to address the previous challenges, we often are faced with conflicting design goals. Thus, for a given problem, there are often many possible solutions, each representing different engineering tradeoffs. In this section we attempt to illustrate this process of engineering tradeoffs by considering five dimensions (Figure 1.14):

- Capacity – such as disk I/O throughput and network bandwidth utilization;
- Time – such as start-up delay and response time;
- Space – such as storage requirement and buffer requirement;
- Quality – such as media quality and service quality; and
- Complexity – such as computational complexity and implementation complexity.

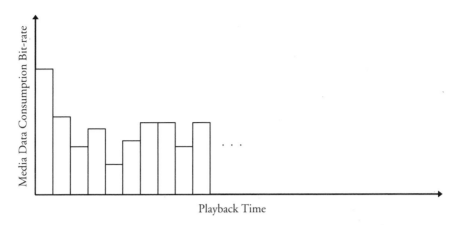

Figure 1.15   A media stream with time-varying playback bit-rates

These five dimensions serve to illustrate the complexities and possibilities in engineering a media streaming system and thus are not meant to be exhaustive nor the most important.

For illustrative purposes, we will consider the streaming of a media stream with time-varying playback bit-rates as shown in Figure 1.15. We assume that the media stream is divided into constant-duration segments of T seconds, i.e., the playback duration for each segment is the same. The size of each segment, however, is variable. Let r_i be the rate at which segment i (0, 1, ...) is consumed in playback. Our goal is to deliver this media stream from a media server to a media client over the network.

1.6.1 Trade-off in Capacity

For simplicity, we will assume that the media stream is the only traffic in the network, which has a finite and fixed bandwidth available. We first consider a simple solution by trading off network capacity. Specifically, knowing the bit-rates of all the media segments, we can simply allocate network bandwidth according to the maximum bit-rate of all segments, i.e., we allocate a network bandwidth of $C = \max\{r_i \mid \forall i\}$, as illustrated in Figure 1.16.

The obvious shortcoming of this simple solution is that except for the segment(s) with peak bit-rate, some of the allocated network bandwidth will be unused and thus wasted. With a fixed amount of total network bandwidth available, this solution results in guaranteed delivery but reduced usable streaming capacity.

1.6.2 Trade-off in Time

Observing the inefficiency of the previous solution we proceed to consider another dimension of trade-off – time. Specifically, the media stream Figure 1.15 has its peak bit-rate in the first segment. Thus, another strategy to stream the media is to send the initial segment at a bit-rate lower than the playback bit-rate. Now this implies that the first media segment will take more than T seconds to arrive at the receiver and so the playback must be delayed accordingly,

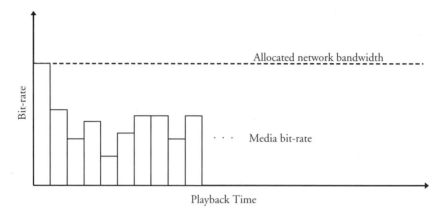

Figure 1.16 Streaming a variable-bit-rate media by allocating network bandwidth according to the peak bit-rate (a solution with trade-off in network capacity)

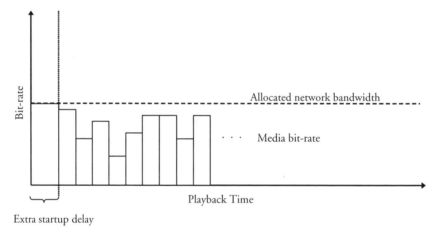

Figure 1.17 Reducing the network bandwidth allocation by sending the first segment with peak bit-rate at a lower bit-rate (a solution with trade-off in start-up delay)

as shown in Figure 1.17. Depending on the rate variability of the media stream, this strategy can result in substantial savings in the network bandwidth allocated, at the expense of longer start-up delay.

1.6.3 Trade-off in Space

In the two solutions discussed above we can observe that from time to time some network bandwidth can still be unused (e.g., when a media segment has a very low bit-rate). To take advantage of this we can perform work-ahead at the media server to send media data ahead of their playback schedule to exploit the otherwise unused network bandwidth. This is illustrated

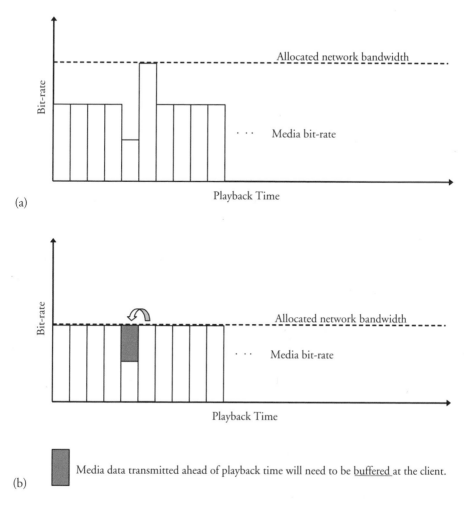

Figure 1.18 Reducing the peak bit-rate by sending media data ahead of their playback schedule (a solution with trade-off in client buffer space)

in Figure 1.18 with an admittedly contrived bit-rate pattern where the fifth media segment is of lower than average bit-rate and the sixth media segment is of higher than average bit-rate.

Applying the work-ahead principle we simply start sending the sixth media segment immediately after sending the fifth media segment. As there is more network bandwidth than is needed to send the fifth segment, this work-ahead transmission can be done without affecting the arrival time of the fifth segment. On the other hand, as the sixth media segment is now transmitted earlier, there is more time for the transmission and so the average transmission rate can be reduced as well. This can lower the peak bit-rate if the sixth media segment happens to be the one with the highest bit-rate. Note that for sake of illustration, the example given here is necessarily simplistic. In general, the media bit-rate profile can vary substantially in both short and long time scales. However, by combining the principles of trading off time and space, the

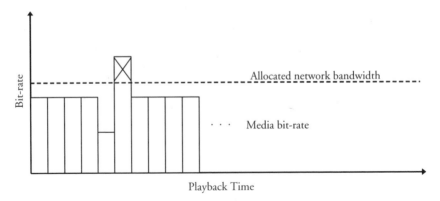

Figure 1.19   Reducing the peak bit-rate by skipping the transmission of some media data (a solution with trade-off in media quality)

media bit-rate variability can be reduced very effectively. We will revisit this topic in Chapter 7 and investigate the streaming of variable bit-rate media in mixed-traffic networks.

1.6.4 Trade-off in Quality

So far we have assumed that the media server must send all media data to the media client. If we remove this constraint, i.e., allowing some media data to be skipped, then we can also lower the network bandwidth required in delivering the media stream. This is illustrated in Figure 1.19 where some of the media data in the segment with peak bit-rate are not transmitted.

With incomplete media data, however, the media client obviously will not be able to reconstruct the original compressed media stream for playback. The amount of quality degradation incurred depends heavily on the media encoding algorithm employed, the data skipping algorithm used, the amount of data skipped, as well as the type of the skipped data (e.g., headers, video data, audio data, etc.).

In addition to simply skipping data for transmission, another approach is to dynamically reshape the media segments to a lower bit-rate before transmission. For example, knowing that the sixth media segment in Figure 1.19 requires a bit-rate higher than the network bandwidth available, the media server could re-encode the media segment to a lower bit-rate that is within bandwidth limit. This re-encoding can be done, for example, by decoding the media segment and then re-encoding it at a lower bit-rate (by discarding more information); or it can be done using a media transcoder than can reduce the media bit-rate without going through the complete decoding/encoding processes, resulting in less quality degradation and possibly lower processing complexity as well. We further explore media transcoding in Section 2.5 and adaptive streaming in Chapter 8.

1.6.5 Trade-off in Complexity

The final trade-off to be discussed is complexity. In multimedia streaming media data are almost always compressed before transmitted over the network. Thus the choice of compression algorithm used will affect the resultant media bit-rate. Take video compression as an example. There are many video compression algorithms available, such as MPEG2, MPEG4, and H.264. In terms of compression efficiency, i.e., achievable compression ratio at a given video quality level, it is generally agreed that MPEG2 < MPEG4 < H.264. However, the more efficient video codec such as H.264 also demands substantially more computations in both the encoding and decoding processes. Thus, choosing a more efficient compression algorithm such as H.264 will lower the network bandwidth required but will increase the computation complexity at both the encoder and decoder.

Note that for stored media the encoder complexity is less of an issue as the encoding process can be performed offline. However, if the media stream is encoded from a live source in real-time, then the encoding complexity will become a significant constraint. Decoding, on the other hand, is usually less complex than the encoder and thus presents less of an issue in the choice of codec in a media streaming system.

1.7 Performance Guarantee

In the previous discussions we illustrated five common dimensions for engineering trade-off. Regardless of the approach taken, the goal is to deliver the media data to the client in time for playback, i.e., to provide performance guarantee. Common to all these different approaches is the need to consider the worst-case scenario – in this case the peak bit-rate among all media segments. This is typical in designs that provide deterministic performance guarantee, i.e., the performance is met under all valid scenarios.

It is easy to see that deterministic performance guarantee often result in poor resource utilization, especially if the worst-case scenario rarely occurs. Alternatively, if we relax the requirement to meet performance under all valid scenarios, and instead guarantee that performance is met most of the time, then we can often reduce the resource requirements substantially. This approach is often known as probabilistic performance guarantee or statistical performance guarantee. The trade-off then is between resource utilization and the probability of failing the performance guarantee.

Finally, we should also mention a third type of performance guarantee (or lack thereof) – best effort. By best effort it implies that the system will attempt to meet the performance requirements using the available resources but there is no guarantee at all. Note that best effort does not necessary mean poor service. Rather, it simply implies that the probability that any given performance requirements is met is not known or controllable.

1.8 Admission Control

An issue closely related to providing performance guarantee is the need for admission control. Specifically, when we consider the different engineering tradeoffs in Section 1.6 we have always assumed that a given network bandwidth needs to be allocated before media streaming can start. In other words, if the network utilization is so high that the required bandwidth is not available, the media server will then reject the request for a new media streaming session (Figure 1.20). This process is known as admission control and it is one of the key elements in providing performance guarantee.

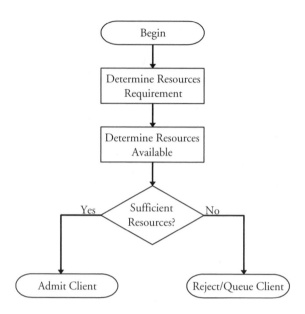

Figure 1.20   Flow chart for a general admission control procedure

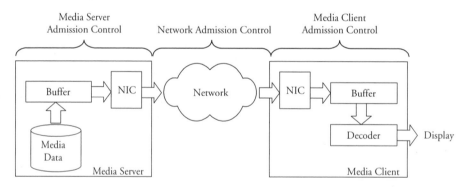

Figure 1.21   Providing end-to-end performance guarantee requires performance guarantee to be available in all the intermediate system components

Now let us examine the media streaming system in more detail as shown in Figure 1.21. The first observation is that only performing admission control in the network may not be sufficient to provide performance guarantee. In particular, the media server as well as the media client also has finite resources available for media streaming. Moreover, at the media server these resources are often shared among multiple concurrent media streams and possibly with other services as well. At the client, on the other hand, there are likely multiple applications running (in a multi-tasked operating system) that compete for the resources such as processor time, memory, or even network bandwidth.

Therefore, if we want to provide absolute end-to-end performance guarantee it will be necessary to investigate all the intermediate system components from the data storage all the way to the client's display, and employ resource allocation together with admission control in these system components to guarantee performance.

To support resource allocation and admission control, we first need to establish performance models of the underlying system components that can relate the workload to the desired performance metric (e.g., delay, buffer requirement, etc.). Using the performance models the admission controller can then compute the performance metric using the new workload (assuming a new media session is admitted, for example) to see if the required performance is still met. If so, then the resources are allocated and the new media stream is admitted. Otherwise the new media stream is rejected for service, or put into a queue to wait for sufficient resources to become available.

In practice, however, modeling the system components is often far from simple, especially if the system component has dynamic interactions with other system components. Thus, another approach to admission control called observational admission control has been proposed in the literature [3]. In observational admission control the system performance model is not known (or not accurately known). The admission controller simply measures the resource utilization as well as various performance metrics of the system, and uses these measurement results to estimate whether admitting a new media stream will result in overload. Clearly, this observational approach will not be able to provide a deterministic guarantee. Nevertheless, for complex system components that cannot be modeled accurately, this observational approach can still improve an otherwise entirely best effort service.

1.9 Summary

In this chapter we have reviewed the basic concepts of continuous media streaming, from the types of media data, media delivery, to the essential system components. Moreover, we presented the many new challenges in the design and implementation of continuous media streaming systems, as well as the common dimensions for engineering trade-offs. Finally, we discussed three types of performance guarantees and their relation to resource allocation and admission control. In the rest of the chapters in Part I we will present more detailed discussions on several key topics in continuous media streaming, namely, media data coding, compression, and adaptation (Chapter 2), storage and retrieval (Chapters 3, 4), fault tolerance (Chapter 5), and media data streaming (Chapters 6, 7, 8).

References

[1] M. Claypool and J. Tanner, The Effects of Jitter on the Perceptual Quality of Video, *Proceedings of the Seventh ACM International Conference on Multimedia*, vol. 2, Florida, USA, Nov. 1999, pp. 115–118.

[2] B. Goode, Voice over Internet Protocol (VoIP), *Proceedings of the IEEE*, vol. 90, no. 9, Sept. 2002, pp. 1495–1517.

[3] D.D. Vleeschauer, J. Janssen, and G.H. Petit, Delay and Distortion Bounds for Packetized Voice Calls of Traditional PSTN Quality, *Proceedings of the 1st IP-Telephony Workshop (IPTel 2000)*, GMD Report 95, pp. 105–110. Berlin, Germany, 12–13 April 2000.

[4] D.A. Patterson, G.A. Gibson, and R.H. Katz, A Case for Redundant Array of Inexpensive Disks (RAID), *Proceedings of the ACM Conference on Management of Data*, Chicago, IL, USA, June 1988, pp. 109–116.

[5] H.M. Vin, A. Goyal, and P. Goyal, An Observation-based Admission Control Algorithm for Multimedia Servers, *Proceedings of the International Conference on Multimedia Computing and Systems*, 15–19 May 1994, pp. 234–243.

2

Media Compression

In multimedia streaming systems media are almost always compressed to reduce their data rate. Thus, the choice and properties of the media codec will have a significant impact on the media streaming process. In this chapter we first introduce the basic concepts in media coding and compression, and then discuss the issues related to media streaming. In particular, we investigate the issue of matching the media bit-rate to the network bandwidth available, and study in some detail video transcoding techniques that can be used to reshape the media bit-rate to fit within the varying available network bandwidth. Readers interested in a more general introduction to the area of media coding and compression are referred to the many excellent texts available [1–5].

2.1 Introduction

Compression of media data is a highly developed research area with many successful applications. The prime examples are the development of the various MPEG compression standards for audio and video compression – MPEG1 [6–10], MPEG2 [11–19], and MPEG4 [20–35]. These compression standards have since formed an essential component in the development of Video CD, Digital Versatile Disc (DVD), HDTV broadcasting, and so on. In the following, we briefly review the basic concepts in audio and video compression.

2.1.1 Digital Audio

Audio data are typically encoded in PCM at a certain data width and sampling frequency. For example, Figure 2.1 illustrates the audio data sequence for CD audio, which employs a sampling frequency of 44.1 kHz, with each sample represented by a 16–bit word. Thus, a two-channel audio recording will generate data at a rate of 1.4 Mbps, not an insignificant number even in today's broadband access networks. Thus, researchers have since developed many audio compression algorithms to reduce the audio bit-rate, with the MP3 [36, 37] being the best known and most widely adopted.

In addition to the CD audio format, the industry has recently introduced (two actually) new formats for next-generation audio distribution. The first one, called DVD audio [38], is

Scalable Continuous Media Streaming Systems   Jack Y. B. Lee
© 2005 John Wiley & Sons, Ltd.

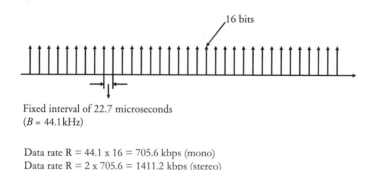

Fixed interval of 22.7 microseconds
($B = 44.1$ kHz)

Data rate R = 44.1 x 16 = 705.6 kbps (mono)
Data rate R = 2 x 705.6 = 1411.2 kbps (stereo)

Figure 2.1   A sequence of data samples in CD-quality digital audio

a natural extension of the PCM format but with a word length of up to 24 bits and sampling frequency of up to 192 kHz. The medium is also capable of encoding surround sound with up to 6 discrete channels of audio. The resultant data generation rate therefore can reach up to 27 Mbps. Due to the extremely high data rate, the DVD audio standard supports the use of a lossless compression algorithm called Meridian Lossless Packing (MLP) to reduce the data rate by approximately a factor of 2.

Apart from the DVD audio standard, another industry group has introduced a second standard targeted to replace the CD format – Super Audio CD (SACD) [38]. Unlike CD and DVD audio, SACD employs 1-bit delta sigma modulation at a sampling rate of 2.8 Mhz. Extensive noise shaping is then employed to achieve very high signal-to-noise ratio and wide frequency response performance. Again the standard supports surround sound of up to 6 channels, resulting in a raw data rate of up to 16.9 Mbps. The standard also includes a lossless compression algorithm called Direct Stream Transfer (DST) to reduce the data rate by approximately a factor of 2.

It is worth noting that both of the emergent audio formats employ only lossless compression that does not discard any audio information in the compression process. This is clearly done to preserve the highest audio quality, which is one of the major advantages of the new formats. Nevertheless, the resultant data rate even at compressed format is still very substantial, and thus whether lossy compression can be employed to reduce the bit-rate further and yet achieve audio quality superior to the standard CD will be an interesting area of research.

2.1.2 Digital Video

In addition to audio, video is the other most commonly used continuous media. A digital video sequence is divided into many video frames, each frame capturing one snapshot of the video scene. The video frames are typically captured at periodic time intervals, such as 25 frames per second (fps) in PAL-standard video and 29.9 fps in NTSC-standard video. In continuous media system, however, there is more flexibility in the frame rate used, predominately due to the often limited network/server bandwidth for delivering the video data. In these applications it is common to use frame rates lower than the previous broadcast standards if network bandwidth is limited. Clearly, the reduction in frame rate will generate fewer data but at the expense of reduced visual quality (e.g., fast motion becomes jerky or jumpy).

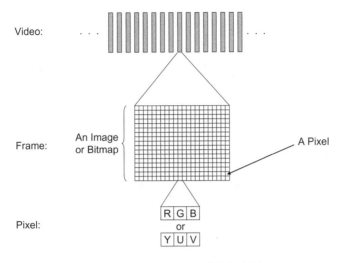

Figure 2.2 Compositions of digital video

Each video frame is further sub-divided into a two-dimensional grid of pixels (Figure 2.2). The numbers of horizontal and vertical pixels are defined by the resolution of the video. Each pixel further sub-divides into a number of color channels. The common color models employed in digital video are RGB, YUV, and YCrBr. In compressed video such as MPEG, the YCrBr color model is employed to exploit different properties of the human visual system (e.g., the human eye is more sensitive to intensity changes, i.e., the Y component, than color changes, i.e., the Cr and Br components).

The following are some common video standards and their basic properties:

- Common Interchange Format (CIF), (ITU-TS H.261):
  - 352 × 288 for luminance (Y)
  - 176 × 144 for chrominances (U, V)
  - Raw data rate = 36 Mbps.
- Quarter-Common Interchange Format (QCIF):
  - 176 × 144 for luminance (Y)
  - 176 × 144 for chrominances (U, V)
  - Raw data rate = 18 Mbps.
- Digitizing NTSC Video Signal:
  - Sampling rate: Y(13.5 Mhz), U (6.75 Mhz), V (6.75 Mhz)
  - Digitizing NTSC video signal
  - Raw data rate = (13.5 + 6.75 + 6,75) × 8 = 216 Mbps
  - Raw pixel resolution = 864 × 525 pixels (removing retrace, etc.)
  - Active video area = 720 × 486 pixels
  - Sub-sampling (4:2:2) (reduce bit-rate by 33%)
    Y (720 × 486), U (360 × 486), V (360 × 486)
  - 8-bits per sample per signal channel
  - Net raw data rate after sub-sampling = 168 Mbps.

- High-Definition Video:
  - Resolutions up to 1920×1080 pixels
  - 30 fps
  - 24 bits per pixel
  - Raw data rate = 1.5 Gbps.
- Emerging Ultra-High-Definition Video:
  - 4 times the resolution (and data rate) of HDTV.

2.1.3 Media Compression

From the previous discussions we can clearly observe that the raw data rates generated by high-resolution audio and video are extremely high, even compared to today's broadband network technologies. Therefore, it is essential to apply compression to the media data to reduce the storage and bandwidth requirements to economical levels. Many extremely successful media compression standards have been developed over the years and some of them are summarized below:

- MPEG-1 [6–10]:
  - VCR-quality video up to 8 Mbps
  - Used in video-CD, CD-I and video-on-demand systems.
- MPEG-2 [11–19]:
  - Broadcast quality video from 3 to >10 Mbps
  - Used in DVD, HDTV, and video-on-demand systems.
- MPEG-3:
  - Originally slated for HDTV but later dropped due to the incorporation of HDTV into MPEG-2.
- MPEG-4 [20–35]:
  - Originally targeted at low-bit-rate video for video telephony systems.
  - Now expanded to a broad range of bit-rates up to high-definition video.
  - The advanced video coding standard in Part 10 [29] of the MPEG-4 standard is also known as H.264. This advanced codec can achieve even greater compression ratio using more sophisticated compression algorithms.

For audio and video compression a vast body of work has been conducted in the past several decades and there are also many excellent texts and chapters available in the literature [1–5]. Thus, instead of repeating materials available elsewhere, we will focus on the impact of media compression on media streaming.

2.2 Media Multiplexing

Figure 2.3 depicts the typical structure of an MPEG encoder. The encoder in fact comprises two independent encoders, one for audio stream and one for video stream. These two media streams are first encoded independently to produce the corresponding compressed audio and compressed video streams, and then multiplexed together by the system encoder to a system stream. The multiplexer serves the important function of adding presentation time stamps to the

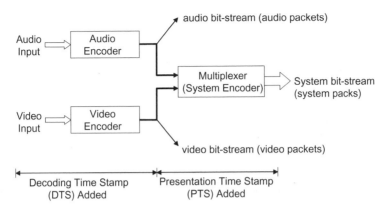

Figure 2.3 The encoding process in a typical MPEG media encoder

system stream and multiplexes the audio and video data streams according to their timing correlations. This system stream can then be delivered over the network to the clients for playback, where the reverse decoding process will occur, i.e., it first demultiplexes the system stream into separate audio and video streams, and then decompresses them for synchronized playback.

In media streaming, the media multiplexer can reduce the complexity of the media server. In fact, the media server can simply treat the multiplexed system stream as a binary bit stream encoded at a combined system stream bit-rate, irrespective of the detail compression algorithms and data format employed. This is a significant advantage as it decouples the media server implementation from the media compression standard employed. In other words, we can reuse without modification the same media server to stream media data compressed using new compression algorithms when they become available. By the same token, we can also stream multiple types of media streams encoded with different compression standards using the same media server, thereby reducing cost and operational complexity.

Alternatively, a media server can also send the compressed audio and compressed video data streams separately over the network (cf. Chapter 6), bypassing the MPEG multiplexer altogether. In this case, the media server will then need to send the data streams in such a way that audio and video data will arrive in time for synchronized playback. This usually requires the media server not only to perform I/O, but also to inspect and interpret the contents of the media streams to extract timing information to schedule data transmissions, and to construct packet headers with presentation timing information.

In this model the media server implementation will be coupled to the media compression standards employed, and it will also consume more processor cycles in processing the media data. Nevertheless, this approach does give the media server more control over the data delivery process. For example, the service provider could produce multiple versions of the media streams at different bit-rates to cater for varying network bandwidth availability. In this case the media server can begin streaming the highest quality media streams to a client, and then dynamically switch to a lower quality (and thus lower bit-rate) media stream if the available network bandwidth drops below a threshold to ensure continuous media playback. Network bandwidth variations are common in the current Internet and thus the capability to dynamically adapt the video content to avoid playback interruptions is an important and useful feature to the end users (cf. Section 2.5).

2.3 Temporal Dependencies in Compressed Video

Video encoders typically exploit three types of redundancies to reduce the compressed video bit-rate, namely, spatial redundancy, temporal redundancy, and entropy. Spatial redundancy refers to the correlation between pixels within the same video frame. This is also known as intra-frame coding as only pixels within the same video frame are used in the encoding process. The resultant encoded video frame, commonly called the I frame, can be decoded independently.

Temporal dependency refers to correlations between adjacent frames. As the video captures a snapshot of a video scene periodically at, say, 25 to 30 fps, adjacent frames will likely contain very similar visual objects, often with some displacements due to motion of the objects or the camera. Thus, the encoder can exploit this correlation by predicting a video frame from the neighboring frames. In MPEG, for example, this is done through the use of predictive frames (P frames) and bi-directional predictive frames (B frames) as shown in Figure 2.4.

Specifically, beginning with an intra-coded I frame, the encoder will first predict the P frame using a process called motion estimation. In motion estimation the encoder will search for similar blocks of pixels in the I frame and the to-be-encoded P frame. After the search is completed, only the displacement of the block (due to motion) and the prediction errors are encoded to form the data for the P frame. Thus, P frames can be encoded using substantially fewer bits than an I frame. This encoded P frame will then be used to predict the next P frame and so on until another I frame is introduced.

In addition to P frames, a number of B frames are also introduced between a pair of anchor frames (I or P frame). These B frames, as shown in Figure 2.4, are predicted from both anchor frames to further reduce the resultant bit-rate. Therefore, B frames usually consume the fewest bits compared to P frames and I frames in the same video stream. Both P and B frames are called inter-coded frames. Note that, unlike I frames, P and B frames cannot be decoded independently. Instead, the required anchor frames must first be decoded and then used in decoding the inter-coded frames. This has two implications to media streaming.

First, as shown in Figure 2.4, the temporal dependencies dictate that the B frames cannot be decoded for playback unless all two anchor frames are received and decoded. Thus, if the media server streams out the video data according to their temporal order, the client will need to buffer up B frames to wait for the second anchor frame to arrive before decoding for the B frames can proceed. In practice, the video encoder often re-orders the frame sequence according to the decoding order as shown in Figure 2.5 to reduce the client buffer requirement.

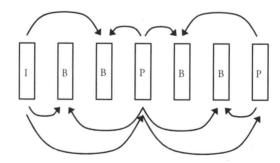

Figure 2.4   Temporal dependencies in compressed video

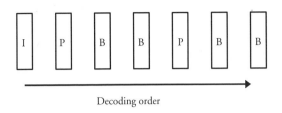

Decoding order

Figure 2.5   Decoding order of I, P, and B frames

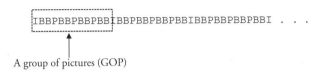

A group of pictures (GOP)

Figure 2.6   Intra-coded I frames are inserted into the compressed video stream to limit error propagation as well as to support random access

Second, when streaming video data over an unreliable network, some packets may be lost in transit. If the lost packets cannot be recovered, then the affected video frame will suffer from quality degradation. Worse still, if the packet losses affect anchor frames (I and P frames), then the predicted frames will be affected as well. For data loss in a P frame, the predicted B frames and the subsequent P frames will all be affected. Data loss in an I frame is worse as practically all subsequent predicted frames up to the next I frame are affected.

Thus, to limit this error propagation problem, a video encoder will periodically introduce I frames to break the temporal dependencies. As an example, the frame sequence in Figure 2.6 reintroduces an I frame, for every 11 predicted frames. Thus, in the worst case of losing some data in the I frame, only up to 12 frames (including the I frame) are affected. With a video frame rate of 25 fps, this translates into slightly less than half a second of quality degradation.

An I frame together with the predicted frames is called a group of pictures (GOP). Depending on the encoder, the GOP structure, such as the number of B frames between anchor frames and the number of P frames, is usually configurable by the user during encoding. Moreover, the GOP structure is not necessarily fixed across the whole video stream. More advanced encoders will attempt to align the GOP boundary with scene changes in the video content (e.g., when switching camera) to improve visual quality and/or to further reduce the encoded video bit-rate.

2.4 Bit-rate Variations

Another side effect of media compression is bit-rate variations. As illustrated in the previous section, different types of video frames (I, P, B) generally consume different amount of bits after encoding. Thus, if we compute the average video bit-rate on a frame-by-frame basis, then the video bit-rate will vary quite substantially across different frames. Figure 2.7 plots the frame size versus frame number for a video encoded in MPEG1.

Not surprisingly, the I frames are generally larger than the P and B frames, evident in Figure 2.7. For a media server this bit-rate variation creates a problem. Specifically, if the media server is to transmit the video data in the exact bit-rate on a frame-by-frame basis,

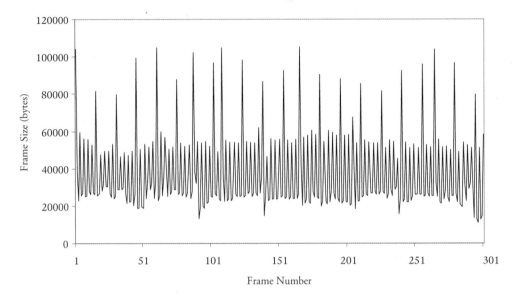

Figure 2.7 Small time scale bit-rate variations in a constant bit-rate encoded video

the outgoing data traffic will vary substantially. This will certainly make network resource planning and/or allocation more complex. The varying traffic may even lead to occasional packet loss due to instantaneous congestion in the network routers or at the client. The extent of bit-rate variations is determined by the rate control algorithm in the video encoder, which can be classified into two types – constant bit-rate encoding and constant quality encoding.

In constant bit-rate (CBR) encoding, the encoder will attempt to compress the video stream to a constant average bit-rate for the whole duration of the video content. However, as discussed earlier, different frame types inherently consume different amount of bits. Thus, in practice, the average bit-rate is not computed in a frame-by-frame basis. Instead, the average bit-rate is computed over a fixed number of frames, such as a GOP, and then the encoder will try to keep the average bit-rate the same at the GOP level.

Alternatively, an encoder can also operate in a constant quality mode. Unlike the CBR encoding mode where constant bit-rate is the goal, a constant quality encoder will attempt to adjust the bit-rate to maintain a consistent visual quality over the whole video content. Recall that a video encoder achieves some of its compression from removing temporal redundancies, which is closely correlated to the amount of motions in the video content. Thus, for fast motion scenes there will be less temporal redundancies and thus more errors in the predicted frame, which then consumes more bits to encode. In CBR encoder, the available bit budget is fixed due to the need to keep the rate constant, and so more information will be discarded, leading to lower visual quality for fast-motion scenes.

By contrast, constant quality encoder will simply use more bits to encode the information to maintain the same quality level, thus resulting in higher bit-rate for fast-motion scenes. Consequently, constant quality encoders will produce compressed video stream of varying bit-rates. Note that the rate variations can be of far longer range (in seconds or even tens of minutes) than the duration of a GOP (typically less than 1 second). Streaming a variable bit-rate media stream will present additional complexities in terms of scheduling, I/O, and resource allocation. We will return to this issue in Chapter 6 when we introduce bit-rate smoothing.

2.5 Media Adaptation

Once a media stream is compressed, it is typically stored in a media server awaiting user requests for streaming. The media server typically will not further process the media data prior to streaming. In some cases, however, the ability to reshape the media stream can be very desirable.

Take media streaming in the current Internet as an example. As the Internet is a best-effort network, it cannot provide any guarantee on bandwidth availability. Thus if the available network bandwidth drops below the encoded media bit-rate, then the media streaming session will often be disrupted – not uncommon in today's Internet.

In another scenario, a public media server may need to serve the same content to clients with different bandwidth availability, e.g., some may be connected via ADSL (1.5 Mbps) while others may have high-speed connections. The service provider commonly will either encode the media according to the lowest bit-rate of their target users (thus sacrificing quality for users with better connections) or encode multiple versions of the same content at different bit-rates for the users to choose. This latter approach is costly as the encoding process is often labor-intensive and takes considerable time.

The previous two scenarios in fact are due to the same fundamental problem – matching the media bit-rate to the bandwidth available. One emerging solution is to use layered video coding where a video is encoded into one base layer and a number of enhancement layers [39–42]. The base layer provides the lowest bit-rate video (with the lowest visual quality) while adding each enhancement layer will progressively improve the visual quality. With layered video coding, the server can then adjust the number of layers to transmit according to the bandwidth available or according to the user's connection speed.

Another solution is video transcoding. A video transcoder can convert a compressed video from a high bit-rate stream to a low bit-rate stream by selectively dropping information in the process. Unlike layered video coding, the output bit-rate is continuously adjustable, and it does not require any modification to the decoder for playback. Common transcoding techniques are requantization [43–44], spatial downscaling [45–48], and temporal downscaling [49]. Each technique has different ranges of achievable bit-rate reductions (see Section 2.5.2) and hence the choice of transcoding techniques will depend on the amount of bit-rate reduction required [50].

In the following, we investigate the achievable rate reductions using requantization and spatial downscaling based on the MPEG1 compression standard. Experiments reveal that neither requantization nor spatial downscaling alone can achieve a sufficiently wide range of bit-rate reductions. Instead, by combining them and selecting the appropriate transcoding techniques we can achieve a wide range of output transcoded bit-rate from 100% down to 20% of the original media bit-rate.

2.5.1 Transcoding Techniques

In this session, we review requantization and spatial downscaling, two existing algorithms for performing video transcoding.

2.5.1.1 Requantization

The first way to transcode video to a lower bit-rate is by increasing the quantization step size. One example is the Cascaded Pixel-Domain Transcoder (CPDT) [51], in which encoded video is fully decoded back into its original pixel-domain representation and then re-encoded using larger quantizers to reduce bit-rate. The computational complexity is high as a complete decode–encode cycle, including the time-consuming motion estimation process, is needed.

Another approach proposed by Assunção and Mohammed [43] reuses the original motion vectors in re-encoding, and thus eliminates the need for motion estimation. Additionally, by computing motion compensation in the DCT domain using the MC-DCT function proposed by Chang and Messerschmitt [52], the IDCT and DCT operations can also be eliminated. Their results showed that it can achieve results comparable to CPDT with significantly lower computational complexity.

2.5.1.2 Spatial Downscaling

In spatial downscaling, we reduce the video's spatial resolution, e.g., from X by Y pixels to $X/2$ by $Y/2$ pixels. Compared to requantization, spatial downscaling is far more complicated because of two reasons. First, given four DCT macroblocks, we have to synthesize a downscaled DCT macroblock. Second, as four macroblocks are combined into one macroblock, we cannot simply reuse the original motion vector for the new combined macroblock.

Downscaling can be performed either in the pixel domain or in the DCT domain. In the pixel domain approach, the four DCT blocks are decoded back to their original pixel representation, downscaled by pixel averaging, and then re-encoded. In the DCT domain approach, four DCT matrices can be downscaled to one DCT matrix without IDCT using the algorithm proposed by Natarajan and Vasudev [53].

For the second problem, instead of performing motion estimation for the downscaled frames, we can also reduce the computations by confining the search window to a few candidate motion vectors [46, 48] or compute the new motion vector directly from the four original motion vectors [45, 47].

2.5.2 Transcoder Design

As the goal of the transcoder is to adjust the transcoded video bit-rate to match the available network bandwidth, we first consider the feasible operating ranges for requantization and spatial downscaling, shown in Figure 2.8 for two different MPEG-1 CIF (352 × 288) video streams *Education* and *Onthestrip*.

There are two observations. First, spatial downscaling can achieve a wider range of transcoded bit-rate compared to requantization, especially at the lower bit-rates. Second, there are regions where both techniques are feasible, and requantization achieves higher PSNR, especially at the higher bit-rates.

These observations suggest that to maximize visual quality, we should use requantization at higher bit-rates and adopt spatial downscaling at lower bit-rates when requantization cannot be used. These two observations are incorporated into the integrated transcoder presented in this chapter.

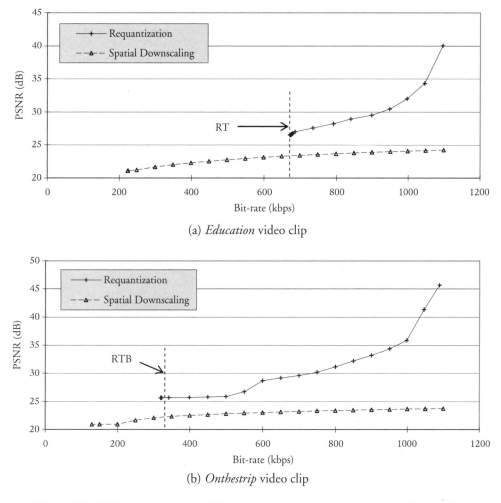

Figure 2.8 PSNR versus transcoded bit-rates using requantization and spatial downscaling

Specifically, the integrated transcoder comprises two parts. First, the original compressed video is transcoded once offline using the largest quantizer value (e.g., 31 in MPEG-1) to find the lowest bit-rate achievable by requantization. This is shown as the requantization threshold bit-rate (RTB) in Figure 2.8.

Second, during online transcoding, this RTB parameter is used to determine the transcoding technique to employ. In particular, if the desired output bit-rate is lower than RTB, spatial downscaling is used, otherwise requantization is used to achieve better visual quality (i.e., PSNR).

Using this two-part design, we can eliminate the need to compute the PSNR versus bit-rate curves, which is very computationally expensive, to determine the transcoding method to use given a target bit-rate. Instead, we only need to perform transcoding once to obtain the RTB for such purpose.

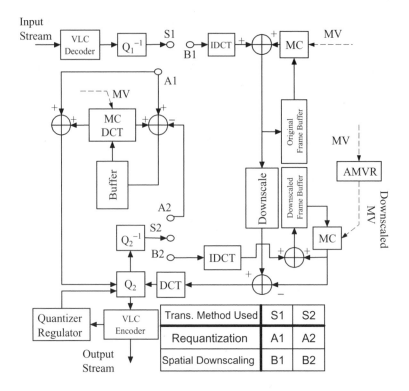

Figure 2.9   Logical design of the integrated transcoder

2.5.3 Implementation Issues

Figure 2.9 depicts the logical design of the integrated transcoder. The switches S1 and S2 control whether requantization or spatial downscaling is activated. If requantization is used, switches S1 and S2 will be connected to A1 and A2 respectively, otherwise they will be connected to B1 and B2 respectively to activate spatial downscaling. The following sections present in more detail the algorithms used in the integrated transcoder.

2.5.3.1 Drift Compensation Loop

Both requantization and spatial downscaling introduce noises in the transcoded video frames. As MPEG-1 makes use of predictive coding for B frames and P frames, these noises can accumulate along frames in the same group of pictures (GOP), further degrading the visual quality in the predicted frames. To prevent this problem, known as drift, we can feed back the noises to the predictor to compensate for the prediction errors. This is known as drift compensation.

For requantization, error is introduced during the requantization process where larger quantization step sizes are used. To compensate for the quantization errors, we can dequantize (Q_2^{-1}) the quantized DCT coefficients and compute the quantization error by comparing to the original DCT coefficients. The quantization error is then stored in the corresponding error buffer for the macroblock, which is then extracted using MC-DCT [52] and added to the next predicted frame to compensate for the quantization error.

For spatial downscaling, as error is introduced in both quantization and DCT operations, we need to perform both dequantization and inverse DCT to compute the aggregate errors. We can then incorporate these errors into the residual errors in the motion compensation process to prevent drifting.

2.5.3.2 Reconstruction of the Downscaled Motion Vectors

In spatial downscaling, instead of performing motion estimation again on the downscaled frames, which is computationally expensive, we can use the AMVR algorithm proposed by Bo *et al.* [45] to reconstruct the motion vectors for the downscaled frame. In AMVR, the downscaled motion vector, denoted by MV', is computed from

$$MV' = \frac{1}{2} \frac{\sum_{i=1}^{4} mv_i A_i}{\sum_{i=1}^{4} A_i} \tag{2.1}$$

where mv_i is the original motion vector of the macroblocks i in the original $N \times N$ video, and A_i is the activity measurement associated with macroblock i. For simplicity, A_i is taken to be the number of non-zero AC coefficients in the macroblock i which can be readily obtained when the coefficients are parsed [45].

2.5.3.3 Quantizer Regulation

In order to match the output video bit-rate to the target bit-rate, a quantizer regulator similar to the quantization control scheme in the Test Model 5 (TM5) [54] is incorporated to control the quantizer scale in each macroblock.

Instead of using the original TM5 quantization control scheme directly in requantization, we need to modify the quantizer regulation scheme so that the new quantizer scale will not be smaller than the original one. Specifically, let q' be the quantizer scale computed using the TM5 algorithm. Then the new quantizer scale, denoted by Q', is obtained from $\max\{q', Q\}$, where Q is the original quantizer scale. This modification reduces bit-rate consumption as using a quantizer smaller than the original one cannot improve visual quality and yet increases the number of bits required to encode the quantized coefficient. Note that the same modification is not required in spatial downscaling because DCT coefficients of the downscaled macroblocks are reconstructed during downscaling.

2.5.4 Experimental Results

To evaluate experimentally the performance of the presented integrated transcoder, we implemented the transcoder depicted in Figure 2.9 in software running under the Microsoft Windows platform. We tested the transcoder using a variety of different videos. The results of two video segments, referred to as *Onthestrip* and *Education*, are presented in this section. Both segments are originally MPEG-1 encoded in CIF resolution with a frame-rate of 30 fps at 1.15 Mbps.

2.5.4.1 PSNR Performance

Figure 2.10 compares the PSNR performance (in terms of PSNR versus output bit-rate) of the integrated transcoder with requantization and spatial downscaling. The PSNR of the video segments generated with integrated transcoding at bit-rates below RTB and above RTB is essentially identical to that obtained by spatial downscaling and requantization individually. The results confirm that the integrated transcoder can always generate video with PSNR higher than or equal to that obtained from using either requantization or spatial downscaling alone.

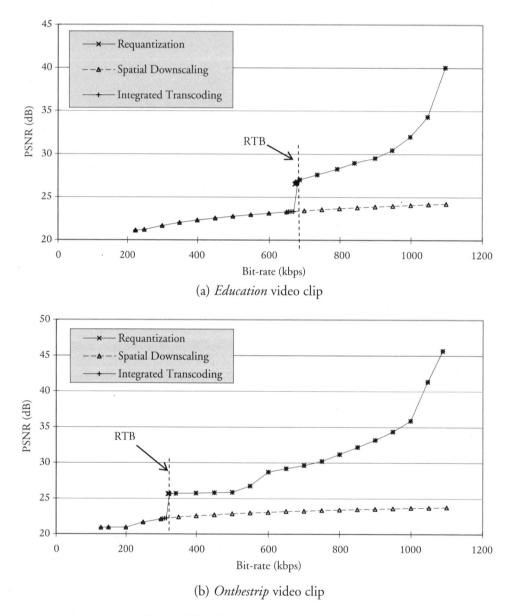

(a) *Education* video clip

(b) *Onthestrip* video clip

Figure 2.10   PSNR versus transcoded bit-rate

2.5.4.2 Bit-rate Conformance

Bit-rate conformance, defined as the ratio between the output bit-rate and the target bit-rate, measures how well a transcoder can control the transcoding process to achieve the desired output bit-rate.

Figure 2.11 plots the bit-rate conformance versus target bit-rate ranging from 100 kbps to 1.1 Mbps. The results show that requantization cannot adapt to the lower bit-rates (e.g., below

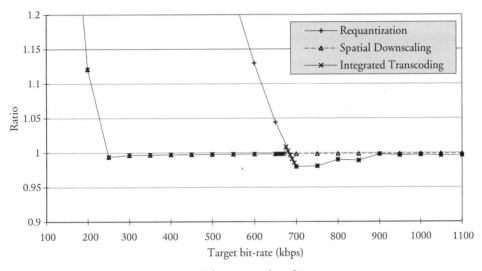

(a) *Education* video clip

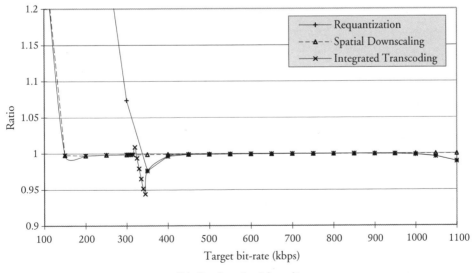

(b) *Onthestrip* video clip

Figure 2.11   Bit-rate conformance versus target bit-rate

670 kbps in Figure 2.11a). Both spatial downscaling and the integrated transcoder, on the other hand, can cover the whole spectrum of target bit-rates. However, as results in the previous section show, the integrated transcoder can achieve significantly better visual quality at target bit-rates higher than the requantization threshold bit-rate discussed in Section 2.5.2.

With the integrated transcoder, we can control the output bit-rate to within 5% of the target bit-rate, which in turn can range from 100% down to 20% of the original video bit-rate. This provides ample flexibility for video streaming servers to adapt to and sustain video playback under the ever-changing network traffic conditions in the Internet. More importantly, the integrated transcoder does not require modification to the decoder and thus is compatible with the vast installed base of MPEG-1 decoders and video players, thereby greatly simplifying deployment of the integrated transcoder.

2.6 Summary

In this chapter we have briefly reviewed the basic concepts in media compression and reviewed the common media compression standards. This area is vast and we only touch upon the surface of the many interesting and challenging issues in media compression. With a view to streaming the compressed media over a network to a client for playback, our focus is on the impact of media compression on media storage, retrieval, and transmission. In these aspects the bit-rate variations introduced by VBR encoding techniques substantially complicate the delivery process, especially if the underlying network does not have some form of quality of service control and guarantee.

The latest developments in layered video coding and video transcoding are promising solutions to tackle this challenge. For example, the MPEG-1 transcoder presented in Section 2.5 can be used to adapt the video bit-rate anywhere from 100% to 20% of the original bit-rate, with trade-offs in visual quality. Moreover, this transcoding process can be performed entirely in software in real time using commodity processors, thus opening a new way to deliver compressed media over networks with unpredictable bandwidths.

References

[1] M. Bosi and R.E. Goldberg, *Introduction to Digital Audio Coding and Standards*, Kluwer Academic Publishers, 2002.
[2] C.A. Poynton, *A Technical Introduction to Digital Video*, John Wiley & Sons, 1996.
[3] J.L. Mitchell, D.L. Gall, and C. Fogg, *MPEG Video Compression Standard*, Chapman & Hall, 1996.
[4] F. Pereira and T. Ebrahimi, *The MPEG4 Book*, Pearson Education; 2002.
[5] Y. Wang, J. Ostermann, and Y.Q. Zhang, *Video Processing and Communications*, Prentice Hall, 2001.
[6] ISO/IEC 11172-1:1993, Information Technology – Coding of Moving Pictures and Associated Audio for Digital Storage Media at up to about 1.5 Mbit/s – Part 1: Systems, International Organization for Standardization (ISO).
[7] ISO/IEC 11172-2:1993, Information Technology – Coding of Moving Pictures and Associated Audio for Digital Storage Media at up to about 1.5 Mbit/s – Part 2: Video, International Organization for Standardization (ISO).
[8] ISO/IEC 11172-3:1993, Information Technology – Coding of Moving Pictures and Associated Audio for Digital Storage Media at up to about 1.5 Mbit/s – Part 3: Audio, International Organization for Standardization (ISO).
[9] ISO/IEC 11172-4:1995, Information Technology – Coding of Moving Pictures and Associated Audio for Digital Storage Media at up to about 1.5 Mbit/s – Part 4: Compliance Testing, International Organization for Standardization (ISO).
[10] ISO/IEC TR 11172-5:1998, Information Technology – Coding of Moving Pictures and Associated Audio for Digital Storage Media at up to about 1.5 Mbit/s – Part 5: Software Simulation, International Organization for Standardization (ISO).

[11] ISO/IEC 13818-1:2000, Information Technology – Generic Coding of Moving Pictures and Associated Audio Information – Part 1: Systems, International Organization for Standardization (ISO).

[12] ISO/IEC 13818-2:2000, Information Technology – Generic Coding of Moving Pictures and Associated Audio Information – Part 2: Video, International Organization for Standardization (ISO).

[13] ISO/IEC 13818-3:1998, Information Technology – Generic Coding of Moving Pictures and Associated Audio Information – Part 3: Audio, International Organization for Standardization (ISO).

[14] ISO/IEC 13818-4:1998, Information Technology – Generic Coding of Moving Pictures and Associated Audio Information – Part 4: Conformance Testing, International Organization for Standardization (ISO).

[15] ISO/IEC TR 13818-5:1997, Information Technology – Generic Coding of Moving Pictures and Associated Audio Information – Part 5: Software Simulation, International Organization for Standardization (ISO).

[16] ISO/IEC 13818-6:1998, Information Technology – Generic Coding of Moving Pictures and Associated Audio Information – Part 6: Extensions for DSM-CC, International Organization for Standardization (ISO).

[17] ISO/IEC 13818-7:1997, Information Technology – Generic Coding of Moving Pictures and Associated Audio Information – Part 7: Advanced Audio Coding (AAC), International Organization for Standardization (ISO).

[18] ISO/IEC 13818-9:1996, Information Technology – Generic Coding of Moving Pictures and Associated Audio Information – Part 9: Extension for Real Time Interface for Systems Decoders, International Organization for Standardization (ISO).

[19] ISO/IEC 13818-10:1999, Information Technology – Generic Coding of Moving Pictures and Associated Audio Information – Part 10: Conformance Extensions for Digital Storage Media Command and Control (DSM-CC), International Organization for Standardization (ISO).

[20] ISO/IEC 14496-1:2001, Information Technology – Coding of Audio-Visual Objects – Part 1: Systems, International Organization for Standardization (ISO).

[21] ISO/IEC 14496-2:2004, Information Technology – Coding of Audio-Visual Objects – Part 2: Visual, International Organization for Standardization (ISO).

[22] ISO/IEC 14496-3:2001, Information Technology – Coding of Audio-Visual Objects – Part 3: Audio, International Organization for Standardization (ISO).

[23] ISO/IEC 14496-4:2000, Information Technology – Coding of Audio-Visual Objects – Part 4: Conformance Testing, International Organization for Standardization (ISO).

[24] ISO/IEC 14496-5:2001, Information Technology – Coding of Audio-Visual Objects – Part 5: Reference Software, International Organization for Standardization (ISO).

[25] ISO/IEC 14496-6:2000, Information Technology – Coding of Audio-Visual Objects – Part 6: Delivery Multimedia Integration Framework (DMIF), International Organization for Standardization (ISO).

[26] ISO/IEC TR 14496-7:2002, Information Technology – Coding of Audio-Visual Objects – Part 7: Optimized Reference Software for Coding of Audio-Visual Objects, International Organization for Standardization (ISO).

[27] ISO/IEC 14496-8:2004, Information Technology – Coding of Audio-Visual Objects – Part 8: Carriage of ISO/IEC 14496 Contents over IP Networks, International Organization for Standardization (ISO).

[28] ISO/IEC TR 14496-9:2004, Information Technology – Coding of Audio-Visual Objects – Part 9: Reference Hardware Description, International Organization for Standardization (ISO).

[29] ISO/IEC 14496-10:2003, Information Technology – Coding of Audio-Visual Objects – Part 10: Advanced Video Coding, International Organization for Standardization (ISO).

[30] ISO/IEC 14496-12:2004, Information Technology – Coding of Audio-Visual Objects – Part 12: ISO Base Media File Format, International Organization for Standardization (ISO).

[31] ISO/IEC 14496-14:2003, Information Technology – Coding of Audio-Visual Objects – Part 14: MP4 File Format, International Organization for Standardization (ISO).

[32] ISO/IEC 14496-15:2004, Information Technology – Coding of Audio-Visual Objects – Part 15: Advanced Video Coding (AVC) File Format, International Organization for Standardization (ISO).

[33] ISO/IEC 14496-16:2004, Information Technology – Coding of Audio-Visual Objects – Part 16: Animation Framework eXtension (AFX), International Organization for Standardization (ISO).

[34] ISO/IEC 14496-18:2004, Information Technology – Coding of Audio-Visual Objects – Part 18: Font Compression and Streaming, International Organization for Standardization (ISO).

[35] ISO/IEC 14496-19:2004, Information Technology – Coding of Audio-Visual Objects – Part 19: Synthesized Texture Stream, International Organization for Standardization (ISO).

[36] B. Ponce, The Impact of MP3 and the Future of Digital Entertainment Products, *IEEE Communications Magazine*, vol. 37, no. 9, Sept. 1999, pp. 68–70.

[37] M. McCandless, The MP3 Revolution, *IEEE Intelligent Systems*, vol. 14, no. 3, May–June 1999, pp. 8–9.

[38] K. Konstantinides, An Introduction to Super Audio CD and DVD-Audio, *IEEE Signal Processing Magazine*, vol. 20, no. 4, July 2003, pp. 71–82.

[39] Y.C. Chang, C.S. Li, and D.G. Messerschmitt, Adapting Network Video to Multi-time Scale Bandwidth Fluctuations, *Proceedings of IEEE International Conference on Multimedia and Expo*, vol. 2, 30 July–2 Aug. 2000, pp. 999–1002.

[40] C. Kuhmanch and C. Schremmer, Empirical Evaluation of Layered Video Coding Schemes, *Proceedings of 2001 International Conference on Image Processing*, vol. 2, 7–10 Oct. 2001, pp. 1013–1016.

[41] R. Atta and M. Ghanbari, A Layered Video Coding Scheme with Its Optimum Bit Allocation, *Proceedings of 2003 IEEE International Conference on Acoustics, Speech, and Signal Processing*, vol. 3, 6–10 April 2003, pp. 77–88.

[42] W. Li, Overview of Fine Granularity Scalability in MPEG-4 Video Standard, *IEEE Trans. of Circuits and Systems for Video Tech.* vol. 11, no. 3, March 2001, pp. 301–317.

[43] P.A.A. Assunção and G. Mohammed, A Frequency-Domain Video Transcoder for Dynamic Bit-Rate Reduction of MPEG-2 Bit Streams, *IEEE Trans. of Circuits and Systems for Video Tech.*, vol. 8, no. 8, Dec. 1998, pp. 923–967.

[44] C.W. Lin and Y.R. Lee, Fast Algorithms for DCT-Domain Video Transcoding, *Proceedings of IEEE Int. Conf. Image Processing*, vol. 1, Oct. 2001, pp. 421–424.

[45] S. Bo, L.K. Sethi, and B. Vasudev, Adaptive Motion-Vector Resampling for Compressed Video Downscaling, *IEEE Trans. of Circuits and Systems for Video Tech*, vol. 9, no. 6, Sept. 1999, pp. 929–963.

[46] J.W.C. Wong, O.C. Au, P.H.W. Wong, and A. Tourapis, Predictive Motion Estimation for Reduced-Resolution Video from High-resolution Compressed Video, *Proc. of IEEE Int. Symposium on Circuits and Systems*, vol. 4, June 1999, pp. 524–527.

[47] M.J. Chen, M.C. Chu, and S.Y. Lo, Motion Vector Composition Algorithm for Spatial Scalability in Compressed Video, *IEEE Trans. of Consumer Electronics*, vol. 47, no. 3, Aug. 2001, pp. 319–325.

[48] Z. Lei and N.D. Georganas, H.263 Video Transcoding for Spatial Resolution Downscaling, *Proc. of IEEE Int. Conf. Information Technology: Coding and Computing*, April 2002, pp. 425–430.

[49] K.T. Fung, Y.L. Chan, and W.C. Siu, Dynamic Frame Skipping for High-Performance Transcoding, *IEEE Int. Symposium Circuits on Circuits and Systems*, vol. 3, May 2002, pp. 389–392.

[50] Y.Q. Liang and Y.P. Tan, A New Content-Based Hybrid Video Transcoding Method, *Proceedings of IEEE Int. Conf. Image Processing*, vol. 1, Oct. 2001, pp. 429–432.

[51] H. Sun, W. Kwok, and J.W. Zdepski, Architecture for MPEG Compressed Bitstream Scaling, *IEEE Transactions of Circuits and Systems for Video Technology*, vol. 6, no. 2, April 1996, pp. 191–199.

[52] S.F. Chang and D.G. Messerschmitt, Manipulation and Compositing of MC-DCT Compressed Video, *IEEE Journal on Selected Areas in Communications*, vol. 13, no. 1, Jan. 1995, pp. 1–11.

[53] B.K. Natarajan and B. Vasudev, A Fast Approximate Algorithm for Scaling Down Digital Images in the DCT Domain, *Proceedings of IEEE Int. Conf. Image Processing*, vol. 2, Oct. 1995, pp. 241–243.

[54] International Organization for Standardization, Test Model Editing Committee, *ISO-IEC/JTC1/SC29/WG11/ N0400 – Test Model 5*, April 1993.

3

Continuous Media Storage and Retrieval

Continuous media data streams generate vast amount of data. Even after compression, the data rate is often still substantial, especially if audio and video are encoded. In terms of storage, the amount of storage space required in a, say, hard disk-based media server can easily be estimated. Moreover, the recent advances in hard disk technology have significantly reduced the price-per-byte storage cost. At the time of writing the cost for 100GB of hard disk storage is less than US$100, and hard disk models of size up to 300GB are already widely available. Thus, although it is still an issue to be considered in the design of media servers, the storage capacity issue is becoming less of a problem in recent years.

By contrast, the I/O capacity of hard disk, while it has improved steadily over the years, has not progressed at the same pace as storage capacity. While we can store many media contents in a large hard disk, it may not have sufficient I/O capacity to satisfy the streaming demand. Therefore, the challenge of improving disk I/O efficiency is still a relevant and important research issue.

In this chapter, we review the hard disk technology with an emphasis on the impact to media streaming. We present ways to model the hard disk to derive performance results useful in designing media servers. Of particular emphasis are the design and engineering of the disk scheduler, which has significant impact on disk I/O efficiency. We analyze the trade-offs in disk scheduler design in both single-disk and multi-disk media servers.

3.1 Structure and Model of Hard Disk

Figure 3.1 depicts the basic mechanical structure of a hard disk. There are a number of disk platters mounted on the same spindle rotating at a constant speed, ranging from 3,600 rounds-per-minute (rpm) to over 15,000 rpm. Each disk platter is sub-divided into many concentric tracks, which are further sub-divided into many sectors. A sector is the smallest unit for data storage and retrieval. To read and write to the disk, a disk head is mounted on a disk arm which

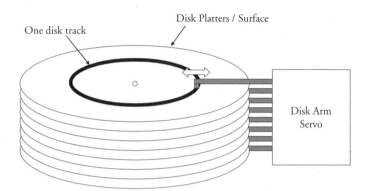

One disk track

Disk Platters / Surface

Disk Arm
Servo

Figure 3.1   The mechanical structure of a hard disk

can position the disk head to any of the tracks in the disk surface. Note that each disk platter has its own disk head and all these disk heads are moved in a synchronized manner.

To read data from a sector, the disk controller will first locate the position of the sector in terms of the track number and platter number. Then it repositions the disk arm to move the disk head to the destination track and then waits for the sector to rotate underneath the disk head to begin transferring the data to memory.

To model this data retrieval process we consider two types of processing delay. First, there are fixed delays that are independent of the amount of data to be retrieved, such as processing delay at the disk controller, delay in acquiring the data bus for data transfer, and so on. For simplicity we lump together these fixed delays and represent it by the disk parameter α.

Next we consider the variable delays. First, the time it takes to position the disk head to the disk track containing the data to retrieve is called seek time, and is represented by the random variable t_{seek}. Note that seek time depends on two factors – the current location of the disk head, and the destination track to move to. Second, once the disk head is in the right track, it will need to wait for the disk sector containing the data to rotate underneath the disk head. This is known as the rotational latency, and is represented by the random variable $t_{latency}$. Note that rotational latency depends on the current sector the disk head happens to be in after seeking, and the sector containing the data to be retrieved. The rotational latency also depends on the speed at which the disk platter rotates. For example, if the disk platter is rotating at a speed of W rounds per second, then the latency will range from 0 to W^{-1} seconds, with an average and worst-case latency of $0.5W^{-1}$ and W^{-1} seconds respectively. This is also why hard disk manufacturers strive to increase the disk platter rotation speed to improve performance. Finally, once the disk sector is reached, the actual data transfer will begin. The delay clearly depends on the rate at which data is transferred – called the disk transfer rate, and the amount of data to read (assuming it is placed in consecutive disk sectors). In practice, there are additional complications such as the data to read may span more than one track, or span more than one disk platter. We will ignore these complications for now and revisit them in Chapter 5.

Therefore, we can model the delay to retrieve a Q-byte data block from the disk using the following equation:

$$T_{read}(n) = \alpha + \beta\sqrt{n} + T_{latency} + \frac{Q}{R_{disk}} \tag{3.1}$$

where n is the number of tracks to seek, α is the parameter for fixed overhead, and β is the proportionality constant for seek time. Seek time is proportional to the square root of n to approximate the effect of physically accelerating the disk head.

3.2 Disk Scheduling

The previous disk model quantifies the time to retrieve one block of media data. A media server often serves multiple media streams concurrently, with each stream issuing retrieval requests periodically. In the simplest case, the system can process retrieval requests in the order they are issued, i.e., first-come-first-serve (FCFS) scheduling. This simple disk scheduler, however, can incur excessive overheads in seeking, as illustrated in Figure 3.2.

In this example three retrieval requests identified as 1, 2, and 3 are issued in that order by on-going media streams. The data for streams 1 and 3 are located in the outer tracks while the data for stream 2 are located in the inner track. As a result, the disk head will need to first travel to the outer track to retrieve data for stream 1, and then travel all the way across the disk surface to the inner track to retrieve data for stream 2, and finally travel all the way back to the outer track to retrieve data for stream 3. In the worst case the service time to retrieve one data block becomes

$$T_{fcfs} = T_{read}\,(N_{track} - 1) \tag{3.2}$$

This is extremely inefficient as the data for stream 1 and 3 are in fact located in nearby tracks. This contrived example illustrates the inefficiency of the FCFS scheduler, which is seldom employed in media servers.

Most media servers instead employ round-based disk schedulers to reduce the disk seek overhead. Figure 3.3 depicts a common retrieval and transmission scheduler serving requests from multiple concurrent media streams, denoted by numeric 1 to 5. For simplicity, we assume the media streams are homogeneous and have the same average data rate, denoted by R. Therefore, in the simplest case, the scheduler will retrieve one fixed-size block of data, say, of Q bytes, for each of the active streams in a service round of duration Q/R seconds.

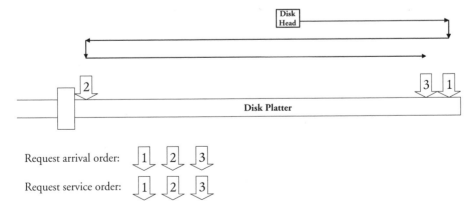

Figure 3.2   First-come-first-serve scheduling of disk retrieval requests

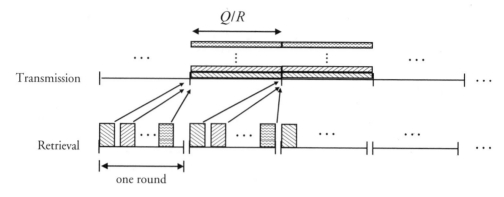

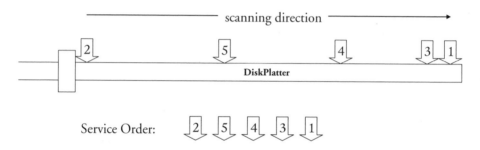

Figure 3.3   Retrieval and transmission scheduling in a media server

Figure 3.4   Using SCAN scheduler to reduce disk seek overhead

Note that the order of data retrievals within a disk service round is not fixed, and can vary from round to round. Specifically, the service order is selected according to the scanning direction of the disk head, which goes alternatively from the innermost track to the outermost track in a round, and then back from the outermost track to the innermost track in the next round (see Figure 3.4). This enables the retrievals to be performed without back-and-forth seeking within a service round, thus reducing seeking overhead. This is known as the Circular-SCAN (CSCAN) scheduler or elevator seeking.

Note that to use CSCAN we need to know which data blocks to retrieve before the beginning of a round. This is possible in a media server because most of the time a media stream retrieves media data sequentially from the disk for transmission to the client, thus enabling the server to know the future data retrievals given the media bit-rate and retrieval block size.

3.2.1 Performance Modeling

Using the disk model in equation (3.1), we can formulate the time required to complete a disk service round using CSCAN. Assuming the disk serves k requests in a round, then the service round length, denoted by $t_{round}(k)$, is given by

$$t_{round}(k) = k\alpha + \sum_{i=1}^{k}\left(t_{seek}^{i} + t_{latency}^{i} + \frac{Q}{R_{disk}}\right) + t_{seek}^{end} \quad (3.3)$$

where i denotes the ith request in a service round; t_{seek}^i and t_{seek}^i are the seek time and rotational latency incurred in serving request i; and t_{seek}^{end} is the time to position the disk head to the last track to prepare for the next scan. We will use this generic disk model for capacity dimensioning in the next section.

3.2.2 Capacity Dimensioning

The goal of capacity dimensioning is to determine the maximum number of concurrent media streams that can be sustained with deterministic performance guarantee so that proper admission control can be performed to prevent system overload. Consider a system with a homogeneous media bit-rate of R bytes per second and a constant request size of Q bytes. Using double buffering as shown in Figure 3.3, data blocks retrieved in a disk service round will be transmitted in the next round at the media bit-rate R. In other words, the retrievals in a service round must be completed within a duration of Q/R seconds or else the transmission will be delayed, possibly leading to playback jitter at the client. This is also known as the continuity condition in the literature.

Formally, this condition can be expressed as

$$t_{round}(k) \leq \frac{Q}{R} \tag{3.4}$$

which must be met for all disk service rounds. In other words, the worst-case disk service round length must not exceed the duration of one transmission round of Q/R seconds.

Now consider equation (3.3) again. The worst-case rotational latency can be computed from the disk's rate of rotation. If the disk spins at a rate of W cycles per second, then the worst-case rotational latency is just one complete rotation, i.e., W^{-1} seconds. For seek time, it can be shown that worst-case seek overheads are incurred when requests are evenly spaced across the disk surface, provided that the seek function is concave.

Modifying equation (3.3) with the previous worst-case values, we can compute the worst-case service round duration:

$$t_{round}^{max}(k) = \max \{t_{round}(k)\}$$
$$= (k+1)\left[\alpha + \beta\sqrt{\frac{N_{track} - 1}{k+1}}\right] + k\left(T_{latency} + \frac{Q}{R_{disk}}\right) \tag{3.5}$$

and then dimension the disk streaming capacity accordingly:

$$C = \max\left\{k | t_{round}^{max}(k) \leq \frac{Q}{R}, k = 1, 2, \ldots\right\} \tag{3.6}$$

where C denotes the dimensioned worst-case disk capacity in number of concurrent media streams.

Note that by using the worst-case values we guarantee that the disk will be able to sustain C concurrent media streams regardless of the actual placement of the requested data blocks. In other words, it does not matter whether a media stream's data are stored sequentially from sector to sector, track to track on the disk, or simply placed randomly over the disk surface.

This is an advantage in storage and data management as the system operator can replace/update the stored media data without worry of adversely affecting the disk's streaming capacity.

Using the Seagate ST12400N SCSI-2 hard disk as an example, the disk transfer rate in the disk specification is 3.35 MBps. If we only consider the disk transfer rate, then with a media stream bit-rate of 1.2 Mbps, the disk will be able to support up to 22 concurrent streams. However, if we account for worst-case disk seek and other overheads, then the resultant capacity is only 12 streams. This serves to illustrate the impact of disk seek overhead on streaming capacity.

3.3 Improving Disk Throughput

Knowing the performance impacts of disk seek, the natural question then is whether we can reduce the disk seek overhead to achieve higher disk throughput. Let us revisit the equation for computing the disk round time and normalize it by the number of data blocks retrieved:

$$t_{round}^{max}(k) = \frac{(k+1)}{k} \left[\alpha + \beta \sqrt{\frac{N_{track} - 1}{k + 1}} \right] + \left(T_{latency} + \frac{Q}{R_{disk}} \right) \tag{3.7}$$

which represents the per-request service time under the worst-case disk seek scenario.

If we examine the system parameters in equation (3.7), we will find that there are three non-configurable system parameters, namely the constant overhead α, disk raw transfer rate R_{disk}, and rotational latency $T_{latency}$. These parameters are properties of the physical disk and thus cannot be controlled by the server application. It is possible to eliminate rotational latency by reading one full track of data at a time. However, this track-based retrieval technique has its own problems such as large buffer/delay (see below) and disk zoning (Section 3.6) will make track-based retrieval very complicated.

By contrast, the remaining components in the equation, namely, k – the number of data blocks to retrieve in a service round, and Q – the size of the data block to retrieve, both can be controlled by the server application. As the disk overheads are relatively fixed, we can improve disk throughput simply by increasing the retrieval block size Q and/or retrieving more data blocks in a service round (i.e., increasing k). Indeed, this is a simple yet effective method to improve the disk throughput, illustrated in Figure 3.5 for two disk models.

But then there are also trade-offs. First, increasing Q and k will consume more memory for buffering. Using the double buffering scheme, the total buffer size is equal to $2kQ$ and, thus, the buffer requirement is proportional to the retrieval block size. Nevertheless, given the decreasing cost of physical memory, this may not be the limiting factor in practice.

Besides buffer size, increasing Q and k will also increase the disk service round length as the disk transfer time (i.e., Q/R_{disk}) in equation (3.7) will increase proportionally. While this does not affect on-going streams, the admission delay experienced by new users will be increased.

To see why, consider a user who initiates a new streaming session by sending a request to the server using some control protocol. Upon receiving the request, the server will first verify the data availability, allocate the system resources (e.g., buffers, state variables, etc.), and then start retrieving data from the disk for transmission. Now as the user request can arrive at any time, it will likely arrive in the middle of a disk service round. In this case the server cannot serve the request in the currently on-going service round as this could lead to additional disk

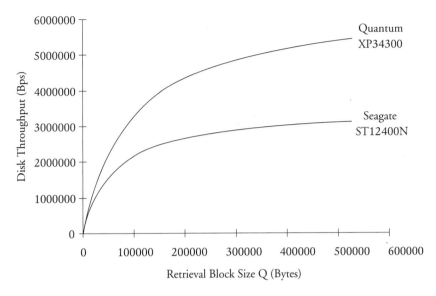

Figure 3.5   Effect of retrieval block size on the disk throughput

seeks that may violate the continuity requirement in equation (3.4). Consequently, the new
request can only start receiving service in the next service round and transmission will follow
after that as shown in Figure 3.3. Therefore, the total delay from receiving the new stream
request to the instant transmission begins will on the average amount to one and a half service
round, or $1.5t_{round}^{max}(k)$.

While the buffer cost will likely reduce as memory cost goes down, the admission delay
is a key performance metric experienced by the end user and a user will only have so much
patience. Therefore, ultimately the admission delay rather than buffer cost will be the limiting
factor in stretching the service round to improve disk efficiency.

3.4 Grouped Sweeping Scheme

In the previous discussion we came to the conclusion that delay/buffer and disk throughput
work against each other and thus it appears that disk throughout can only be increased so far
as limited by the admission delay. However, this analysis was based on the SCAN/CSCAN
disk schedulers and, in particular, due to the constraint that a new request can only be served
in the next service round.

To tackle this problem, Yu *et al.* [1] proposed a more general disk scheduler called the
Grouped Sweeping Scheme (GSS). It shares the common principle with SCAN/CSCAN in
that retrievals are also served in a scanning fashion in fixed-duration service rounds to reduce
disk seek overheads. However, instead of serving all streams in a service round, GSS divides
active streams into groups, say, *g* groups. Streams within the same group are served using
SCAN/CSCAN, while the groups are scheduled for service in a round-robin manner. Thus, the
disk schedule is effectively divided into two levels of service rounds – a micro service round
that serves a group of streams, and a macro service round that comprises the micro service

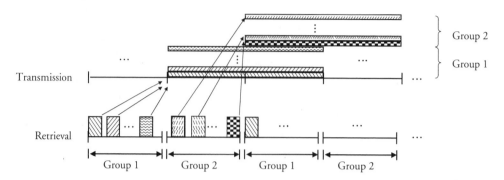

Figure 3.6 Illustration of the Grouped Sweeping Scheme with two groups

rounds of the g different groups. Figure 3.6 illustrates this GSS scheduler with two groups of media streams.

There are two important characteristics in GSS. First, a media stream can join any one of the groups as long as the group is not already running at full capacity. Second, transmission of a data block lasts for a macro service round (i.e., g micro service rounds) to sustain a continuous transmission of media data to the client while waiting for the next media block to be retrieved.

Using GSS, the retrieval of a data block can be completed in a shorter time than in the case of SCAN/CSCAN. For example, suppose at full capacity the server can serve k concurrent media streams. The service round length in SCAN will then be equal to $t_{round}^{max}(k)$. However, if we employ GSS with two groups, then each group will serve up to $k/2$ requests and so the micro-round length will be equal to $t_{round}^{max}(k/2)$ while the macro-round length will be equal to $2t_{round}^{max}(k/2)$.

This shorter service round length is desirable in terms of buffer requirement. A buffer is occupied from the time a service round begins to the time the data retrieval is completed. Thus, comparing GSS to CSCAN, it is clear that the buffer-holding time is reduced from $t_{round}^{max}(k)$ to $t_{round}^{max}(k/2)$, thus allowing the buffer to be reused more quickly for the next service round. With g groups, it can be shown [1] that the buffer requirement of GSS is given by

$$B_{GSS} = \left(k + \left\lceil \frac{k}{g} \right\rceil \right) Q \tag{3.8}$$

Comparing GSS to CSCAN's buffer requirement $2kQ$, it is clear that we can reduce the buffer requirement by using GSS with more groups (i.e., increasing g).

On the other hand, a new user joining the system running GSS may also experience shorter admission delay. This is possible because the new stream can join *any* of the g groups to receive service. If the new stream arrives in the middle of, say, micro-round serving group i, for example, then it can simply join the next group $((i + 1) \bmod g)$ provided that it is not already fully utilized. In this case the average delay is only $1.5t_{round}^{max}(k/g)$, which is shorter than the corresponding delay under CSCAN (i.e., $1.5t_{round}^{max}(k)$).

From the previous discussion it seems that we should use GSS with as many groups as possible to reduce buffer requirement and admission delay. Taking it to the limit for a server with capacity of k concurrent streams we can at most divide them into $g = k$ groups under GSS. However, in this case the disk will serve the streams one by one in a round-robin manner, effectively reducing to FCFS!

Knowing that FCFS has poor disk throughput, it raises the first trade-off in using more groups under GSS. Specifically, with k requests divided into g groups, each micro-round will retrieve only k/g data blocks. Since we are using worst-case analysis to dimension disk capacity, we will assume that the k/g data blocks are evenly spaced across the full span of the disk surface. Thus, with more groups, fewer requests are served in each micro-round, which in turn increases the worst-case per-request seek distance. Obviously this added overhead will result in longer service round length and thus reduces disk throughput.

The second trade-off is more subtle and only occurs under heavy system load. Specifically, if a new stream arrives to find the next micro-round already fully occupied, then it will have to wait for a group with available capacity to join. Note that this scenario will not occur if the number of active streams in service is smaller than k/g, as none of the groups will be fully occupied. As more and more streams join the system, more and more groups will be fully occupied. In the worst case a new stream could arrive in the middle of the only group with available capacity, and thus will need to wait until that particular group cycles back again in the next macro-round. In this worst-case scenario the admission delay will become $(g + 1) \cdot t_{round}^{max}(k/g)$.

Nevertheless, the beauty of the GSS scheduler is that we can control the trade-off between throughput and buffer requirement/admission delay by dividing the streams into different number of groups. In fact, the GSS scheduler subsumes as special cases both the FCFS scheduler (when $g = k$) and the CSCAN scheduler (when $g = 1$). The same principle will also find applications in multi-disk (Section 3.5.3) and multi-server streaming systems (Part II of the book).

3.5 Multi-Disk Storage And Retrieval

So far, we have only considered media storage and retrieval in a single hard disk. In practice, the capacity of a single disk may not be sufficient to serve a large number of concurrent users. While disk storage capacity is advancing rapidly, improvements in disk throughput are far slower due to mechanical limits like spindle speed, disk seek, etc. Therefore in most, if not all, media servers, an array of disks will be needed to provide sufficient streaming capacity. The challenge then is how to efficiently store and retrieve media data from multiple disks.

3.5.1 Partition and Replication

First, if the disks are operated independently, i.e., each disk is accessed independently from other disks using, say, a separate logical disk drive or filesystem, then we can simply distribute the media objects such as movie files to the disks according to certain placement policy. For example, we may place media object 1 in disk 1, media object 2 in disk 2, and so on. When a user requests the streaming of media object 1, then the server will simply retrieve the media data blocks from disk 1 for transmission to the user. Each disk in this case is independent and thus the same disk schedulers discussed earlier in Sections 3.2 to 3.4 can be applied directly without modification.

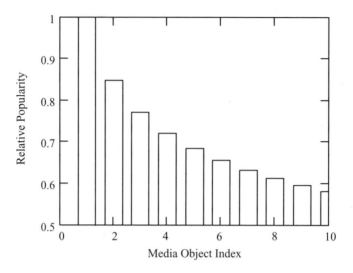

Figure 3.7 Relative popularity of media objects following the Zipf law with $c = 0.237$

However, this *data partition* approach has a significant shortcoming. Specifically, in most applications (e.g., entertainment videos) the popularity of media objects can vary significantly. Previous studies [2–3] using data collected from video rentals, for example, suggested that the popularity in those scenarios follows the Zipf law [4]:

$$P_i \propto \frac{1}{i^c} \tag{3.9}$$

where $\{P_i | i = 1, 2, \ldots, M\}$ is the popularity of media object i, sorted in decreasing order of popularity; and c is a constant characterizing the skewness of the popularities. A value of $c = 0$ represents uniform popularity (i.e., $P_i = P_j \; \forall i, j$), and increasing values of c represent increasing skewness. Figure 3.7 shows the relative popularity for a 10-media system following the Zipf law with $c = 0.237$. The difference between the more popular media objects and the less popular media objects can be quite substantial.

The implication of the Zipf-distributed popularity is that a small portion of the media objects often account for the majority of the user requests. Consequently, the disks storing the popular media objects will have a much higher load than the disks storing unpopular media objects. Researchers have long recognized this load imbalance problem and have proposed various solutions, e.g., careful placement of media objects to even out the load of the disks, or replicating the popular objects on multiple disks for load sharing. Interested readers are referred to the literature for more detail of these approaches [5–7].

At the other extreme, if we replicate all media objects onto all disks in the server, then the load balance problem can be eliminated altogether. However, this approach is only suitable for systems with a small set of media objects, where the extra storage incurred in replication is available and is cost effective. For systems serving a library of, say, hundreds or even thousands of feature-length movies, the multiplied storage costs will quickly become prohibitive.

3.5.2 Disk Striping

An alternative solution to partition/replication is through the use of striped disk array. The principle is to distribute pieces of a media object over all the disks in the array so that the load of each and every media object is equally shared by the disks in the array, illustrated in Figure 3.8. Physically, the disks in a disk array are just ordinary disks, but managed as a single logical storage device either by the disk array controller, by the software disk array module in the operating system, or by the application directly.

If a hardware disk array controller is employed, then the disk array configuration is completely hidden from the operating system and the media server application. If the operating system's software disk array function is employed, then the disk array configuration is visible to and controllable by the operating system, but still hidden from the media server application. In both cases, the disk array will simply appear to the media server application as a single logical disk drive with a large storage capacity and throughput as depicted in Figure 3.9.

These approaches eliminate the need to modify the media server software to support the use of the disk array as it is completely transparent. However, the disk array controller and the operating system's software disk array module may not have been optimized for media streaming applications. By contrast, if the media server implements multi-disk storage and retrieval functions directly (see Figure 3.10), then it can have complete control over the disk array configuration (e.g., interleaving block size, disk scheduling, etc.) to optimize for streaming applications. In the following discussions we assume this third approach, i.e., implementing the disk array within the media server application, and present efficient ways to schedule the media data retrieval process.

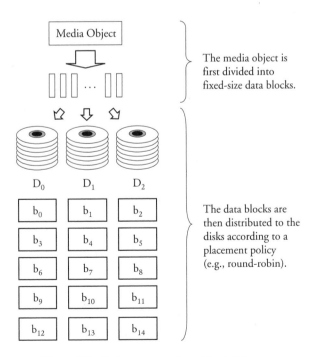

Figure 3.8   Data placement in a striped disk array

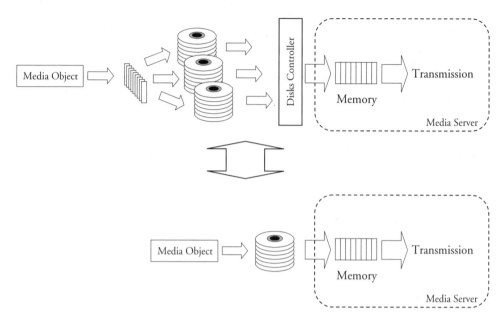

Figure 3.9   A disk array implemented by a disk array controller will be transparent to the media server
application

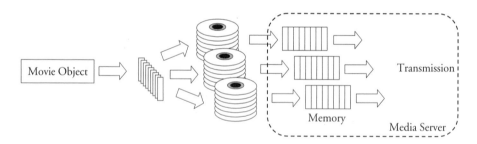

Figure 3.10   The media server implements a striped disk array directly from multiple independent disks

3.5.3 Multi-Disk Scheduling

Now that data blocks of a media object are spread over multiple disks, we can no longer
schedule each disk independently. We present in this section three multi-disk schedulers and
discuss their strengths and weaknesses.

3.5.3.1 Concurrent Schedule

Figure 3.11 depicts the first multi-disk scheduler – *concurrent schedule*, for a striped disk array
comprising three disks. Under the concurrent scheduler all disks in the array are synchronized
to retrieve data blocks for the same set of media objects in fixed-duration rounds. For example,
for a d-disk array with retrieval block size of Q bytes, a total of dQ bytes will be retrieved
for each media stream in a service round. Within each service round each disk performs its

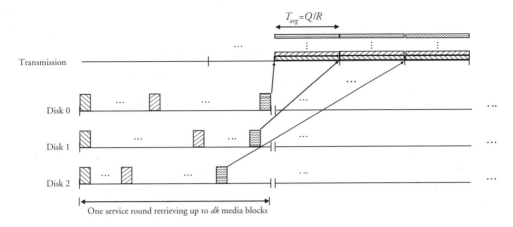

Figure 3.11   Retrieval and transmission under the concurrent scheduler (blocks of the same pattern belongs to the same media stream)

own seek optimization using CSCAN and the data blocks retrieved from all disks are then sequentially transmitted in the next service round as shown in Figure 3.11.

As the disks effectively operate in parallel serving the same set of media streams, it seems that the throughput will then be increased by d times for a d-disk array. Interestingly, the actual achievable throughput under concurrent schedule can in fact be larger than d times the throughput of a single disk.

Recall that in capacity dimensioning the constraint is that the worst-case time spent in a service round must not exceed the transmission time for the data retrieved:

$$t_{round}^{\max}(k) \leq \frac{Q}{R} = T_{avg} \tag{3.10}$$

Extending this to a d-disk striped disk array we have

$$t_{round}^{\max}(k') \leq \frac{dQ}{R} = dT_{avg} \tag{3.11}$$

Now since the number of tracks in the disk is the same no matter a disk is used independently or in a striped disk array, the seek distance on a per-data-block basis must decrease if we increase the number of blocks retrieved in a service round. In other words, we have in general

$$t_{round}^{\max}(dk) \leq dt_{round}^{\max}(k) \leq dT_{avg} \tag{3.12}$$

Thus, by retrieving more data blocks in a service round, the striped disk array can achieve lower seek overhead than operating the disks independently. This performance gain, however, is offset by two trade-offs.

Assume that each disk has a capacity to retrieve k data blocks in a service round of T_{avg} seconds. Ignoring the performance gains due to reduced seek overhead, then in a d-disk array each disk will retrieve dk data blocks in a service round of dT_{avg} seconds, or d^2k data blocks for the entire disk array. Therefore, the buffer requirement is equal to $2d^2kQ$ bytes, half of

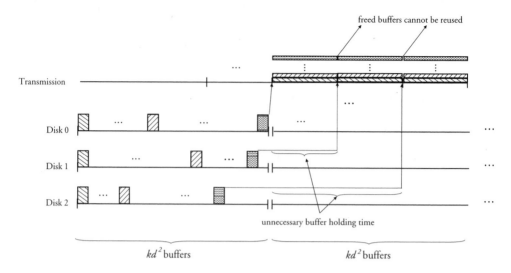

Figure 3.12   Buffer management under concurrent schedule

which are used to store the retrieved data and the other half storing the data for transmission as shown in Figure 3.12.

Now it becomes evident that this concurrent schedule does not scale up well. For example, with $Q = 64$KB and $k = 20$, the *per disk* buffer requirement for 1-disk and 8-disk arrays are 2.5MB and 20MB respectively. This gets worse as the array grows larger. Again, buffer requirement may not be the limiting factor as memory cost has decreased dramatically over the years. The real limiting factor is the start-up delay, which equals 1.5 times the service round length. Thus for concurrent schedule the start-up delay increases linearly with the number of disks in the array (i.e., dT_{avg}).

3.5.3.2 Offset Schedule

As illustrated in Figure 3.12, part of the reason for the increased buffer requirement is the unnecessary buffer holding time after a data block is completely transmitted. The empty buffers cannot be reused until the next disk service round starts. To eliminate this deficiency, we can offset the schedules of the disks' service rounds by T_{avg} seconds as depicted in Figure 3.13 so that a new disk service round is started once every T_{avg} seconds, aligned with the instants when empty buffers become available.

This *offset schedule* has the same buffer requirement for disk retrieval (d^2kQ bytes) but the buffer requirement for transmission is reduced from d^2kQ bytes to dkQ bytes. Therefore, the total buffer requirement is reduced from $2d^2kQ$ to $(d + 1)dkQ$ bytes. Using the same example in the previous section ($Q = 64$KB, $k = 20$) the per-disk buffer requirement is reduced from 20MB to 11.25MB for an 8-disk array.

3.5.3.3 Split Schedule

Despite the reduction, the per-disk buffer requirement under offset schedule still increases with the number of disks in the array. The ultimate solution is to divide the streams into d groups and serve them in separate service rounds of shorter duration – *split schedule*, shown in Figure 3.14.

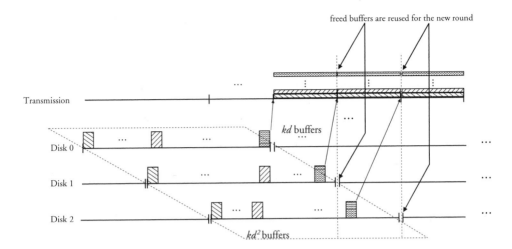

Figure 3.13 Offset schedule enables immediate reuse of empty buffers

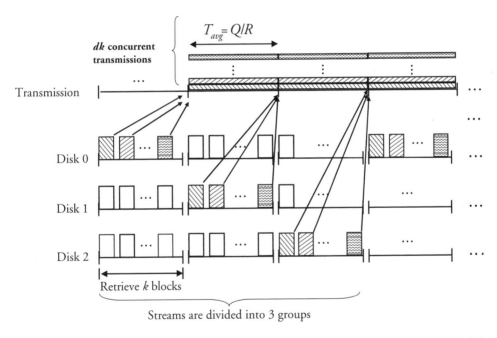

Figure 3.14 The split schedule can achieve constant per-disk buffer requirement regardless of the number of disks in the striped disk array

The key in split schedule is that a media stream retrieves its data blocks from each disk not in parallel as in the case of concurrent schedule and offset schedule, but in turns. For example, if a media stream retrieves data block i from disk 0 in round j for transmission in round $j + 1$, then in the next round $j + 1$ it retrieves data block $i + 1$ from disk 1 for transmission in round $j + 2$, and so on. Compared to the previous two schedulers, this split schedule reduces the service round length from dT_{avg} seconds to T_{avg} seconds. Therefore, in addition to reusing the empty buffers after transmission, the buffer requirement for data retrieval is also reduced from d^2kQ bytes to dkQ bytes, and the total buffer requirement becomes $2dkQ$ bytes, or $2kQ$ bytes per disk – same as the single-disk case.

Therefore from the scalability perspective, a split schedule is linearly scalable in terms of buffer requirement. Start-up delay is more complicated as it also depends on the utilization of the disk array. Similar to the GSS scheduler discussed earlier in Section 3.4, a new user can join any one of the d groups to receive service. Thus, if the disk array is lightly utilized (e.g., with fewer than k on-going streams), then the new stream can simply join the next service round and the average delay will be the same as in the single-disk case, i.e., equal to $1.5T_{avg}$. At higher utilization some of the groups may be fully occupied and in that case the delay will be longer. We will revisit this issue in Chapter 11 where we derive the average admission delay at a given system utilization.

3.6 Disk Zoning

So far in this chapter we have modeled a hard disk to be composed of multiple disk platters, each further divided into tracks, and finally each track sub-divided into a *fixed* number of sectors. This last assumption, however, is not necessarily true in practice. In the race to increase the disk storage capacity, disk drive manufacturers have developed a technique called Zoned-Bit-Recording (ZBR), which breaks the constant-size track assumption.

If we reconsider the physical disk geometry in Figure 3.1 we can easily see that an outer track at the edge of the disk platter will have a larger circumference than the inner track closer to the center of the disk platter. Therefore, if the same number of sectors are used (i.e., same track size), then the recording density of the outer tracks will be lower than the inner tracks. Disk drive manufacturer exploits this by allocating more sectors to outer tracks than the inner tracks. In most cases the tracks are divided into multiple *zones*, with tracks in the same zone having the same number of sectors per track – disk zoning. The number of sectors per track increases as we go from the innermost zone to the outermost zone. This effectively increases the storage capacity of the outer zones/tracks.

The immediate impact of disk zoning is that the disk transfer rate R_{disk} is no longer a constant parameter. As the disk spins at a constant angular velocity (e.g., 10,000 rpm), the transfer rate will be higher for the outer tracks than the inner tracks due to the larger track size. Take Seagate 31200W, as an example. Its disk platter is divided into 23 zones with transfer rates ranging from 2.33 MBps for the innermost zone to 4.17 MBps for the outermost zone. Thus, the difference in transfer rates can be quite substantial.

This creates a problem in disk capacity dimensioning as the disk transfer rate is one of the parameters previously assumed to be a given constant. For worst-case capacity dimensioning the obvious solution is to use the lowest transfer rate among all disk zones as the parameter value in dimensioning the disk capacity. This guarantees that the continuity requirement will be satisfied no matter which zone the requested data happen to be located. However, this obviously will under-utilize the disk as the outer zones, due to their larger storage capacity, account for a larger proportion of the disk's storage. Over the years researchers have come up with clever solutions to tackle this problem, principally by trading off some storage and/or buffer for higher throughput. Interested readers are referred to the works by Birk [8], Mourad [9], Ghandeharizadeh *et al.* [10], and Nerjes *et al.* [11] for more detail.

3.7 Summary

In this chapter we have covered the fundamental principles in continuous media storage and retrieval, including issues in data placement, scheduling, capacity dimensioning, multi-disk storage and retrieval, and disk zoning. In particular, we employed worst-case analysis technique to design the disk schedulers and to dimension the streaming capacity. This worst-case approach is simple to analyze and is able to guarantee the retrieval performance. However, as the worst-case scenario rarely, if ever, occurs in practice, the over-engineering will necessarily result in disk under-utilization. In the next chapter we depart from the worst-case approach to investigate a statistical approach to capacity dimensioning, which can increase disk utilization as well as solving the disk zoning problem.

References

[1] P.S. Yu, M.S. Chen, and D.D. Kandlur, Grouped Sweeping Scheduling for DASD-based Multimedia Storage Management, *ACM Multimedia Systems*, vol. 1, no. 2, 1993, pp. 99–109.

[2] A.L. Cherenal, Tertiary Storage: An Evaluation of New Applications, PhD dissertation, University of California, Berkeley, Dec. 1994.

[3] C. Griwodz, M. Bar, and L.C. Wolf, Long-term Movie Popularity Models in Video-on-Demand Systems or The Life of an on-Demand Movie, *Proceedings of ACM Multimedia Conference*, Seattle Washington, USA, Nov. 1997, pp. 349–357.

[4] Y.S. Chen, Mathematical Modeling of Empirical Laws in Computer Application: A Case Study, *Comput. Math. Applicat.*, Oct. 1992, pp. 77–87.

[5] T.D.C. Little and D. Venkatesh, Popularity-Based Assignment of Movies to Storage Devices in a Video-on-Demand System, *ACM Multimedia Systems*, vol. 2(6), 1995, pp. 280–287.

[6] N. Venkatasubramanian and S. Ramanthan, Load Management in Distributed Video Servers, *Proceedings of 17th International Conference on Distributed Computing Systems*, Baltimore, MD, USA, May 1997, pp. 528–535.

[7] C.C. Bisdikian and B.V. Patel, Issues on Movie Allocation in Distributed Video-on-Demand Systems, *Proceedings of ICC'95*, Seattle, WA, vol. 1, June 1995, pp. 250–255.

[8] Y. Birk, Track-Pairing: A Novel Data Layout for VoD Servers with Multi-Zone-Recording Disks, *Proceedings of International Conference on Multimedia Computing and Systems*, May 1995, pp. 248–255.

[9] A.N. Mourad, Issues in the Design of a Storage Server for Video-on-Demand, *ACM Multimedia Systems*, vol. (4), April 1996, pp. 70–86.

[10] S. Ghandeharizadeh, S.H. Kim, C. Shahabi, and R. Zimmermann, Placement of Continuous Media in Multi-Zone Disks, In *Multimedia Information Storage and Management* (Ed. S.M. Chung), Kluwer Academic Publisher, 1996.

[11] G. Nerjes, P. Muth, and G. Weikum, Stochastic Service Guarantees for Continuous Data on Multi-Zone Disks, *Proceedings of 16th Symposium on Principles of Database Systems (PODS'97)*, May 1997, pp. 154–160.

4

Soft Scheduling

Most of the existing disk schedulers employed in continuous media servers use worst-case analysis in capacity dimensioning. As discussed in Chapter 3, this worst-case approach is relatively simple to implement and yet can provide deterministic performance guarantees, which is particularly desirable in media streaming applications. Nevertheless, the worst-case scenario rarely, if ever, occurs in practice and so the disk ends up under-utilized. In this chapter we investigate a statistical approach to capacity dimensioning – soft scheduling, where performance guarantee is probabilistic rather than deterministic. By tolerating a very small probability of capacity overflow, we can significantly increase the disk utilization. This chapter presents a disk scheduler supporting soft scheduling, a number of performance enhancement techniques, and evaluates the potential performance gains.

4.1 Introduction

For media servers serving stored data (as opposed to data captured in real time), the disk scheduler plays a vital role in providing glitch-free services to the end-users. Many excellent disk schedulers have been proposed in the literature and were reviewed in Chapter 3, including the SCAN [5], Circular-SCAN (CSCAN) [5], and the more general Group Sweeping Scheme (GSS) [4] schedulers. Using worst-case dimensioning techniques, these schedulers are simple to implement and are capable of providing guaranteed performance.

Nevertheless, using worst-case dimensioning techniques – *hard scheduling* – also have shortcomings. As the usable disk capacity (in terms of maximum number of concurrent streams that can be supported) is dimensioned according to worst-case scenarios, the disk will often be under-utilized during actual operation. Moreover, most modern disk drives employ disk zoning to improve disk capacity. Zoning divides the disk surface into multiple zones, where each zone has a number of consecutive cylinders having the same track size (in numbers of sectors per track). As the disk rotates with a constant angular velocity, outer zones can be allocated more sectors per track than inner zones. Therefore, one side-effect is that outer zones will have substantially higher transfer rate than inner zones. If we adopt hard scheduling, then we would need to dimension the disk streaming capacity according to the (lowest) transfer rate of the innermost zone, and hence sacrifice additional disk bandwidth available in the outer zones.

In this chapter, we investigate an alternative approach to disk-scheduler design – soft scheduling. Specifically, by designing the disk scheduler with statistical performance guarantees instead of deterministic performance guarantees, we can use the disk I/O bandwidth more efficiently and at the same time, still be able to satisfy the continuity requirement with high probability. In addition, by placing data randomly instead of sequentially across the disk surface, we can make use of the higher transfer rates of the outer zones in multi-zone disks to achieve higher disk throughput. To further increase the usable disk capacity, we present a Dual-Round Scheduling technique to schedule disk rounds in pairs so that overflows in the next service round can be absorbed in the current service round, and an Early-Admission Scheduling technique to enable the use of larger media block size for better disk efficiency without adversely increasing the system response time. Finally, we present methods for detecting and recovering from service round overflow. Results from a detailed simulation of five disk drives will be used to explore the potential performance gains of the presented techniques over hard-scheduling approaches.

4.2 Statistical Capacity Dimensioning

The worst-case dimensioning technique in hard scheduling enables the disk to provide deterministic performance guarantee. However, as with any worst-case techniques, the trade-off would be lower disk utilization in practice as the worst-case scenario occurs very sparingly. For example, ignoring rotational latency for the moment, the worst-case seek time under CSCAN for a disk with a total of N tracks occurs with probability

$$\Pr\left\{n_i = \frac{N-1}{k+1}, i = 1, 2, \ldots, k\right\} = \left(\frac{1}{N-1}\right)^k \tag{4.1}$$

where n_i denotes seek distance for the ith request (see Table 4.1 for a summary of notations).

For a disk with $N = 5,001$ tracks and $k = 10$, this computes into a probability of 1.024^{-37}. This is clearly negligible in practice and this motivates us to investigate soft scheduling to provide statistical rather than deterministic performance guarantee.

In statistical capacity dimensioning, the objective is to find an operating point that provides higher usable disk capacity than deterministic capacity dimensioning, subject to a given overflow probability constraint. Let $F_{round}(t, k)$ be the cumulative distribution function (CDF) for the disk service round length, i.e.,

$$F_{round}(t, k) = \Pr\{t_{round}(k) \leq t\} \tag{4.2}$$

We can then define an overflow probability constraint ε that specifies the maximum acceptable occurrence probability for violating the continuity condition in equation (3.4). Using this constraint and equation (4.2), we can then compute the usable disk capacity, denoted by $C(\varepsilon)$, from

$$C(\varepsilon) = \max\{k | (1 - F_{round}(T_r, k)) \leq \varepsilon, k = 0, 1, \ldots\} \tag{4.3}$$

where $T_r = Q/R$ is the maximum length of a service round. This is the maximum number of requests that can be served in each service round with an overflow probability no greater than ε.

Note that storage allocation for a media object must be pseudo-randomized under soft scheduling, i.e., available disk blocks are randomly selected to store a media title. This

Table 4.1 Summary of notations

Description	Symbol	Notes
Average media bit-rate	R	Parameter in bytes per second (e.g., 150,000 Bps)
Media block size	Q	Parameter in bytes (e.g., 65,336 bytes)
Length of a transmission cycle	T_r	Computed from Q/R, in seconds
Disk transfer rate	r	Random variable in bytes per second
	r_i	Transfer rate for the ith request
	r_{min}	Minimum transfer rate (i.e. of innermost track)
Fixed overhead in disk read	α	Constant in seconds
Seek time in disk read	t_{seek}	Random variable in seconds
	t_{seek}^i	Seek time for the ith request
Rotational latency in disk read	$t_{latency}$	Random variable in seconds
	$t_{latency}^i$	Rotational latency for the ith request
Head repositioning delay	t_{seek}^{end}	Random variable in seconds
Service time for a request	$t_{request}$	Random variable in seconds
	$t_{request}(i)$	Service time for request i
Length of a service round serving k requests	$t_{round}(k)$	Random variable in seconds
Disk platter rotation rate	W	Constant in cycles per second
Worst-case seek time for serving k requests	$t_{seek}^{max}(k)$	Computed, in seconds
An upper bound for service round length serving k requests	$t_{round}^{max}(k)$	Computed, in seconds
Usable disk capacity under hard scheduling	C	Computed, in number of requests (served in a service round)
Usable disk capacity under soft scheduling	$C(\varepsilon)$	Computed, in number of requests (served in a service round)
Usable disk capacity under Dual-Round Scheduling	$C_{DRS}(\varepsilon)$	Computed, in number of requests (served in a service round)
Number of disk tracks	N	Parameter in numbers
Seek distance for request i	n_i	Variable in number of tracks
Overflow probability constraint	ε	Parameter
Extra number of buffers for Dual-Round Scheduling	B_{early}	Computed, in number of Q-bytes buffers
Probability density function for round length serving k requests	$f_{round}(t, k)$	Parameter
Probability distribution function for round length serving k requests	$F_{round}(t, k)$	Parameter
Overflow probability for serving k requests in a round	$\Omega(k)$	Computed
Size of track i	z_i	Parameter in bytes
Probability of disk head located at track i	p_i	Computed
Track number for request i	v_i	Parameter
	$\mathfrak{R}$	Variable in seconds:

(Continued)

Table 4.1 (*Continued*)

Description	Symbol	Notes
Residual service time of a round	$\mathfrak{R}_{nps}$ $\mathfrak{R}_{ps}$	for non-preemptive scheduling for preemptive scheduling
Arrival time for a new request	t_{new}	Variable, in seconds
End time for the current service round	t_{due}	Variable, in seconds
Seek function, including both seek time and fixed overhead	$f_{seek}(k)$	k is seek distance in number of tracks
Probability of round overflow under First-Block Replication	δ	Computed
Expected scheduling delay	D	Computed, in seconds
Scheduling delay constraint	D_{max}	Parameter in seconds
Retrieval deadline for request i	d_i	Computed, in seconds
Size of partial block retrieval during round overflow	Q_d	Computed, in bytes

randomization is necessary to prevent correlated overflows from one service round to the next. To see why this randomization process is needed, let us assume that media blocks from a media stream are stored sequentially in the disk. Now suppose a new stream joins the system at round i and causes the service round to overflow (i.e., length of service round exceeding T_r). Then due to the sequential data placement, the next round will have requests in similar locations, and hence will likely experience overflow as well. Randomized placement can break up spatial correlation between requests from consecutive service rounds and thus avoid correlated overflows.

In practice, the entire placement information for a media stream can be stored in an index file. Assume 16 bits are used to store the beginning sector number for a media block. Then a 2GB media stream stored in 128KB blocks will consume 32KB to store the index file, which is negligible compared to the size of the media stream. Hence, the server can simply load the entire index file into the memory at the time of stream admission.

4.3 Dual-Round Scheduling

The previous soft-scheduling approach in general can achieve better usable disk capacity than hard scheduling at the expense of a small probability of service round overflow. In this section, we present a Dual-Round Scheduling (DRS) technique to further reduce this overflow probability so that even more streams can be admitted.

4.3.1 Read-Ahead Algorithm

We observe that in practice the majority of the service rounds are shorter than the maximum limit T_r. This is necessary to keep the overflow probability within the given threshold ε. Now during this *slack time* the disk is in fact idle. Thus, instead of wasting the otherwise unused disk time, we can start the retrievals for the *next* service round earlier. In this way the next service round length will be extended beyond T_r seconds and consequently will reduce the overflow probability.

Let the length for round x be t_x and further assume that the next round $x + 1$ is overflowed, i.e., $t_{x+1} > T_r$. If the length of round x is less than or equal to $(2T_r - t_{x+1})$, then we can compensate for the longer service round $x + 1$ simply by starting it $(t_{x+1} - T_r)$ seconds earlier than normal.

Let B_{early} be the number of extra buffers (each Q bytes) available for storing these early-retrieved media blocks. Then the disk scheduler can be modified as follows. If the current round finishes early (i.e., round length shorter than T_r), then the disk will immediately proceed to retrieve data blocks of the next round. This process continues until either the extra buffers are exhausted, or all data blocks in the next round are completely retrieved. On the other hand, no special operation is needed if the current round does not finish early (either round length equals T_r or round overflows). We analyze the capacity gain in Section 4.3.2 next, and derive an upper bound for B_{early} in Section 4.3.3.

4.3.2 Performance Modeling

To quantify the improvement, reconsider the round lengths for any two consecutive disk service rounds. As data placement is random, the service round lengths for any two disk service rounds are independent and identically distributed according to $F_{round}(t, k)$. Let $f_{round}(t, k)$ be the density function of $F_{round}(t, k)$ and let $F_{round}^{(2)}(t, k)$ be the distribution of the sum of two service round lengths, which is the auto-convolution of $F_{round}(t, k)$:

$$F_{round}^{(2)}(t, k) = \int_{-\infty}^{\infty} F_{round}(t - x, k) f_{round}(x, k) dx \tag{4.4}$$

Now consider an arbitrary service round i. With DRS, round i can overflow under two conditions. First, if round $(i - 1)$ does not overflow, then round i will overflow only if the combined round lengths are longer than $2T_r$:

$$\Pr\{(t_i + t_{i-1}) > 2T_r | t_{i-1} \leq T_r\} \leq \Pr\{(t_i + t_{i-1}) > 2T_r\}$$
$$= 1 - F_{round}^{(2)}(2T_r, k) \tag{4.5}$$

Second, if round $(i-1)$ does overflow, then it will be truncated to at most T_r (see Section 4.5.2). In this case, round i overflows if it is longer than T_r:

$$\Pr\{t_i > T_r | t_{i-1} > T_r\} = 1 - F_{round}(T_r, k) \tag{4.6}$$

Hence the overflow probability of round i, denoted by $\Omega(k)$, can be computed from

$$\Omega(k) = \Pr\{t_i > T_r\}$$
$$= \Pr\{(t_i + t_{i-1}) > 2T_r | t_{i-1} \leq T_r\} \Pr\{t_{i-1} \leq T_r\}$$
$$+ \Pr\{t_i > T_r | t_{i-1} > T_r\} \Pr\{t_{i-1} > T_r\} \tag{4.7}$$
$$\leq \left[\left(1 - F_{round}^{(2)}(2T_r, k)\right) F_{round}(T_r, k)\right] + \left[(1 - F_{round}(T_r, k))^2\right]$$

Finally, we can compute the usable disk capacity under DRS from

$$C_{DRS}(\varepsilon) = \max \{k | \Omega(k) \le \varepsilon, k = 0, 1, \ldots\} \tag{4.8}$$

4.3.3 Buffer Requirement

To achieve the capacity gains, there is also a trade-off in DRS – the additional buffers used to store the early-retrieved media blocks. To obtain an upper bound for the extra buffer requirement, we note that in the worst case, the second disk service round will have a length of $t_{round}^{max}(k)$ as given in equation (3.5). To prevent overflow, the server will have to start the service round earlier by

$$T_{early} = t_{round}^{max}(k) - T_r \tag{4.9}$$

Note that DRS cannot compensate for overflowed rounds with length longer than $2T_r$ as the slack time for the previous round is bounded by T_r.

Now the time to retrieve a media block of size Q bytes is bounded from below by

$$t_{read}^{min} = \frac{Q}{r_{max}} \tag{4.10}$$

where r_{max} is the maximum transfer rate (e.g., at the outer-most zone). Hence during the time interval T_{early}, we need at most

$$B_{early} = \min \left\{ \left\lceil \frac{T_{early}}{t_{read}^{min}} \right\rceil, C_{DRS}(\varepsilon) \right\} \tag{4.11}$$

extra buffers to store the early-retrieved media blocks.

To be fair, the extra buffers may also be used to increase the media block size Q, which also increases disk efficiency, instead of using DRS. However, increasing the media block size will result in longer service round length and, consequently, will increase the admission delay for new streams. In practice, a system is likely to have been dimensioned to use the largest media block size for maximum disk efficiency and hence increasing the block size further will not be feasible. By contrast, DRS does not affect the scheduling delay as the media-block size is unchanged and hence we can employ DRS to further increase the usable disk capacity in a system with an already optimized media block size.

4.4 Early-Admission Scheduling

In conventional round-based scheduler such as SCAN and CSCAN, the media block size is one of the key parameters in determining the achievable disk utilization. As current memory costs continue to drop due to rapid increases in memory density, it may appear that one can keep increasing disk utilization simply by choosing larger block sizes. However, in addition to memory cost, the usable block size is also limited by the admission delay, as discussed in Section 3.3.

For example, in conventional round-based scheduler a new request arriving mid-way in round i will receive service beginning in the next round $(i + 1)$. Due to double-buffering, the retrieved block will be transmitted in round $(i + 2)$. Hence, the worst-case admission delay is

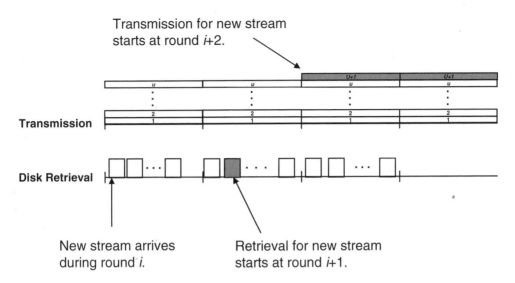

Figure 4.1 Admission scheduling in conventional round-based scheduler

two rounds and the average admission delay is 1.5 rounds (Figure 4.1). This admission delay not only affects the start-up latency, but also affects the system response time when interactive playback controls (e.g., seeking to a new playback point) are performed.

4.4.1 Admission Algorithm

Soft scheduling allows a new way of shortening this admission delay. The principle is to try to admit a new stream into the current round rather than waiting for the next round to begin. If admission to the current round is successful, then we can reduce the admission delay by one service round as shown in Figure 4.2.

Let s_i be the start time for round i. Consider a new stream arriving at time t_a when service round i is in progress. At that instant, the disk head is either moving to the next target track (Case 1) or stationed in a track reading data (Case 2). Let n be the track number for the target track (Case 1) or current track (Case 2). Assuming Poisson arrival and randomized placement, the disk head would be equally probable to be located at any one of the N tracks for non-zoned disks when a new request arrives. For zoned disks, the probability for the disk head to be located at track n when a new stream arrives, denoted by p_n, would be skewed by the track size:

$$p_n = \frac{z_n}{\sum_{j=0}^{N-1} z_j} \tag{4.12}$$

where z_j is the size of track j.

We consider the case where the disk head is scanning in the forward direction with increasing track number. The case for reverse direction scanning is similar and thus is omitted here. To

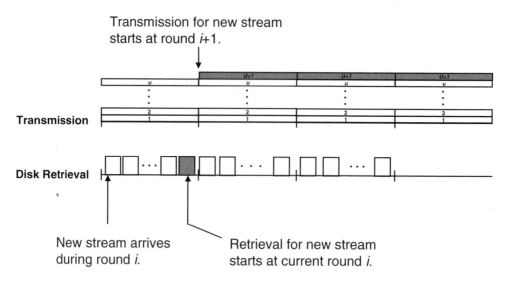

Figure 4.2 Early-Admission Scheduling in soft scheduling

simplify notations, let $f_{seek}(k)$, $k = 1, 2, \ldots, N - 1$, be the seek function that includes both seek time and fixed overhead. Let there be u existing streams (i.e., retrieving u data blocks per round) when the new request arrives. Let t_{new} be the time the new request arrives and t_{due} be the latest time when the current service round must end.

Suppose the disk is retrieving data block for stream w when the new request arrives and the data block for the new request is located in the scanning direction of the disk head, i.e., the track number of the data block for stream $u + 1$ is equal to or larger than the track number of data block for stream w. In this case, we can simply insert the new data block into the list of remaining data blocks to be retrieved, sort it by their track number $\{v_w, v_{w+1}, \ldots, v_{u+1}\}$, and then compute the residual service time, denoted by $\Re$, to complete the service round:

$$\Re = \sum_{j=w}^{u} \left(f_{seek}(v_{j+1} - v_j) + \frac{Q}{r_{j+1}} + t_{latency}^{j+1} \right)$$
$$+ f_{seek}(N - v_{u+1}) + t_{residual} \tag{4.13}$$

where the first summation term is the service time to retrieve the remaining data blocks; the second term is the head-repositioning time; and the last term is the time to complete retrieving the current data block w.

In the worst case equation (4.13) is bounded by

$$\max\{\Re\} = \sum_{j=w-1}^{u} \left(f_{seek}(v_{j+1} - v_j) + \frac{Q}{r_{j+1}} + W^{-1} \right)$$
$$+ f_{seek}(N - v_{u+1}) \tag{4.14}$$

If the worst-case residual service time does not exceed the service round end time t_{due}, then this new request can be admitted into this round as if it had arrived before the round starts. We can check for the overflow condition simply by

$$\max \{\Re\} + t_{new} \leq t_{due} \tag{4.15}$$

Otherwise, the new request will have to wait for service in the next round.

On the other hand, if the media block for the new stream is located in an upstream location where the disk head has already scanned past, then the disk scheduler has two options: it can proceed to retrieve data blocks for existing streams first and then come back to retrieve data block for the new request – *non-preemptive schedule*; or it can backtrack to retrieve data block for the new request first before proceeding with the rest of the existing streams – *preemptive schedule*.

We can compute the residual service time for non-preemptive schedule from

$$
\begin{aligned}
\Re_{nps} = \sum_{j=w}^{u-1} \left(f_{seek}(v_{j+1} - v_j) + \frac{Q}{r_{j+1}} + t_{latency}^{j+1} \right) \\
+ f_{seek}(v_u - v_{u+1}) + \frac{Q}{r_{u+1}} + t_{latency}^{u+1} \\
+ f_{seek}(N - v_{u+1}) + t_{residual}
\end{aligned}
\tag{4.16}
$$

where the first term is the time to retrieve data blocks for the existing streams; the second term is the time to retrieve data block for the new stream; the third term is the head-repositioning time; and the last term is the time to complete retrieving the current data block w.

Similarly, we can compute the residual service time for preemptive schedule from

$$
\begin{aligned}
\Re_{ps} = f_{seek}(v_w - v_{u+1}) + \frac{Q}{r_{u+1}} + t_{latency}^{u+1} \\
+ f_{seek}(v_{w+1} - v_{u+1}) + \frac{Q}{r_{w+1}} + t_{latency}^{w+1} \\
+ \sum_{j=w+1}^{u-1} \left(f_{seek}(v_{j+1} - v_j) + \frac{Q}{r_{j+1}} + t_{latency}^{j+1} \right) \\
+ f_{seek}(N - v_u) + t_{residual}
\end{aligned}
\tag{4.17}
$$

where the first term is the time to backtrack and retrieve data block for the new stream; the second term is the time to retrieve data block for stream $(w + 1)$; the third term is the time to retrieve data blocks for the remaining streams; the fourth term is the head-repositioning time; and the last term is the time to complete retrieving the current data block w.

By comparing $\max\{\Re_{nps}\}$ with $\max\{\Re_{ps}\}$, the scheduler can choose the method with shorter delay and then check for overflow using method similar to Eqs. (4.14) and (4.15). If the round does not overflow, then the new stream can be admitted into the current round, thereby shortening the scheduling delay by one complete service round of T_r seconds.

4.4.2 First-Block Replication

To further improve the chance of successfully admitting a new stream into the current round and to simplify the disk scheduler, we can use a technique called First-Block Replication (FBR) which replicates the first data block of a media stream to both the innermost track and the outermost track. With this technique, we can guarantee that request for a new stream will always be located downstream in the disk head scanning direction.

Moreover, as the disk head has to be repositioned to the platter edge at the end of a service round, seek time for the new request is eliminated as well. The residual service time with FBR can be computed from

$$\Re = \sum_{j=w}^{u-1} \left(f_{seek}(v_{j+1} - v_j) + \frac{Q}{r_{j+1}} + t_{latency}^{j+1} \right)$$

$$+ f_{seek}(N - v_u) + t_{residual} \qquad (4.18)$$

$$+ \frac{Q}{r_{u+1}} + t_{latency}^{u+1}$$

and the scheduler can admit the new stream immediately if the round does not overflow.

The significance of FBR is that it guarantees that the data block of the new stream will be located downstream in the scanning direction. Consequently, rather than using the worst-case to estimate round overflow as in equation (4.14), we can approximate the probability of round overflow by the round-length distribution

$$\delta = \Pr\{overflow\}$$
$$= 1 - F_{round}(T_r, u + 1) \qquad (4.19)$$

which is a conservative measure as the disk head has already scanned past some tracks.

Using equation (4.19) we can then compute the average scheduling delay under EAS+FBR from

$$D = 1.5T_r\delta + 0.5T_r(1 - \delta)$$
$$\approx 0.5T_r \qquad (4.20)$$

for small δ.

Therefore, together Early-Admission Scheduling and First-Block Replication can reduce the scheduling delay by two-thirds for small δ. The trade-off is the additional storage required to replicate the first data block of each media stream, which is relatively insignificant, given the size of typical media streams. For example, the overhead for a one-hour MPEG1 video of bit-rate 150KB/s and block size 128KB is only 0.024%. Note that we can extend this FBR technique to replicate additional blocks within a media stream, such as the first block in the beginning of a seekable chapter to achieve similar admission-delay reduction when performing interactive playback controls (e.g., chapter selection).

To quantify the gain of FBR, we assume that an admission delay constraint D_{max} is given as part of the service requirement. Then we must ensure that the expected admission delay is smaller than D_{max}:

$$\frac{Q}{R}(0.5 + \delta) \leq D_{\max} \tag{4.21}$$

Rearranging gives the largest block size that can be used:

$$Q \leq \frac{R D_{\max}}{0.5 + \delta} \tag{4.22}$$

Note that we also need to round Q computed from equation (4.22) to integral multiples of disk sector size.

4.5 Overflow Management

The previous sections focus on the cases when there is no overflow. In this section, we turn the focus to operational issues in implementing soft scheduling and tackle the issue of overflow detection and recovery.

4.5.1 Deadline-Driven Detection

The analytical models in Sections 4.2 and 4.3 give the probability of experiencing a round overflow during system operation. This means that overflow can eventually occur and so the system must detect any overflow condition and take corrective actions. To illustrate, consider the scenario in Figure 4.3 where overflow occurs in round i. Consequently, transmission for the last block retrieved in round i cannot proceed normally as it has missed the transmission cycle. Moreover, schedule for the next round is also delayed, further increasing the likelihood of overflow in subsequent rounds. Clearly, we need to contain the problems caused by overflow to prevent overflow propagation.

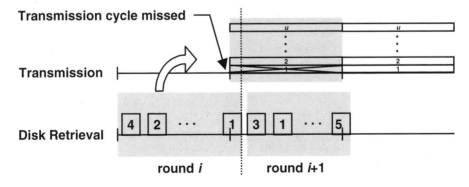

Figure 4.3   An overflow in round i can delay the start of the next round, increasing the likelihood of overflow in round $i + 1$

To detect round overflow, we need to compare the round length with the limit T_r. It may appear that the exact length of any disk round can be computed from

$$t_{round} = \sum_{j=0}^{u-1} \left(f_{seek}(v_{j+1} - v_j) + \frac{Q}{r_{j+1}} + t_{latency}^{j+1} \right) + f_{seek}(N - v_u) \tag{4.23}$$

given that we know *a priori* the locations of all the data blocks to be retrieved.

Unfortunately, the exact rotational latency cannot be computed in advance and consequently the round length remains a random variable. Therefore, rather than computing the exact round length, we can employ a Deadline-Driven Detection algorithm to detect overflow. Specifically, the time to retrieve the ith block in a round, denoted by $t_{request}(i)$, is given by

$$t_{request}(i) = f_{seek}(v_i - v_{i-1}) + \frac{Q}{r_i} + t_{latency}^i \tag{4.24}$$

Note that except for the latency term $t_{latency}^i$, all the other terms in equation (4.24) are known. Let t_{due} be the time when the current service round must end, and let d_i be the time to start retrieving block i. Then, to prevent overflow in reading block i, we must ensure that:

$$t_{due} \geq d_i + t_{request}(i) + f_{seek}(N - v_i) \quad \forall i \tag{4.25}$$

where the last term is head-positioning time under CSCAN. Hence, we can guarantee that overflow will not occur if

$$d_i \leq t_{due} - \left(f_{seek}(v_i - v_{i-1}) + f_{seek}(N - v_i) + \frac{Q}{r_i} + \frac{1}{W} \right) \tag{4.26}$$

where we replaced the rotational latency by the worst-case of one complete round of rotation.

The retrieval deadlines d_i ($i = 1, 2 \ldots, u$) can be computed at the beginning of a service round. Before retrieving media block i, the system compares the current time t against d_i. If the deadline is exceeded (i.e., $t > d_i$), then it raises an overflow exception and proceeds with corrective action, discussed in the next section.

4.5.2 Overflow Recovery

The goal of overflow recovery is to truncate the overflowed round to within T_r seconds. One straightforward solution is to skip retrieving all those media blocks that have missed their retrieval deadlines specified in equation (4.26). To reduce data loss, instead of dropping a data block altogether, we can also retrieve a partial data block by reducing the amount of data to read from Q to Q_d such that

$$t + \left(f_{seek}(v_i - v_{i-1}) + f_{seek}(N - v_i) + \frac{Q_d}{r^i} + \frac{1}{W} \right) \leq t_{due} \tag{4.27}$$

is satisfied. Rearranging, we can then obtain the reduced block size:

$$Q_d \leq r^i \left(t_{due} - t - f_{seek}(v_i - v_{i-1}) - f_{seek}(N - v_i) - \frac{1}{W} \right) \tag{4.28}$$

Depending on the system design, the affected clients can handle data loss in several ways. First, the client could attempt recovery by means of retransmission if that is supported by the system. The effectiveness of this approach will depend on the amount of client buffers as well as the time required in performing the retransmission. Second, the client could reduce the effect of data loss by error concealment. The effectiveness will depend on the coding algorithm employed as well as the type of data lost. In practice, round overflow will be relatively rare as not only is the overflow probability small, a service operator also rarely runs the media server at 100% utilization.

4.6 Performance Evaluation

To compare the performance of soft scheduling and hard scheduling, we conducted extensive simulations using detailed disk models obtained from the DiskSim simulator project [14, 15]. We simulated five modern disk drives from three manufacturers (Quantum Atlas-III, Quantum Atlas-10K, Seagate Barracuda, Seagate Cheetah, IBM 9ES). The disk models include parameters such as seek time, rotational latency, number of disk zones, number of cylinders per zone, number of sectors per track in each zone, etc. Block sizes of 64KB, 128KB, 256KB, and 512KB are simulated for each of the five disk models.

4.6.1 Service Round Length Distribution

Figure 4.4 shows the round length distribution for the Quantum Atlas-10K disk model for round sizes of $K = 10$, 20, and 30 respectively. A remarkable observation is that the

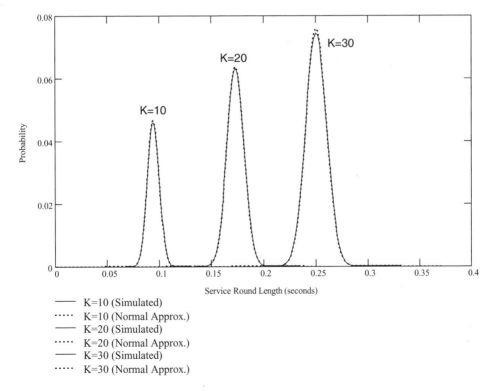

Figure 4.4 Service round length distributions and the corresponding normal approximations

distributions all resemble the normal distribution. The same observation holds for all five disk models simulated. In retrospect, this is expected since the round length is a summation of multiple random variables and hence would approach normal according to the central limit theorem.

We take advantage of this observation and use the normal distribution in place of $F_{round}(t,k)$ to compute numerical results in the following sections. As shown in Figure 4.4, the normal approximation curves closely overlap with their simulated counterparts and hence justify their use for computing numerical results.

4.6.2 Statistical Streaming Capacity

Once the round length distribution is known, we can compute the usable disk capacity from equation (4.3). The first set of results is obtained from simulation with media bit-rates of 150KB/s (e.g., MPEG-1 video). Figures 4.5a and 4.5b show the normalized gains in disk capacity versus overflow probability constraint for bit-rate of 150KB/s and block sizes of 64KB and 128KB respectively. Figures 4.6a and 4.6b show a similar set of results for bit-rate of 600KB/s (e.g., MPEG-2 video) and block size of 256KB and 512KB respectively. Note that the normalized capacity gain is defined as

$$G = \frac{C(\varepsilon) - C}{C} \tag{4.29}$$

where $C(\varepsilon)$ is the usable disk capacity under soft scheduling with overflow constraint ε and C is the usable disk capacity under hard scheduling. The lowest overflow probability constraint is set to 1×10^{-10}, equivalent to a mean-time-between-overflow of 138.5 years assuming the disk is operated continuously at full capacity 24 hours a day. Depending on the overflow probability constraint, the block size, and the particular disk model, the capacity gains ranges from around 20% to over 40%.

To further investigate the effect of media block size on capacity gains, we plot in Figure 4.7 the capacity gains versus media block sizes for media bit-rate of 150KB/s and overflow probability constraint of 10^{-6}. We observe that while the capacity gains vary according to the chosen block size, the gains remain substantial for all block sizes, with all but one case exceeding 25%.

4.6.3 Dual-Round Scheduling

To investigate the additional gains achievable using Dual-Round Scheduling, we compute the normalized additional capacity gain from

$$= \frac{C_{DRS}(\varepsilon) - C(\varepsilon)}{C(\varepsilon) - C} \tag{4.30}$$

and plot the results in Figure 4.8 for block size of 64KB and bit-rate of 150KB/s. The results clearly show that DRS can further improve capacity gains over single-round scheduling. Note that there are ups and downs in the curves due to variations of the factor $C(\varepsilon)$ in equation (4.30). We also observe that, in general, DRS is more effective for smaller overflow probability

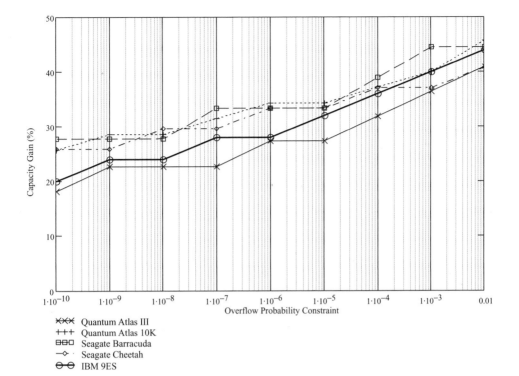

Figure 4.5a Capacity gain in soft scheduling ($Q = 64$KB, $R = 150$KB/s)

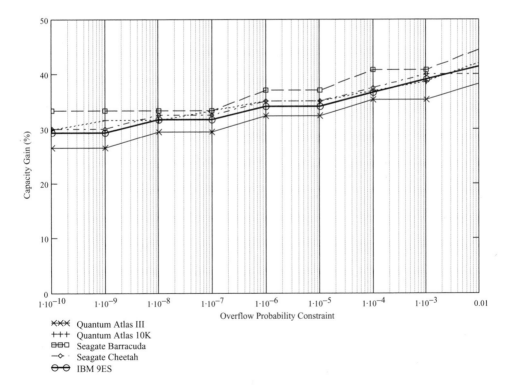

Figure 4.5b Capacity gain in soft scheduling ($Q = 128$KB, $R = 150$KB/s)

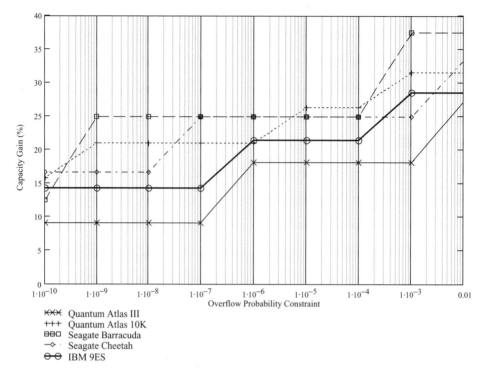

Figure 4.6a Capacity gain in soft scheduling ($Q = 256$KB, $R = 600$KB/s)

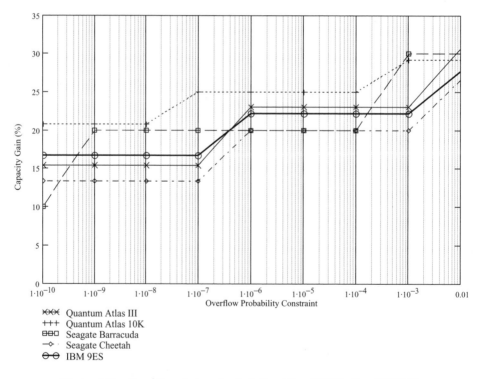

Figure 4.6b Capacity gain in soft scheduling ($Q = 512$KB, $R = 600$KB/s)

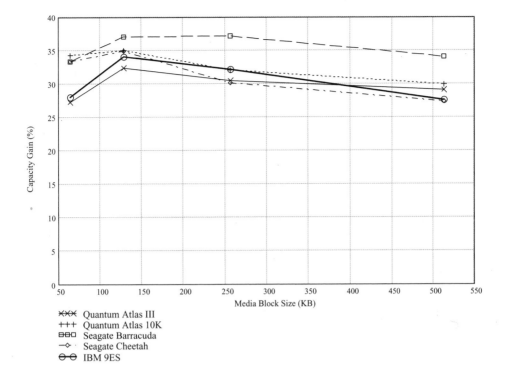

Figure 4.7 Capacity gain versus media block size ($R = 150\text{KB/s}, \varepsilon = 10^{-6}$)

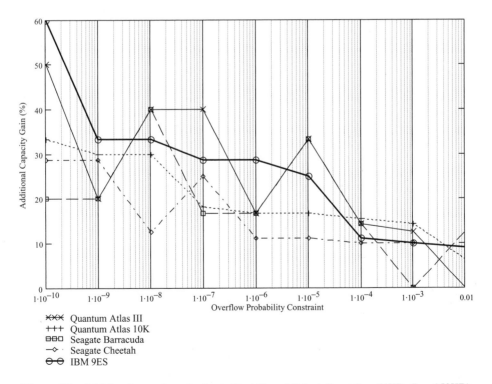

Figure 4.8 Additional capacity gain due to Dual-Round Scheduling ($Q = 64\text{KB}, R = 150\text{KB}$)

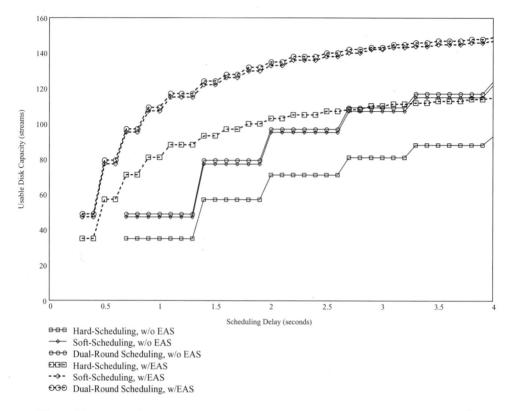

Usable Disk Capacity (streams)

Scheduling Delay (seconds)

⊟⊟⊟   Hard-Scheduling, w/o EAS
─◆─   Soft-Scheduling, w/o EAS
⊖⊖⊖   Dual-Round Scheduling, w/o EAS
⊟⊞⊟   Hard-Scheduling, w/EAS
-◇-   Soft-Scheduling, w/EAS
⊖⊝⊖   Dual-Round Scheduling, w/EAS

Figure 4.9   Usable disk capacity versus scheduling-delay constraint ($R = 150$KB, $\varepsilon = 10^{-6}$)

constraints. This is explained by the fact that it is more likely to have sufficient slack time in a round to absorb overflow when the overflow probability constraint is small. Given today's low memory cost, DRS is an attractive option for achieving better capacity at the expense of modest increase in buffer requirement.

4.6.4 Early-Admission Scheduling

To study the capacity gains from Early-Admission Scheduling, we compute the media block size according to equation (4.22), and round it down to multiples of 64KB. The usable disk capacities for various combinations are shown in Figure 4.9. The horizontal axis is the admission delay constraint used for computing the block size.

We observe that in all cases EAS can substantially increase the capacity, including both hard-scheduling and soft-scheduling cases. For example, with an admission delay constraint of one second, the usable disk capacity increases from 35 to 81 (131% increase) for hard scheduling and increases from 47 to 107 (128% increase) for soft scheduling. The improvement in DRS is similar. These dramatic increases in usable disk capacity are explained by the fact that the admission delay is reduced by two-thirds under EAS. Therefore, substantially larger block size can be used to improve disk efficiency.

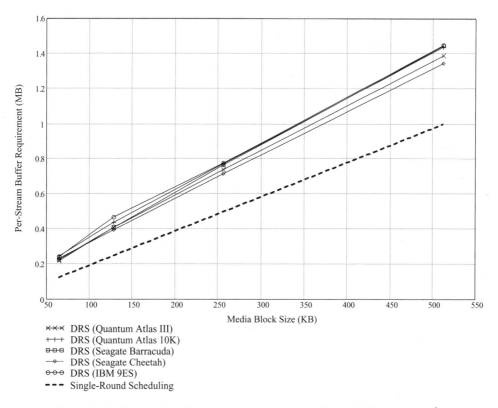

Figure 4.10   Per-stream buffer requirement comparison ($R = 150$KB/s, $\varepsilon = 10^{-6}$)

4.6.5 Buffer Requirement

Soft scheduling does not modify the way in which buffer is managed and hence has the same buffer requirement as hard scheduling. For disk scheduling algorithms such as SCAN and CSCAN, the buffer requirement will be two buffers per stream, one for disk retrieval and one for transmission. If DRS is employed, then additional buffers will be required to stage early-retrieved media blocks. Using the formulae in Section 4.3.3, we computed the per-stream buffer requirement for various media block sizes and summarized the results in Figure 4.10.

Here we have two observations. The first observation is that buffer requirement in all cases increases with larger media block size as expected. For single-round scheduling (SRS), buffer requirement is independent of disk models. Differences between disk models in the DRS case are due to differences in the service round length. The second observation is that the buffer requirement of DRS is only modestly higher than SRS. This is because additional buffers are only needed to stage retrievals whereas buffer requirement for transmission is not affected by DRS. As an illustration, a PC media server with 512MB available memory will have sufficient buffers for more than 1,000 concurrent streams (Quantum Atlas-10K, $Q = 128$KB, $R = 150$KB/s) using DRS. Hence, whether DRS should be employed would simply become a cost-effectiveness issue (i.e., memory cost versus disk costs) to be worked out by the system designer.

4.7 Related Work

The principle of statistical multiplexing is not new and has been applied in many different areas. In this section, we briefly review the related literature.

Vin *et al.* [7] proposed a statistical admission control algorithm for multimedia servers. Their admission control algorithm used disk round-length distribution and request size distribution to admit more streams than deterministic admission control algorithms. Their simulation results showed that substantially more streams can be admitted compared to the deterministic case. Instead of using empirical distributions, Chen and Little [8] took an analytical approach to derive the disk round length function and then use Chebychev's Inequality to obtain statistical bounds. By allowing a small probability of overflow, their results also showed performance gains over conventional hard-scheduling algorithms. Note that both studies assumed disks without zoning. A more recent work by Nerjes *et al.* [9] incorporated zoning into their analytical disk model and used the method of Chernoff bounds to obtain tighter statistical bounds for the SCAN scheduling algorithm. These pioneering works on statistical admission control all focused on achieving better usable disk capacity by exploiting the statistical behavior of the disk's service-round length. The key differences are in the way the disk's service-round length is modeled and in the way statistical bounds are obtained.

Other researchers have studied the disk zoning problem in isolation of the admission control problem. For example, Birk [10] proposed a data layout technique called Track-Pairing for media servers with multi-zone disks. Under Track-Pairing, a media stream is placed alternatively between tracks in the outer zones and matching tracks in the inner zones. During retrieval, tracks from both zones will be retrieved in a cycle so that a more uniform disk throughput can be obtained. Another study by Ghandeharizadeh *et al.* [11] proposed a placement algorithm called FIXB where media blocks are striped across all disk zones in a round-robin manner. Media blocks are then retrieved from every zone in a disk cycle so all zones will be utilized to contribute to the average throughput. There are also other methods such as Logical Track [12] and deadline-driven techniques [13] for tackling this disk zoning problem but these studies did not investigate statistical admission control issues in the context of continuous-media servers.

4.8 Summary

In this chapter, we presented a soft-scheduling approach to increase disk efficiency in continuous-media servers. Our results obtained from detailed simulations demonstrated that existing hard-scheduling approaches sacrifice substantial disk efficiency for scheduling simplicity. This is particularly significant for disk drives with zoning. Capacity gains in the range of 20% to 40% are achievable by soft scheduling. With the additional dual-round scheduling technique, usable disk capacity can be further increased by another 10–20% depending on disk models and system parameters. In addition, the Early-Admission Scheduling technique allows the use of much larger block size without adversely increasing the scheduling delay, thus further increasing disk efficiency. Finally, we also presented procedures for detecting and recovering from round overflow, which are necessary for practical implementations.

References

[1] J. Gemmell and S. Christodoulakis, Principles of Delay Sensitive Multimedia Data Storage and Retrieval, *ACM Transactions on Information Systems*, vol. 10(1), Jan. 1992, pp. 51–90.

[2] H.M. Vin and P. V. Rangan, Designing a Multi-User HDTV Storage Server, *IEEE Journal on Selected Areas in Communications*, vol. 11, no. 3, Jan 1993, pp. 153–164.

[3] A.N. Mourad, Issues in the Design of a Storage Server for Video-on-Demand, *ACM Multimedia Systems*, vol. (4), April 1996, pp. 70–86.

[4] P.S. Yu, M.S. Chen, and D.D. Kandlur, Grouped Sweeping Scheduling for DASD-based Multimedia Storage Management, *ACM Multimedia Systems*, vol. 1, no. 2, Dec. 1993, pp. 99–109.

[5] A.L.N. Reddy and J.C. Wyllie, I/O Issues in a Multimedia System, *IEEE Computer*, vol. 27, no. 3, Mar. 1994, pp. 69–74.

[6] D.J. Gemmell, H.M. Vin, D.D. Kandlur, P.V. Rangan, and L.A. Rowe, Multimedia Storage Servers: A Tutorial, *IEEE Computer*, vol. 28, no. 5, May 1995, pp. 40–49.

[7] H. Vin, P. Goyal, and A. Goyal, A Statistical Admission Control Algorithm for Multimedia Servers, *Proceedings of* 2nd *ACM International Conference on Multimedia '94*, San Francisco, CA, USA, Oct. 1994, pp. 33–40.

[8] H.J. Chen and T.D.C. Little, Storage Allocation Policies for Time-Dependent Multimedia Data, *IEEE Transactions on Knowledge and Data Engineering*, vol. 8, no. 5, Oct. 1996, pp. 855–864.

[9] G. Nerjes, P. Muth, and G. Weikum, Stochastic Service Guarantees for Continuous Data on Multi-Zone Disks, *Proceedings of* 16th *Symposium on Principles of Database Systems (PODS'97)*, May 1997, pp. 154–160.

[10] Y. Birk, Track-Pairing: A Novel Data Layout for VoD Servers with Multi-Zone-Recording Disks, *Proceedings of International Conference on Multimedia Computing and Systems*, May 1995, pp. 248–255.

[11] S. Gandeharizadeh, S.H. Kim, C. Shahabi, and R. Zimmermann, Placement of Continuous Media in Multi-Zone Disks, In *Multimedia Information Storage and Management* (ed. S.M. Chung), Kluwer Academic Publisher, 1996.

[12] M.F. Mitoma, S.R. Heltzer, and J.M. Menon, *Logical Data Tracks Extending Among a Plurality of Zones of Physical Tracks of One or More Disk Devices*, US Patent No. 5,202,799, April 1993.

[13] S. Ghandeharizadeh and S.H. Kim, Design of Multi-User Editing Servers for Continuous Media, *Multimedia Tools and Applications*, vol. 11, no. 1, May 2000, pp. 339–365.

[14] G.R. Ganger, B.L. Worthington, and Y.N. Patt, *The DiskSim Simulation Environment Version 2.0*. http://www.ece.cmu.edu/~ganger/disksim.

[15] B.L. Worthington, G.R. Ganger, Y.N. Patt, and J. Wilkes, On-Line Extraction of SCSI Disk Drive Parameters, *Proceedings of the ACM Sigmetrics Conference*, May 1995, pp. 146–156.

5

Reliable and Fault-Tolerant Storage Systems

In addition to streaming capacity, reliability is also an important issue in the deployment of media streaming services. In particular, a high-capacity media server will likely be equipped with a large disk array comprising many disks. Failure in any one of these disks, however, will cripple the entire media server. This is why a RAID is often employed to enable the media server to sustain the rare but possible disk failures.

Nevertheless, even though the media server can continue operation after a disk failure, the failed disk and the data it contain will eventually need to be replaced. Otherwise the media server will be susceptible to data loss in case of additional disk failures. In this chapter we address this issue and investigate rebuild algorithms to automatically rebuild data stored in a failed disk into a stand-by spare disk. The rebuild process is automatic, i.e., does not require human intervention, and is transparent to the on-going streaming service. We investigate both block-based and track-based rebuild algorithms and present buffer sharing techniques to reduce the buffer requirement. Our results show that automatic rebuild of a failed disk can be completed in a reasonable amount of time even at relatively high server utilization (e.g., less than 1.5 hours at 90% utilization), thus contributing to improve the availability of the media server.

5.1 Introduction

Since the introduction of media servers, a large number of researchers have investigated ways to improve server capacity to cope with the bandwidth requirement in delivering high-quality audio-visual contents to a large number of users. Apart from the challenge of capacity, another challenge – reliability – readily comes into the picture when companies deploy paid services to a large user population.

Specifically, a media server usually employs multiple disks in the form of a disk array for media data storage and retrieval. Media data are then distributed evenly across all the disks in small units so that data retrieval for a media stream will spread across all disks for load balancing. However, one downside of this disk-array-based storage is reliability. In particular, failure of any one of the disks in the array will render the server inoperable due to data lost. Worst still, the reliability will decrease further when one adds more and more disks to scale up the system capacity, thereby limiting the system's scalability.

This reliability problem has been investigated by many researchers in the last decade [1–8] and a number of innovative solutions have been proposed and studied. While the exact method varies, the basic principle is similar, i.e., by adding redundant data to the disks so that data lost in a failed disk can be reconstructed in real time for delivery to the client.

A media server operates in *normal mode* when there is no disk failure, and switches into *degraded mode* operation once a disk has failed. While existing solutions (e.g., using RAID) can sustain disk failure without service interruption, operating the media server under degraded mode is still a temporary measure because additional disk failures will result in system failure and permanent data loss. Therefore, the media server needs to initiate a *rebuild mode* to reconstruct data lost in the failed disk and store them on a spare disk to bring the server back to normal mode operation. Once the rebuild process is complete, the media server can sustain another disk failure without total system failure or permanent data loss. This gives the system operator more time to repair or replace the failed disk with a new spare disk.

It is worth noting that today's hard disks generally have fairly long mean-time-between-failure (MTBF) ratings, ranging from 300,000 hours to nearly 1,000,000 hours depending on the disk model. Consider a media server with 16 disks (including one parity disk) plus a spare disk. The MTBF for the disk array computed using a formula derived by Chen *et al.* [9] is over 42,000 years if the rebuild time is one hour and 4,200 years if the rebuild time is ten hours. While an MTBF of 4,200 years may appear to be sufficient, Chen *et al.* [9] also pointed out that the computed MTBF should be taken conservatively because disk failures in practice are not necessarily independent and hence the likelihood of a second disk failure could be much higher after the first disk failure. As the disk array MTBF is inversely proportional to the rebuild time, it is therefore important to quickly rebuild the failed disk to prevent total system failure.

This chapter addresses this problem by investigating efficient rebuild algorithms to rebuild the failed disk *automatically* and *transparently* in a media server serving constant-bit-rate (CBR) media streams. Automatic refers to the fact that the rebuild process does not require human intervention such as locating and loading a back-up tape to restore data. Transparent refers to the fact that the rebuild process itself can operate without any adverse effect on existing users.

The rest of this chapter is organized as follows. Section 5.2 reviews some previous works; Section 5.3 presents and formulates the system model studied in this chapter; Section 5.4 presents and analyzes a block-based rebuild algorithm; Section 5.5 presents and analyzes a track-based rebuild algorithm; Section 5.6 presents a pipelined rebuild algorithm to reduce buffer requirement in track-based rebuild; Section 5.7 compares the presented algorithms quantitatively using numerical results; and Section 5.8 summarizes the chapter and discusses some future works.

5.2 Background

The problem of supporting degraded mode of operation in media servers has been investigated by a number of researchers [1–8]. One approach makes use of data replications such as mirroring to sustain disk failure. The idea is to place two or more replicas in different disks so that at least one copy is available after a disk failure. Examples include the rotational mirrored declustering scheme proposed by Chen *et al.* [4], the doubly striped mirroring scheme proposed by Mourad [6], and the random duplicated assignment proposed by Korst [8].

Another approach makes use of parity encoding for data redundancy. A parity block together with a number of data blocks forms a parity group. The entire parity group can be reconstructed even if one of the blocks in the parity group is lost in a disk failure. Compared to replication, parity encoding generally requires less redundancy overhead, but higher buffer requirement for data reconstruction. This approach has been investigated by Tobagi *et al.* [1] in their Streaming RAID architecture, by Cohen *et al.* [3] in their pipelined disk array, by Berson *et al.* [2] in their non-clustered scheme, and by Özden *et al.* [5] in their declustered parity scheme and prefetch scheme.

In another work by Cohen and Burkhard [7], a segmented information dispersal (SID) scheme was proposed to allow fine grain trade-off between the two extremes of mirroring and RAID-5 parity encoding. Reconstruction reads under SID are contiguous, leading to better disk efficiency. The authors showed that the SID schemes match the performance of RAID-5 and schemes based on balanced incomplete block designs under normal mode, and outperforms them under degraded mode of operation.

The previous studies all focus on the normal mode and degraded mode of operation. The problem of rebuilding data in a failed disk to a spare disk in a media server has received little attention. While there are many existing studies on disk rebuild, they have all focused on data applications such as online transaction processing (OLTP) servers. Some examples are the work by Menon and Mattson [10–11], Hou *et al.* [12–13], Thomasian and Menon [14–16], Mogi and Kitsuregawa [17], and so on.

Disk rebuild in media server applications, however, differs from that of OLTP applications in two major ways. First, OLTP applications generally do not have the stringent performance requirement of a media server. In particular, performance of OLTP applications is commonly measured using response time. While shorter response time is desirable, it is not a condition for correct operation. Therefore, in disk rebuild, the focus is to balance service response time with rebuild time. For example, one can use priority scheduling in OLTP applications to give higher priority to normal requests to minimize their response time and to serve rebuild requests with the unused disk time.

By contrast, a media server has to *guarantee* the retrieval of media data according to a fixed schedule. Even a small delay beyond the schedule will result in service disruption. Consequently, the rebuild process can take place only if normal media data retrievals can still be completed on time. This requires detailed disk modeling and the use of worst-case analysis to determine exactly how much disk time can be spent on the rebuild process. Unlike rebuild algorithms for OLTP applications, the amount of disk time to spend on rebuild is determined *a priori*, given the disk parameters. Moreover, retrievals for playback data and rebuild data are scheduled to minimize disk-seek time instead of according to priority as in the OLTP case.

Second, OLTP applications commonly employ the RAID-5 striping scheme to maximize I/O concurrency [9]. On the other hand, media server applications commonly employ the RAID-3 striping scheme for reasons to be discussed in Section 5.3.1. This fundamental difference in the striping scheme, together with the inherently round-based disk scheduling algorithm employed in media servers, requires different designs for the rebuild algorithm.

5.3 System Model

In this section, we present the system model used throughout the chapter. In particular, we discuss the rationale for adopting the RAID-3 striping scheme instead of the RAID-5 striping scheme; we present a storage allocation policy and I/O scheduling algorithm based on the RAID-3 striping scheme; we present a detailed disk model; and explain a capacity dimensioning procedure.

5.3.1 Disk Redundancy

In the pioneering study by Patterson *et al.* [18], a number of striping schemes are proposed to support disk-level fault tolerance. Among them, RAID-5 is the most widely used in general data and OLTP applications. RAID-5 is designed to maximize I/O concurrency by allowing individual disks in a disk array to serve different requests simultaneously. This design choice is particularly suitable for OLTP applications as request size is generally small and performing I/Os in parallel can reduce response time.

For media servers, however, RAID-5 is less suitable for two reasons. First, media server generally uses large data block size to maximize disk throughput. Moreover, minimizing response time is less important as the disk schedulers typically operate in fixed-duration cycles. As long as a block can be retrieved within the cycle, the exact response time is irrelevant.

Second, when a RAID-5 disk array operates in degraded mode with a failed disk, significant overheads (up to 100% depending on the placement policy) will be incurred in the remaining disks because reconstructing an unavailable block requires reading corresponding blocks from the same parity group from all the remaining disks. While increases in response time due to this overhead are not critical in OLTP applications, the same overhead will destroy the performance guarantee required in a media server. This problem can be solved by performing striping at the application level instead of at the storage level (cf. Section 3.5.2), with all blocks in a parity group storing data from the same media stream. However, it is still necessary to retrieve the entire parity group to reconstruct the unavailable block, and by doing so, the RAID-5 disk array is practically operated as a RAID-3 disk array with block interleaving instead of bit interleaving. In this case, the algorithms presented in this chapter will also be applicable.

For the previous two reasons, it is more common to employ the RAID-3 striping scheme for use in media server applications. Unlike RAID-5, where each disk can serve a different I/O request, all the disks in RAID-3 participates in serving an I/O request, thereby maximizing disk throughput. More importantly, no additional overhead will be incurred by the remaining disks during degraded mode. This is because each I/O always retrieves the entire parity group from all the remaining disks and hence reconstruction can take place immediately. We will focus on this RAID-3 striping scheme in the rest of the chapter.

5.3.2 Storage Allocation and I/O Scheduling

Table 5.1 summarizes the notations used in this chapter. Let N_D be the number of disks in the server where $(N_D - 1)$ of them store data while the remaining one stores parity. The storage is divided into blocks of Q bytes as shown in Figure 5.1. Assuming the disks are homogeneous, then a parity group comprises one block at the same location from each one of the N_D disks.

Table 5.1 Summary of notations

Description	Symbol	Notes
Average media bit-rate	R_v	Parameter in bytes per second (e.g., 150,000 Bps)
Media block size	Q	Parameter in bytes (e.g. 65,336 bytes)
Length of a transmission cycle	T_r	Computed from Q/R, in seconds
Number of disks	N_D	Parameter
Disk storage capacity	G	Parameter in bytes
Number of recording surfaces in a disk	N_{suf}	Parameter in numbers
Number of tracks per recording surface in a disk	N_{trk}	Parameter in numbers
Size of a disk sector	S	Parameter in bytes
Number of zones in a disk	N_{zone}	Parameter in numbers
Number of sectors per track in zone i	Y_i	Parameter in numbers
Minimum number of sectors per track	Y_{min}	Parameter in numbers
Maximum number of sectors per track	Y_{max}	Parameter in numbers
Transfer rate in zone i	X_i	Computed, in bytes per second
Minimum transfer rate	X_{min}	Computed, in bytes per second
Disk platter rotation rate	W	Parameter in cycles per second
Fixed overhead in disk read	α	Parameter in seconds
Seek time in disk read	t_{seek}	Random variable in seconds
Rotational latency in disk read	t_{rot}	Random variable in seconds
Transfer time in disk read	t_{xfr}	Random variable in seconds
Disk seek function	$f_{seek}(n)$	Seek time for a seek distance of n tracks
Maximum number of requests that can be served in a service round	K	Computed, in number of requests
Head-switching time in disk	t_{hsw}	Parameter in seconds
Total track-to-track seek time in serving a request	t_{track}	Random variable, in seconds
Earliest completion time for retrieving track i	e_i	Random variable, in seconds
Latest completion time for retrieving track i	l_i	Random variable, in seconds
Minimum time for retrieving one track	Δ	Parameter in seconds
Deviation bound for track-retrieval completion times	D_{asyn}	Parameter in seconds
Server load/utilization	u	Variable, in number of streams
Number of active streams for a server	U	Variable, in number of streams
Server utilization	ρ	Variable, real number ranging from 0 to 1
Number of rebuild blocks retrieved by a working disk in a round	n_b	Computed, in number of media blocks
Rate at which rebuild data are retrieved from the working disks	R_{rb}	Computed, in bytes per second
Rate at which lost data are rebuilt	$R_{rebuild}$	Computed, in bytes per second
Buffer requirement for playback process	B_p	Computed, in bytes
Buffer requirement for rebuild process	B_r	Computed, in bytes
Total buffer requirement without buffer sharing	B_{sum}	Computed, in bytes
Total buffer requirement with buffer sharing	B_{share}	Computed, in bytes

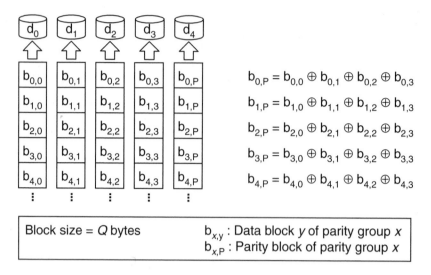

Figure 5.1   Storage organization under RAID-3 striping scheme

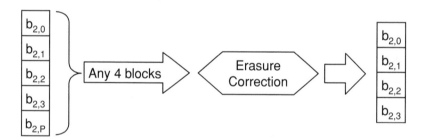

Figure 5.2   Reconstructing media data with erasure correction

The parity block is computed from the $(N_D - 1)$ data blocks using exclusive-or computation. The storage is then allocated in whole parity groups instead of individual blocks to ensure that data blocks within a parity group always store data from the same media stream. This RAID-3 striping scheme enables the system to mask the failure of any one of the N_D disks to continue operating through erasure correction processing (Figure 5.2).

Retrievals and transmissions are organized into rounds as shown in Figure 5.3. We assume that all media streams are encoded with constant-bit-rate encoding at the same bit-rate. Short-term bit-rate variations (e.g., due to I, P, B frame differences in MPEG) are assumed to be absorbed by client buffers and hence the disk can simply retrieve exactly one media block from each of the N_D disks for each active media stream in each round. Note that this can also be extended to support other bit-rates which are multiples of a base rate. These higher-rate streams can be treated as multiple base-rate streams and hence we will ignore this minor complication in the rest of the chapter.

With the previous disk scheduler, a complete parity group, including the $(N_D - 1)$ media blocks and the associated parity block are all retrieved for each stream in a service round. This

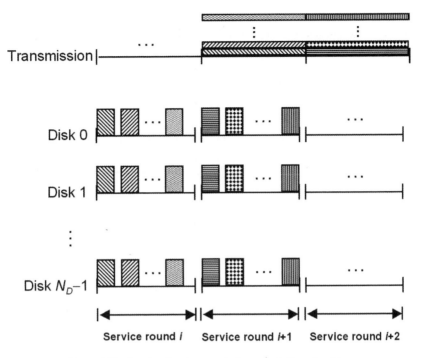

Figure 5.3   Retrieval and transmission scheduling algorithm

enables the server to sustain non-stop service even when one of the disks fails by computing the unavailable media block using erasure-correction computation over the remaining blocks in the parity group.

Let R_v be the media bit-rate. Then the retrieved $(N_D - 1)$ media blocks will be transmitted in the next service round and the service round length is thus given by

$$T_r = \frac{(N_D - 1)Q}{R_v} \tag{5.1}$$

Under this scheduling algorithm, the total number of buffers required is given by

$$B_p = KN_DQ + K(N_D - 1)Q \tag{5.2}$$

where the first term is the buffer requirement for retrieval, the second term is the buffer requirement for transmission, and K is the maximum number of requests that can be served in a service round (see Section 5.3.4). Transmission requires fewer buffers because the retrieved parity block is not transmitted and hence the buffer can be reused.

For a server with a large number of disks, the single parity disk may not provide sufficient redundancy to maintain acceptable reliability. This problem can be solved by dividing the disks into clusters where each cluster has its own parity disk (e.g., Streaming RAID [1]). Multiple disk failures can be sustained as long as no more than one disk fails in a cluster. Results discussed in this chapter can be directly extended to these clustered schemes and hence we will focus on single-cluster disk arrays in the rest of the chapter.

5.3.3 Disk Performance Model

To model the disk, we extend the disk model introduced in Chapter 3 by incorporating additional details such as head-switching time and retrieval across track boundary. Let N_{trk} be the number of tracks per recording surface (or number of cylinders), N_{suf} be the number recording surfaces, W be the disk rotation speed in rounds per second, S be the sector size in bytes, N_{zone} be the number of zones, Y_i ($i = 1, 2, \ldots, N_{zone}$) be the number of sectors per track in zone i. Note the disk transfer rate, denoted by X_i ($i = 1, 2, \ldots, N_{zone}$), is also zone dependent and is given by

$$X_i = SY_i W \tag{5.3}$$

To simplify notations in later sections, we define

$$X_{min} = \min\{X_i | i = 1, 2, \ldots, N_{zone}\} \tag{5.4}$$

$$Y_{min} = \min\{Y_i | i = 1, 2, \ldots, N_{zone}\} \tag{5.5}$$

$$Y_{max} = \max\{Y_i | i = 1, 2, \ldots, N_{zone}\} \tag{5.6}$$

and we shall leave out the subscript i in X_i and Y_i when representing random variables (i.e., X, Y) instead of system parameters.

To model disk performance, we first consider the time it takes to serve a request. Specifically, disk time for retrieving a single request can be broken down into four components, namely, fixed overhead (e.g., head-switching time, settling time, etc.) denoted by α, seek time denoted by t_{seek}, rotational latency denoted by t_{rot}, and transfer time denoted by t_{xfr}:

$$t_{req} = \alpha + t_{seek} + t_{rot} + t_{xfr} \tag{5.7}$$

Seek time depends on the seek distance and can be modeled by a seek function $f_{seek}(n)$ where n is the number of tracks to seek. For rotational latency, the random variable t_{rot} will be uniformly distributed between 0 and W^{-1}. Finally, the transfer time t_{xfr} comprises three components:

$$t_{xfr} = \frac{Q}{X} + t_{hsw} + t_{track} \tag{5.8}$$

where the first term is the time it takes to read the media block of Q bytes from the disk surface; the second term is the total head-switching time incurred if the media block spans more than one track in the cylinder; the last term is the total track-to-track seek time incurred if the media block spans more than one consecutive cylinder.

Take the Quantum Atlas-10K disk model as an example, transfer time for retrieving a 64KB media block ranges from 4.59ms to 6.69ms depending on the zone, head-switching time is 0.176ms, and track-to-track seek time is 1.25ms. Therefore, unless one sacrifices some storage (7–10% depending on zone) to prevent a media block spanning two tracks, the effect of track-crossing should not be ignored.

We note that the previous disk model is only an approximation and is chosen for the sake of simplicity. The results can be extended to more detailed and complex models for more accurate performance predictions.

5.3.4 Capacity Dimensioning

The goal of capacity dimensioning is to determine the maximum number of concurrent media streams that can be sustained with deterministic performance guarantee. Based on the previous disk model, we first obtain an upper bound for the length of a service round. Rewriting equation (5.7) by replacing t_{xfr} with the transfer time formula in equation (5.8), we have

$$t_{req} = \alpha + t_{seek} + t_{hsw} + t_{track} + t_{rot} + \frac{Q}{X} \tag{5.9}$$

Assuming the use of CSCAN serving k requests per round and the seek time function is concave, then the seek time overhead is maximized when the requests are evenly spaced along the disk surface, i.e.,

$$t_{seek}^{max} = \max \{t_{seek}\} = f_{seek}(N_{trk}/(k+1)) \tag{5.10}$$

Note the use of $(k+1)$ instead of k to account for the effect of head-repositioning delay in CSCAN [19–20].

To determine an upper bound for t_{hsw} and t_{track}, we note that a media block of size Q bytes can span at most

$$n_{hsw} = \lceil Q/(SY_{min}) \rceil \tag{5.11}$$

recording surfaces. Therefore the worst-case total head-switching time can be computed from

$$\max \{t_{head}\} = \left\lceil \frac{Q}{SY_{min}} \right\rceil t_{hsw} \tag{5.12}$$

Similarly, the same request can span at most

$$n_{cyl} = \left\lceil \frac{Q}{SY_{min} N_{suf}} \right\rceil \tag{5.13}$$

cylinders. Hence, the worst-case total track-to-track seek time can be computed from

$$\max \{t_{track}\} = \left\lceil \frac{Q}{SY_{min} N_{suf}} \right\rceil f_{seek}(1) \tag{5.14}$$

Finally, the maximum rotational latency is simply given by W^{-1} while the maximum reading time can be obtained from

$$\max \left\{ \frac{Q}{X} \right\} = \frac{Q}{X_{min}} \tag{5.15}$$

While summing the previous upper bounds will also bound the request service time, we note that most, if not all, modern disk drives support a read-on-arrival feature [3] (all five disk drive models studied in Section 5.7 support read-on-arrival). Specifically, the service time model assumes that if the disk head arrives to the track to find it passing through the middle of the media block, it will wait until the first sector of the media block rotates back to the head position before commencing reading. The read-on-arrival feature removes this restriction and allows the disk to start reading data immediately even in the middle of the requested media block. This avoids the worst-case scenario of waiting one complete rotation before reading.

Therefore, for a service round of k requests, the round length will be bounded from above by

$$t_r(k) = k \left(\alpha + t_{seek}^{max} + \left\lceil \frac{Q}{SY_{min}} \right\rceil t_{hsw} + \left\lceil \frac{Q}{SY_{min} N_{suf}} \right\rceil f_{seek}(1) + \left\lceil \frac{Q}{SY_{min}} \right\rceil W^{-1} \right) + t_{seek}^{max}$$

(5.16)

where the first term represents the worst-case time to read k requests using CSCAN and the second term is the head-repositioning time.

Reconsider the scheduling algorithm in Section 5.3.2, the server needs to ensure that a complete parity group must be retrieved within time T_r to maintain continuous transmission:

$$t_r(k) \leq T_r$$

(5.17)

This timing constraint, also known as the continuity condition, determines the maximum number of concurrent media streams, denoted by K, that can be supported by the server:

$$K = \max \left\{ k | t_r(k) \leq \frac{(N_D - 1)Q}{R_v}, k = 1, 2, \dots \right\}$$

(5.18)

5.4 Automatic Data Rebuild

A system is said to operate under normal mode when there is no failure. The system switches to degraded mode of operation once a disk failure occurs. Under this degraded mode of operation, unavailable data are recomputed in real time from the remaining disks to sustain service. Although still operational, the system must return to normal mode of operation as soon as possible because any further disk failure will cripple the entire system. The goal of data rebuild is to bring the system back to normal mode of operation by reconstructing data lost in the failed disk into spare storage hardware.

A number of modern disks and disk controllers not only can detect a disk failure, but can also predict a disk failure in advance. This early-warning signal can be used to initiate data rebuild even before the actual failure occurs. However, there are also complications that must be handled properly. First, if the data disks are updated (i.e., being written to) during the rebuild process, then the spare disk will have to be updated accordingly as well. This is less of a problem in a media server as the disks primary serve read requests. One can also disallow updating until the rebuild process completes. Second, if the actual failure occurs in a disk other than the predicted one, then the rebuild process will have to be aborted and then restarted to rebuild the disk that failed. Nevertheless, this is equivalent to the case without failure prediction and hence will not degrade rebuild performance. For simplicity, we will not make use of this failure prediction feature in the rest of the chapter.

5.4.1 Sparing Scheme

To support automatic data rebuild, a dedicated spare disk is reserved to store data reconstructed in the rebuild process. The spare disk is connected to the server at all times but is not used during normal mode and degraded mode of operation. In this sparing scheme, the recomputed data will be stored in the spare disk, which will replace the failed disk once the rebuild process is completed. Note that human intervention is still required to replace the failed disk with another spare disk to cater for another disk failure but this is less time-critical.

5.4.2 Rebuild Algorithm

The challenge of automatic rebuild is to proceed with the rebuild process without interrupting user services. Specifically, all retrievals in a disk service round must finish within T_r seconds and the addition of rebuild requests must not violate this limit. Clearly, we can only utilize unused disk capacity to serve rebuild requests. Once rebuild blocks from the surviving disks are retrieved into memory, the server can then perform an erasure-correction computation to reconstruct the lost media blocks and store them to the spare disk. This process repeats until all the media blocks lost in the failed disk are reconstructed to the spare disk, which then simply replaces the failed disk to bring the system back into normal mode of operation. The failed disk will later be replaced or repaired manually and a new spare disk will be reinserted into the system to prepare for the next rebuild cycle.

5.4.3 Analysis of Rebuild Time

A key performance metric in evaluating automatic data rebuild algorithms is rebuild time, defined as the time required to completely rebuild data in the failed disk to the spare disk. For a server with N_D disks (one of which has failed) and one spare disk, the rebuild process consists of reading $(N_D - 1)$ blocks for each parity group from the surviving $(N_D - 1)$ disks and reconstructing the lost media block for storage in the spare disk. Note that this is true even if the failed disk happens to be the parity disk because all $(N_D - 1)$ data blocks in a parity group are required to recompute the parity block for storage in the spare disk.

Let $u, 0 \le u \le K$, be the number of active streams in the server. We define a server utilization $\rho, 0 \le \rho \le 1$, as follows:

$$\rho = \frac{u}{K} \tag{5.19}$$

Now the number of rebuild blocks retrieved by a working disk in a service round, denoted by n_b, will be given by

$$n_b = K - u \tag{5.20}$$

which is the same for all disks. Given that there are $(N_D - 1)$ working disks, the rate at which rebuild data are retrieved, denoted by R_{rb}, is then given by

$$R_{rb} = \frac{(N_D - 1)n_b Q}{T_r} = \frac{(N_D - 1)n_b Q}{(N_D - 1)(Q/R_v)}$$

$$= n_b R_v \tag{5.21}$$

where the numerator is the total amount of rebuild data retrieved in a service round, and the denominator is the length of a service round. Note that if R_{rb} is only the rate at which rebuild data are retrieved, the reconstruction process will consume $(N_D - 1)$ rebuild blocks to reconstruct one lost media block. Therefore, the rebuild rate $R_{rebuild}$, defined as the rate at which lost data are reconstructed, can be computed from R_{rb}:

$$R_{rebuild} = \frac{R_{rb}}{(N_D - 1)} = \frac{n_b R_v}{(N_D - 1)} \tag{5.22}$$

Using equations (5.19) and (5.20), we can simplify equation (5.22) into

$$R_{rebuild} = \frac{(K - u)R_v}{(N_D - 1)}$$

$$= \frac{(1 - \rho)}{(N_D - 1)} K R_v \tag{5.23}$$

Equation (5.23) computes the achievable rebuild rate under a given server utilization. Let G be a disk's storage capacity. Assuming storage in the entire disk array is fully utilized, we can then calculate the rebuild time from

$$T_{rebuild} = \frac{G}{R_{rebuild}} = \frac{G(N_D - 1)}{K R_v (1 - \rho)} \tag{5.24}$$

5.4.4 Buffer Requirement

Two types of buffers are required for a media server supporting automatic rebuild. First, the server needs buffers for the normal retrieval and transmission of media blocks for playback (henceforth called *playback buffers*). Second, the server also needs buffers to support the rebuild process (henceforth called *rebuild buffers*).

The playback buffer requirement is given by equation (5.2) in Section 5.3. To determine the requirement for rebuild buffers, we consider the rebuild process depicted in Figure 5.4. Note that write operations on the spare disk are scheduled using the same periodic scheduler as read operations on the data disks. Rebuild blocks retrieved in a round are used to reconstruct the lost media blocks for writing to the spare disk in the next round. This algorithm simplifies the server implementation as the same scheduler can be used for both read and write operations.

For every unavailable media blocks reconstructed, $(N_D - 1)$ blocks must be retrieved for erasure-correction computation. Therefore, the buffer requirement for rebuild is given by

$$B_r = K(N_D - 1)Q + KQ \tag{5.25}$$

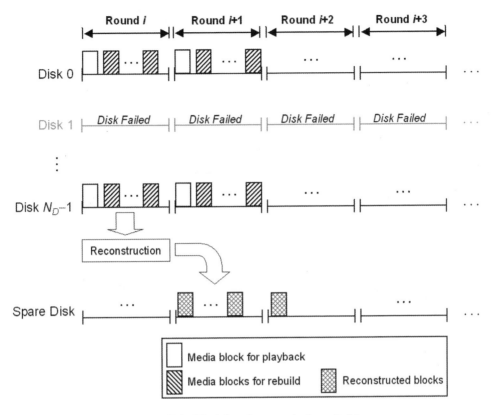

Figure 5.4 Block-based automatic data rebuild.

where the first term is the buffer requirement for retrieval and the second term is the buffer requirement for storing the reconstructed media data blocks. The multiplication factor K represents the maximum number of blocks that can be retrieved in a round for rebuild. The total buffer requirement can then be computed from the sum of equations (5.2) and (5.25):

$$B_{sum} = B_p + B_r$$
$$= K(3N_D - 1)Q \qquad (5.26)$$

It may occur to the reader that while simple in implementation, this periodic write scheduler is inefficient in buffer usage because reconstructed blocks are buffered for up to one cycle before writing to the spare disk. However, we discover that a simple buffer-sharing scheme will completely eliminate the additional buffer requirement.

Specifically, we notice that the server would need fewer buffers for playback and more buffers for rebuild when the utilization is low, and vice versa. This motivates us to investigate buffer-sharing between the retrieval process and the rebuild process. Specifically, the server can allocate a pool of, say, N_{share} buffers (each Q bytes) during initialization and then allocate the buffers to the retrieval process and the rebuild process in an on-demand manner.

Now consider the retrieval process. Given u active streams, the number of playback buffers required is given by

$$B_p(\rho) = u(2N_D - 1)Q \tag{5.27}$$

For the rebuild process, the buffer requirement when there are u active streams is given by

$$B_r = (K - u)N_D Q \tag{5.28}$$

Hence, the combined buffer requirement with buffer sharing is then given by

$$\begin{aligned} B_{share} &= u(2N_D - 1)Q + (K - u)N_D Q \\ &= ((N_D - 1)u + KN_D)\, Q \\ &\leq (2N_D - 1)KQ \quad \forall u \end{aligned} \tag{5.29}$$

which surprisingly just equals the buffer requirement for the retrieval process. Therefore, with the proposed buffer-sharing scheme one can use the same round-based disk scheduler for both reading and writing without any additional buffer.

5.5 Track-Based Rebuild

Most, if not all, modern disk drives employ zoning to increase disk storage capacity as discussed in Section 3.6. A side-effect of zoning is the variation in track size. In particular, inner tracks have less storage capacity than outer tracks. Due to this uneven track-size problem, the disk scheduler in most continuous-media server designs retrieves media units in fixed-size blocks instead of tracks. While reading the entire track can eliminate the rotational latency, the amount of buffer required to maintain a balanced disk schedule is often prohibitively large [6].

Unlike the retrieval process, the rebuild process is a non-real-time process that does not require data retrieval at a constant rate. Consequently, we can employ track-based retrieval for the rebuild process to improve rebuild performance and keep using block-based retrieval for the streaming process to maintain low buffer requirement.

5.5.1 Rebuild Algorithm

Figure 5.5 depicts the track-based rebuild algorithm. In reading data from the data disks, playback data are still retrieved in fixed-size blocks but rebuild data are retrieved in tracks. This allows the elimination of rotational latency during rebuild data retrieval.

Specifically, in block-based retrieval, the disk head must wait for the required disk sector to rotate to beneath the disk head before data transfer can begin. In the worst case where the required sector has just passed over the disk head after searching is complete, the disk will have to wait for one complete round of rotation before beginning data transfer.

By contrast, under track-based retrieval, the disk head can start data transfer as soon as seeking is completed because the entire track is to be retrieved. Clearly, the reading time is

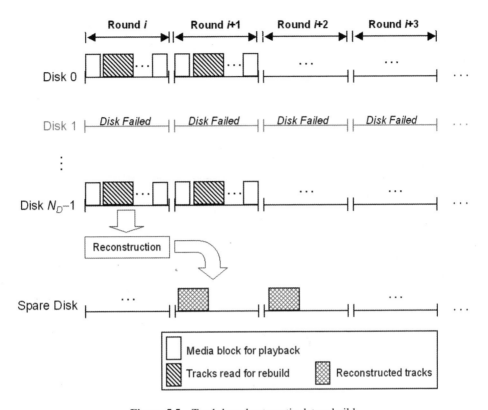

Figure 5.5   Track-based automatic data rebuild.

simply the time for one disk rotation, i.e., W^{-1}. After reading the corresponding tracks from all ($N_D - 1$) disks, the server can then reconstruct the lost track and write it to the spare disk.

Additionally, the rebuild process rebuilds tracks sequentially starting from one end of the disk surface with all track retrievals performed back-to-back in one go. For example, let $y_i, i = 0, 1, \ldots, (u - 1)$ and $y_i \leq y_j$ for $i < j$, be the track numbers for the u data blocks to be retrieved for playout in a round. Suppose that the next track to rebuild is track number x and a total of v tracks are to be rebuilt. Then the order of retrievals will be $y_0, y_1, \ldots, y_i, x,$ $y_j, \ldots, y_{u-1}$, where $y_i \leq x \leq y_j$. In other words, all v tracks are retrieved in one go between the retrievals of block i and j. Consequently, the seek time between track retrievals is reduced to $t_{seek}(1)$. The rebuild process will retrieve as many tracks as possible in a round for rebuild as long as retrieval performance for normal data blocks can still be guaranteed.

5.5.2 Analysis of Rebuild Time

To model the rebuild process, let u ($u \leq K$) be the number of media blocks to retrieve for playback and v be the number of tracks to retrieve for rebuild in a service round. Using the disk model in Section 5.3.3, the modified service round length is bounded from above by

$$
\begin{aligned}
t_r(u, v) = u \left(\alpha + f_{seek}(N_{trk}/(u+2)) + \left\lceil \frac{Q}{SY_{\min}} \right\rceil t_{hsw} + \left\lceil \frac{Q}{SY_{\min}N_{suf}} \right\rceil f_{seek}(1) \right. \\
\left. + \left\lceil \frac{Q}{SY_{\min}} \right\rceil W^{-1} \right) + f_{seek}(N_{trk}/(u+2)) + v \left(\alpha + t_{hsw} + W^{-1} \right) \\
+ \left\lceil \frac{v-1}{N_s} \right\rceil t_{seek}(1) + f_{seek}(N_{trk}/(u+2))
\end{aligned}
\tag{5.30}
$$

The first term is the service time for reading u media blocks; the second term is the additional seek time due to rebuild; the third term is the time for reading v tracks, the fourth term is the track-to-track seek time for reading rebuild tracks, and the last term is the head-repositioning delay.

Now invoking the continuity condition in equation (5.17), we can determine the maximum number of tracks that can be retrieved for rebuild given there are already u data requests in a round, denoted by $V(u)$, from

$$
V(u) = \max \left\{ v | t_r(u, v) \le \frac{(N_D - 1)Q}{R_v}, v = 0, 1, \ldots \right\}
\tag{5.31}
$$

Given a disk with N_{suf} recording surfaces and N_{trk} tracks per surface, the rebuild time can then be computed from

$$
T_{rebuild} = \frac{N_{trk}N_{suf}}{V(u)} \cdot \frac{(N_D - 1)Q}{R_v}
\tag{5.32}
$$

5.5.3 Buffer Requirement

Under track-based rebuild, tracks retrieved in a service round will be consumed by the reconstruction process to compute the lost tracks for writing to the spare disk in the next service round. With a sector size of S bytes and up to Y_{max} sectors per track, the maximum buffer requirement for rebuild can be obtained from

$$
B_r = V(0)(N_D - 1) SY_{max} + V(0) SY_{max}
\tag{5.33}
$$

where the first term is the buffer for reading from the $(N_D - 1)$ working disks and the second term is the buffer for writing to the spare disk.

Without buffer sharing, the total buffer requirement would be the sum of equations (5.2) and (5.33):

$$
\begin{aligned}
B_{sum} &= B_p + B_r \\
&= K(2N_D - 1)Q + V(0)N_DSY_{max}
\end{aligned}
\tag{5.34}
$$

Using buffer-sharing technique, we can compute the combined buffer requirement at a given server utilization from

$$
B_{share}(u) = u(2N_D - 1)Q + V(u)N_DSY_{max}
\tag{5.35}
$$

and the maximum buffer requirement can be computed from

$$B_{share} = \max \{ B_{share}(u) | u = 0, 1, \ldots, K \} \tag{5.36}$$

Intuitively, the larger the track size (i.e., SY_{max}) compared to the block size (i.e., Q), the more likely that the buffer requirement will be dominated by the rebuild process, and vice versa. In the next section, we present a novel pipelined rebuild algorithm to reduce this buffer requirement.

5.6 Pipelined Rebuild

The possibility of track-based rebuild in multi-zone disks stems from the fact that rebuild requests are non-real-time and hence can be served at variable rates. Another observation is that tracks are always retrieved sequentially to avoid seek overhead. This sequential property differs from normal data requests where the order of retrieval can change from round to round due to the CSCAN algorithm. We present in this section a pipelined rebuild algorithm to take advantage of this sequential property to reduce the buffer requirement to insignificant levels.

5.6.1 Buffer Requirement

Figure 5.6 depicts the pipelined rebuild algorithm. The scheduling algorithm for retrieving data from the data disks are the same as track-based rebuild. The difference is in scheduling the write operations to the spare disk. Specifically, tracks reconstructed from track-based rebuild are buffered until all track retrievals are completed before writing to the spare disk. By contrast,

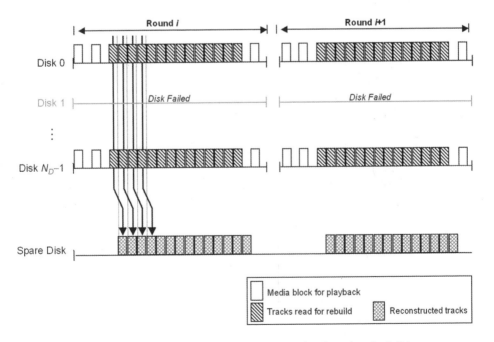

Figure 5.6   Pipelined rebuild under ideal scenario of synchronized disks

in pipelined rebuild as soon as a track is retrieved from each of the $(N_D - 1)$ surviving disks, the server will reconstruct the lost track and store it to the spare disk immediately. In this way, the track reading and writing processes operate simultaneously in a pipelined manner.

Under this pipelined rebuild algorithm, the rebuild buffer requirement is reduced to

$$B_r = (N_D - 1) SY_{\max} + SY_{\max} \tag{5.37}$$

where the first term is the buffer required for reading and the second term is the buffer required for writing.

However, the scenario in Figure 5.6 is idealized with the assumption that track retrievals for all surviving disks complete at the same instant. In practice, this is unlikely to be the case due to variations in disk rotational latencies incurred in reading media blocks prior to reading the rebuild tracks. To account for this disk asynchrony, we introduce a deviation bound D_{syn} defined as the maximum difference between the time the first track retrieval completes and the time the last track retrieval completes.

Mathematically, let $t_{i,j}$ be the retrieval completion time for reading rebuild track $i, i = 0, 1, \ldots, N_{trk}$, by disk $j, j = 0, 1, \ldots, (N_D - 1)$, as shown in Figure 5.7. We define a *track group* as the set of corresponding tracks from all $(N_D - 1)$ surviving disks that forms a parity group. For example, track group i comprises track i from each of the $(N_D - 1)$ disks.

Let e_i and l_i be the earliest completion time and latest completion time respectively for track group i:

$$e_i = \min\{t_{i,j} | \forall j\} \quad \text{and} \quad l_i = \max\{t_{i,j} | \forall j\} \tag{5.38}$$

Then D_{asyn} can be computed from

$$D_{asyn} = \max\{l_i - e_i | \forall i\} \tag{5.39}$$

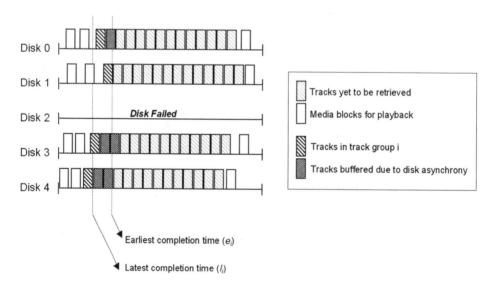

Figure 5.7   A snapshot of track retrievals at time $t = l_i$ with disk asynchrony

Let b_r be the number of buffers (each SY_{max} bytes) allocated for the rebuild process. Then at time $t = l_i$, all $(N_D - 1)$ tracks for track group i are completely retrieved. Due to disk asynchrony, some of the disks may have completed retrieving track i earlier than l_i and have started reading subsequent tracks. In particular, the earliest time for a disk to start reading tracks $i + 1$ will simply be equal to e_i. Let Δ be the minimum time for retrieving a track:

$$\Delta = \alpha + t_{hsw} + W^{-1} \tag{5.40}$$

Then a disk can retrieve up to track

$$i + \left\lceil \frac{l_i - e_i}{\Delta} \right\rceil \tag{5.41}$$

by time $t = l_i$. In the worst case, all but the last disk have performed early retrievals and the buffer requirement (in number of tracks) will be given by

$$b_r = (N_D - 1) + \left\lceil \frac{l_i - e_i}{\Delta} \right\rceil (N_D - 2) + 1 \tag{5.42}$$

where the first term is the buffers for reading track group i; the second term is the buffers for early retrievals; and the last term is the buffer for writing to the spare disk.

Finally, the maximum buffer requirement can be obtained from

$$B_r = \max \left\{ (N_D - 1) + \left\lceil \frac{l_i - e_i}{\Delta} \right\rceil (N_D - 2) + 1 \,\middle|\, \forall i \right\} SY_{max}$$

$$= \left((N_D - 1) + \left\lceil \frac{D_{asyn}}{\Delta} \right\rceil (N_D - 2) + 1 \right) SY_{max} \tag{5.43}$$

of which equation (5.37) becomes a special case of equation (5.43) with $D_{asyn} = 0$.

5.6.2 Active Disk Synchronization

In deriving the buffer requirement in equation (5.43), we assumed that disks that completed reading a track earlier than others will continue reading the subsequent tracks. While this appears to be making efficient use of disk time, it is in fact counter-productive. Unlike transaction processing (OLTP) applications, residual disk time in a continuous-media server will not be used for retrieving additional media blocks due to the periodicity of the disk schedule. Therefore, even if there is residual disk time after reading all media blocks and rebuild tracks, the disk will just sit idle until the next service round.

This observation motivates us to introduce an active disk synchronization (ADS) scheme to further reduce the buffer requirement in equation (5.43). Specifically, track retrievals for the surviving disks under ADS are actively synchronized according to the slowest disk. For example, in reading track group i, all disks will start their retrieval for track i at time $t = l_{i-1}$ instead of $t_{i-1,j}$ for disk j as shown in Figure 5.8. Note that the added delay will not affect the normal retrieval process or the rebuild process as they are dimensioned according to the worst-case scenario.

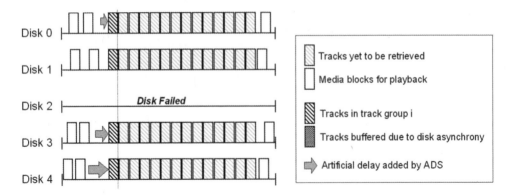

Figure 5.8   A snapshot of track retrievals at time $t = l_i$ with Active Disk Synchronization

Theoretically, with ADS the deviation-bound D_{asyn} will become zero. In practice, small deviations might still exist because the server is likely to send disk commands serially to each of the surviving disks. Assuming this deviation is small compared to Δ, then

$$\left\lceil \frac{D_{asyn}}{\Delta} \right\rceil = 1 \quad \text{for } D_{asyn} \leq \Delta \tag{5.44}$$

and the buffer requirement is reduced to

$$B_r = (2N_D - 2)\, SY_{\max} \tag{5.45}$$

5.7 Performance Evaluation

Using the performance models derived in the previous sections, we present in this section numerical results computed for five disk drive models to quantitatively compare the studied algorithms. The disks' parameters are extracted from the disk specifications in Ganger *et al.* [21] and summarized in Table 5.2. Unless stated otherwise, the results are computed using a disk array configuration of four data disks, one parity disk, and one spare disk.

5.7.1 Comparison of Rebuild Time

Figures 5.9 and 5.10 show the rebuild time versus server utilization for block-based rebuild and track-based rebuild respectively. We observe that the rebuild time increases modestly until around a utilization of 0.8, after which it increases rapidly due to the limited capacity available for rebuild. For example, rebuild time for the Quantum Atlas-10K disk increases from 44.4 minutes at $\rho = 0.5$ to 221.9 minutes at $\rho = 0.9$ for block-based rebuild. Comparing Figure 5.9 with Figure 5.10, it is clear that track-based rebuild significantly outperforms block-based rebuild. With the same disk model, the rebuild time for track-based rebuild is only 12.8 minutes at $\rho = 0.5$ and 87.7 minutes at $\rho = 0.9$.

This result is encouraging, as rebuilding a failed disk requires less than 1.5 hours even at a server utilization of 0.9. Given that a service provider is likely to dimension a system to operate

Table 5.2   Parameters of five disk models

Parameter	Quantum Atlas-III	Quantum Atlas-10K	Seagate Barracuda	Seagate Cheetah	IBM 9ES
No. of tracks	8057	10042	5172	6581	11474
No. of surfaces	10	6	5	8	5
Fixed overhead	0 ms	0 ms	0 ms	0 ms	0 ms
Head-switching time	0.999 ms	0.176 ms	0.1 ms	0.195 ms	0.062 ms
Spinning speed	7200 rpm	10025 rpm	7200 rpm	10033 rpm	7200 rpm
Sector size	512 bytes	512 bytes	512 bytes	512 bytes	512 bytes
Max track size	256 sectors	334 sectors	186 sectors	195 sectors	390 sectors
Min track size	168 sectors	229 sectors	119 sectors	131 sectors	247 sectors
Disk capacity	9.1GB	9.1GB	2GB	4.5GB	9GB
Track-to-track seek	1.663 ms	1.245 ms	1.943 ms	0.636 ms	1.086 ms

Source: Ganger *et al.* [21]

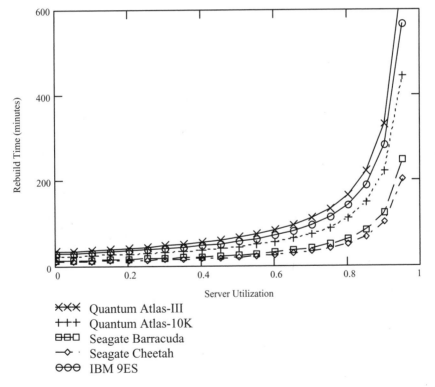

✕✕✕	Quantum Atlas-III
+++	Quantum Atlas-10K
⊟⊟⊟	Seagate Barracuda
–◇·	Seagate Cheetah
⊖⊖⊖	IBM 9ES

Figure 5.9   Rebuild time versus server utilization for block-based rebuild ($Q = 64KB$, $N_D = 5$)

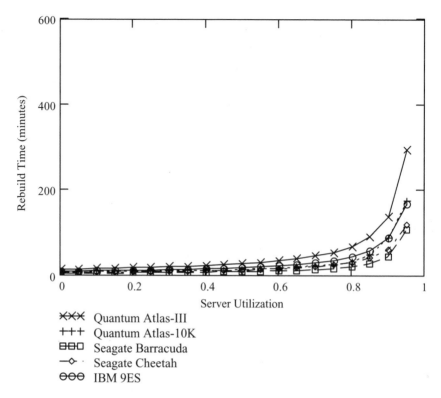

Figure 5.10   Rebuild time versus server utilization for track-based rebuild ($Q = 64KB$, $N_D = 5$)

well below such a high utilization to minimize blocking, the rebuild time in practice is likely to be even shorter.

5.7.2 Sensitivity to Server Utilization

Figure 5.11 plots the reduction in rebuild time by track-based rebuild versus server utilization. We observe that track-based rebuild consistently achieves significant rebuild-time reductions over a wide range of server utilization. This result demonstrates that performance gain of the track-based rebuild algorithm is stable with respect to server utilization.

5.7.3 Sensitivity to Media Block Size

In Figure 5.12, we plot the rebuild time versus the media block size at three server utilizations of 0, 0.25 and 0.5 respectively. We observe that rebuild time for block-based rebuild decreases with increases in media block size as larger block size increases the overall disk efficiency. By contrast, rebuild time for track-based rebuild is relatively insensitive to the media block size used as retrievals are done in whole tracks instead of blocks.

Figure 5.13 plots the reduction in rebuild time versus the media block size. As expected, the reduction decreases for increases in media block size as rebuild performance for block-based

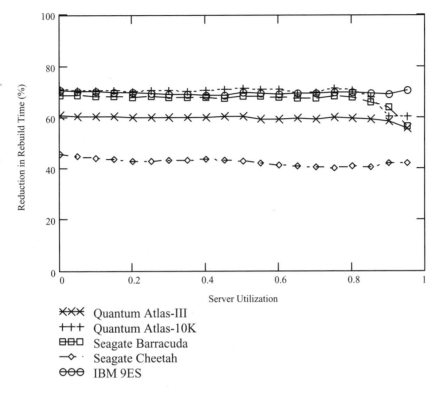

Figure 5.11   Comparison of rebuild time reduction by track-based rebuild ($Q = 64$KB, $N_D = 5$)

rebuild improves. However, even at a very large block size of 640KB, track-based rebuild still outperforms block-based rebuild by about 30%.

5.7.4 Buffer Requirement

We plot the buffer requirement for the studied algorithms versus number of disks in Figure 5.14 for the Quantum Atlas-10K disk model. We observe that track-based rebuild without buffer sharing has the largest buffer requirement as expected. However, even with buffer sharing track-based rebuild still requires more buffers than block-based rebuild. This is due to the fact that the block size (64KB) used is smaller than the track size (varies from 114.5KB to 167KB) and hence the rebuild buffers dominate the buffer requirement.

By contrast, the pipelined rebuild algorithm has only a slightly larger buffer requirement than the best scheme – block-based rebuild with buffer sharing. For a five-disk server, pipelined rebuild requires only 0.7MB to 1.5MB more buffers than block-based rebuild with buffer sharing (see Table 5.3). Note that block-based rebuild with buffer sharing is already optimal because the same server will require just as much buffer without the rebuild option. Therefore, with pipelined rebuild, we can achieve a significant gain in rebuild performance through track rebuild and at the same time avoid the large buffer requirement.

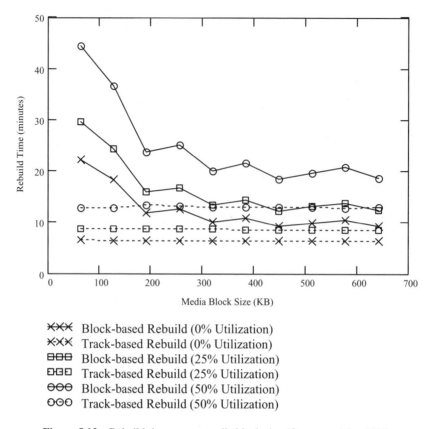

✕✕✕	Block-based Rebuild (0% Utilization)
✕✕✕	Track-based Rebuild (0% Utilization)
⊟⊟⊟	Block-based Rebuild (25% Utilization)
⊟⊟⊟	Track-based Rebuild (25% Utilization)
⊖⊖⊖	Block-based Rebuild (50% Utilization)
⊖⊖⊖	Track-based Rebuild (50% Utilization)

Figure 5.12   Rebuild time versus media block size (Quantum Atlas-10K)

5.8 Summary

In this chapter we have investigated two algorithms for rebuilding data lost in a failed disk to a spare disk automatically and transparently. We first presented a block-based rebuild algorithm derived from the conventional CSCAN disk scheduler and analyzed its performance. A buffer-sharing scheme was then introduced to eliminate the additional buffer requirement during rebuild. Next, we presented a track-based rebuild algorithm that can reduce the rebuild time by 70–80%. The large buffer requirement incurred in track-based rebuild is then reduced to insignificant levels by a novel pipelined rebuild algorithm. Numerical results show that it is feasible to completely rebuild a failed disk using the presented rebuild algorithms in a practical amount of time without causing any performance degradation to the media server.

While the algorithms presented in this chapter are designed for media servers serving CBR media streams, they can also be extended to media servers serving variable-bit-rate (VBR) media streams. One possibility is to replace fixed-size block retrievals with variable-size block retrievals, with the block size corresponding to the video bit-rate. As long as the sizes of the blocks to be retrieved in a disk round is known, we can use the same worst-case analysis as in Section 5.5 to determine how much disk time to allocate for rebuild. The rest of the rebuild process will be similar.

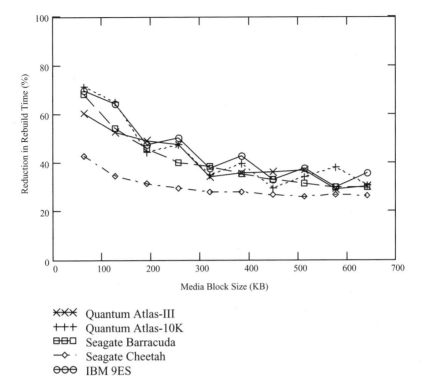

Figure 5.13   Reduction in rebuild time versus media block size ($N_D = 5$, $\rho = 0.5$)

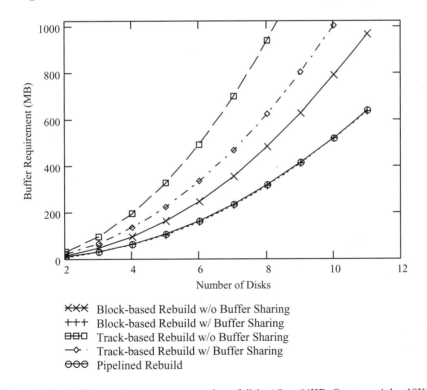

Figure 5.14   Buffer requirement versus number of disks ($Q = 64$KB, Quantum Atlas-10K)

Table 5.3   Buffer requirements ($Q = 64$KB, $N_D = 5$)

Rebuild Algorithm	Quantum Atlas-III	Quantum Atlas-10K	Seagate Barracuda	Seagate Cheetah	IBM 9ES
Block-based rebuild w/o buffer sharing	107 MB	161 MB	67 MB	175 MB	126 MB
Block-based rebuild with buffer sharing	69 MB	104 MB	43 MB	112 MB	81 MB
Track-based rebuild w/o buffer sharing	182 MB	326 MB	132 MB	244 MB	272 MB
Track-based rebuild with buffer sharing	114 MB	222 MB	89 MB	132 MB	191 MB
Pipelined rebuild	70 MB	105 MB	44 MB	113 MB	83 MB

In addition to the rebuild algorithms studied in this chapter, there are also other techniques that may further improve rebuild performance. In particular, when serving active media streams in rebuild mode, the system has to recover lost data blocks for playback purpose. By storing these already reconstructed data blocks to the spare disk, one may be able to further shorten the rebuild time.

However, storing individual data blocks to the spare disk might also adversely affect disk efficiency in track-based rebuild. First, compared to track-based rebuild, more time is spent seeking rather than data transfer in block-based rebuild. Hence, the reduction in reading from the data disks is offset by the loss in disk efficiency in the spare disk. Second, depending on the placement policy, rebuilding individual blocks may also require changes to the track-based rebuild algorithm as some tracks will have some of the blocks already reconstructed. Therefore, the performance impact is not obvious and more work is required to determine the applicability of such technique.

References

[1] F.A. Tobagi, J. Pang, R. Baird, and M. Gang, Streaming RAID^{TM}: A Disk Array Management System for Video Files, *Proceedings of ACM Conference on Multimedia '93*, Anaheium, CA, USA, Aug. 1993, pp. 393–400.

[2] S. Berson, L. Golubchik, and R.R. Muntz, Fault Tolerant Design of Multimedia Servers, *Proceedings of ACM SIGMOD International Conference on Management of Data*, San Jose, CA, USA, May 1995, pp. 364–375.

[3] A. Cohen, W.A. Burkhard, and P.V. Rangan, Pipelined Disk Arrays for Digital Movie Retrieval, *Proc. of IEEE International Conference on Multimedia Computing and Systems*, May 1995, pp. 312–317.

[4] M.S. Chen, H.I. Hsiao, C.S. Li, and P.S. Yu, Using Rotational Mirrored Declustering for Replica Placement in a Disk-Array-Based Video Server, *Proc. of ACM Multimedia '95*, San Francisco, CA, USA, Nov. 1995.

[5] B. Özden, R. Rastogi, P. Shenoy, and A. Silberschatz, Fault-tolerant Architectures for Continuous-Media Servers, *Proc. of ACM SIGMOD International Conference on Management of Data*, Montreal, Canada, June 1996, pp. 79–90.

[6] A.N. Mourad, Issues in the Design of a Storage Server for Video-on-Demand, *ACM Multimedia Systems*, vol. 4, April 1996, pp. 70–86.

[7] A. Cohen and W.A. Burkhard, Segmented Information Dispersal (SID) for Efficient Reconstruction in Fault-tolerant Video Servers, *Proc. of 4th ACM International Multimedia Conference*, Boston, USA, Nov. 1996, pp. 227–286.

[8] J. Korst, Random Duplicated Assignment: An Alternative to Striping in Video Servers, *Proc. of the 5th ACM International Conference on Multimedia*, Nov. 1997, pp. 219–226.

[9] P. Chen, E. Lee, G. Gibson, R. Katz, and D. Patterson, RAID: High-Performance, Reliable Secondary Storage, *ACM Computing Surveys*, vol. 26, no. 2, June 1994, pp. 145–186.

[10] J. Menon and D. Mattson, Performance of Disk Arrays in Transaction Processing Environments, *Proc. of 12th International Conference on Distributed Computing Systems*, June 1992, pp. 302–309.

[11] J. Menon and D. Mattson, Distributed Sparing in Disk Arrays, *Proc. of the 37th IEEE Computer Society International Conference, Compcon Spring 92*, Feb 1992, pp. 410–421.

[12] R.Y. Hou and Y.N. Patt, Comparing Rebuild Algorithms for Mirrored and RAID5 Disk Arrays, *Proc. of the 1993 ACM SIGMOD International Conference on Management of Data*, Washington, DC, USA, May 1993, pp. 317–326.

[13] R.Y. Hou, J. Menon, and Y.N. Patt, Balancing I/O Response Time and Disk Rebuild Time in a RAID5 Disk Array, *Proc. of the 26th Hawaii International Conference on System Sciences*, Jan. 1993, pp. 70–79.

[14] A. Thomasian, Performance Analysis of RAIDS Disk Arrays with a Vacationing Server Model for Rebuild Mode Operation, *Proc. of the 10th International Conference on Data Engineering*, Feb 1994, pp. 111–119.

[15] A. Thomasian, Rebuild Options in RAID5 Disk Arrays, *Proc. of the 7th IEEE Symposium on Parallel and Distributed Processing*, Oct 1995, pp. 511–518.

[16] A. Thomasian and J. Menon, RAID5 Performance with Distributed Sparing, *IEEE Transactions on Parallel and Distributed Systems*, vol. 8, no. 6, June 1997, pp. 640–657.

[17] K. Mogi and M. Kitsuregawa, Hot Mirroring: A Method of Hiding Parity Update Penalty and Degradation During Rebuilds for RAID5, *Proc. of the 1996 ACM SIGMOD International Conference on Management of Data*, Montreal Canada, June 1996, pp. 183–194.

[18] D.A. Patterson, G.A. Gibson, and R.H. Katz, A Case for Redundant Array of Inexpensive Disks (RAID), *Proc. of the ACM Conference on Management of Data*, June 1988, pp. 109–116.

[19] A.L.N. Reddy and J.C. Wyllie, I/O Issues in a Multimedia System, *IEEE Computer*, vol. 27, no. 3, Mar 1994, pp. 69–74.

[20] D.J. Gemmell, H.M. Vin, D.D. Kandlur, P.V. Rangan, and L.A. Rowe, Multimedia Storage Servers: A Tutorial, *IEEE Computer*, vol. 28, no. 5, May 1995, pp. 40–49.

[21] G.R. Ganger, B.L. Worthington, and Y.N. Patt, *The DiskSim Simulation Environment Version 2.0*. http://www.ece.cmu.edu/~ganger/disksim.

6

Media Data Streaming

In the previous chapters the focus was on retrieving media data from the disk storage to the main memory for transmission to the client. In this chapter we follow the data flow to investigate issues in streaming the media data over the network to the client hosts. We can separate the media streaming problem into two aspects: protocol and scheduling. The former covers the issues in the design of the transport/application layer protocols between the media server and the media client. Some key issues include resource identification, playback controls, media data synchronization, authentication, and digital rights management, etc. The latter covers the data transmission issues such as scheduling media data transmission to sustain continuous media playback, transmission of variable-bit-rate media streams, and adaptation of the media stream to changing network conditions.

This chapter addresses the protocol issues, discusses the feasibility of streaming media data using the existing Internet transport protocols (TCP/UDP), and gives a brief overview of the recently standardized Internet streaming protocols RTSP, RTP, and RTCP. The scheduling issues will be addressed in Chapters 7 and 8.

6.1 Streaming over TCP/UDP

Before discussing the specialized streaming protocols, let us first investigate the feasibility of using the existing Internet protocols for streaming applications. If we consider the transport layer protocols, then the Internet already supports the Transmission Control Protocol (TCP) [1] and the User Datagram Protocol (UDP) [2]. TCP is the transport protocol used by most of the Internet applications, including the WWW, FTP, telnet, and so on. It is a connection-oriented protocol that has built-in error control, flow control, and congestion control [3, 4]. In other words, TCP shields the application from much of the complexities in managing traffic flowing through the Internet. This greatly simplifies application development and TCP also possesses a desirable property – it shares network resources with other competing traffic flows in a fair manner [5]. So, given the many desirable features of TCP, the natural question is, can we simply stream media data over TCP as depicted in Figure 6.1?

The answer depends on the bandwidth requirement, network characteristics, and the desired quality of service. For example, if network bandwidth is abundant compared to the media data

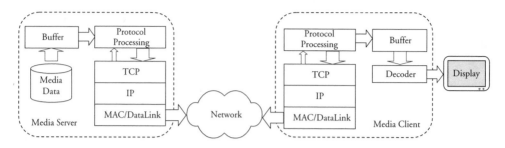

Figure 6.1 Media streaming over the Transmission Control Protocol (TCP)

rate, then streaming media data over TCP will likely work well. In fact, many web portals simply host media contents using ordinary web servers. Thus, when the client application requests the media content using the Hyper-Text Transfer Protocol (HTTP) [6], the web server simply sends back the media content over HTTP, which in turn makes use of TCP for data delivery.

In this case the web server does not explicitly *stream* the media content as it simply sends the media object as fast as TCP will allow, regardless of the media object's intrinsic data rate. The client application after receiving certain amount of data often can begin playback without waiting for the whole media object to be completely received. As long as the media data flow can keep up with the playback data rate, the end result is very much like streaming.

Streaming over HTTP/TCP has many obvious advantages. First, since the web server is used to serve media contents, the service provider will not need to invest in expensive specialized media servers. Second, deployment is simplified as the traffic is treated in the same way as ordinary web traffic, thus enabling them to transparently traverse firewalls and gateways. Third, the wide support for HTTP enhances compatibility with the client applications. Most media player software supports pseudo-streaming over HTTP/TCP protocol in addition to their own proprietary streaming protocols.

The downside of HTTP/TCP streaming, however, is in performance. At the application level, the web server is not designed to deliver time-sensitive media data and thus it may not always be able to sustain a smooth and jitter-free media playback, e.g., when the web server load is high. At the transport level, TCP's features have been developed for generic applications and thus have no provision for time-sensitive and bandwidth-sensitive applications such as streaming media.

For example, TCP's congestion control algorithm ramps up the transmission speed slowly after connection set-up (i.e., the slow-start algorithm), regardless of the bandwidth demand of the application [3, 4]. Moreover, the error control feature in TCP enforces correct and in-sequence data delivery. This means that if a TCP segment is lost, the TCP sender will simply keep retransmitting the lost segment until either an acknowledgement is received from the TCP receiver (see Figure 6.2) or it gives up trying and shuts down the connection. In a media streaming application this may not always be desirable as media data have intrinsic timing information, i.e., they have to be played back at a certain time or the data will become useless. Consequently, if the retransmitted data arrive after the playback deadline, the data will no longer be useful and are discarded by the receiver. In this case the bandwidth consumed in retransmitting the data is simply wasted.

Worst still, TCP's congestion control algorithm will interpret packet loss as an indication of network congestion and thus throttle back its transmission rate by means of reducing the congestion window size [3, 4]. This could end up stalling the sender from sending any more data

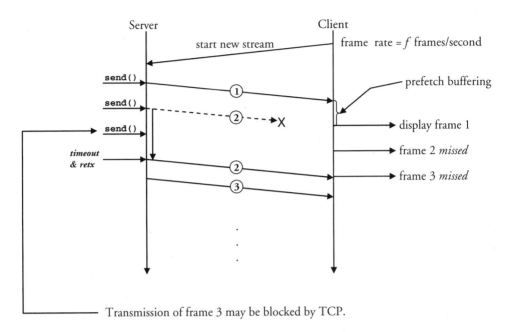

Figure 6.2   Retransmission of lost packet under TCP may cause media data to miss playback deadline at the client

until the congestion window grows back to normal after receiving a number of acknowledgements from the receiver. Again, this sender throttling will cause problems in media streaming as deferring transmission of the media data may cause them to miss the playback deadline, thus rendering them useless even if they are eventually received by the client.

The User Datagram Protocol (UDP), on the other hand, does not suffer from the problems of TCP as it is a relatively simple protocol that transfers datagrams without flow control, congestion control, or any error control at all. Therefore, the protocol itself will not introduce additional delay (ignoring processing time and packetization delay) like the flow control and congestion control algorithm in TCP, making it suitable for delivering time-sensitive media data. Nevertheless, in media streaming it is sometimes still necessary to perform flow control, to react to network congestion, as well as handling packet losses. The key is that when performing these functions the timing and bandwidth requirement of the media data must be taken into account. This can be achieved by implementing another layer of streaming protocol on top of UDP, where the streaming protocol will handle the streaming-specific functions while UDP is simply used to deliver the data and control messages. We review in the next section some of the more popular streaming protocols in the Internet.

6.2 Specialized Streaming Protocols

Over the years a number of streaming protocols have been developed both by commercial companies and the Internet community. On the commercial side, streaming solution companies often develop their own proprietary streaming protocols for use in their streaming products. For example, Microsoft developed a Microsoft Media Services (MMS) for use in its Windows

Media streaming solution. MMS employs TCP for the exchange of control messages and can send the media data over either UDP or TCP. RealNetworks also developed their own Real-Networks Data Transport (RDT) for use in their streaming solution. Because of the proprietary nature of these protocols we will not cover them further in this chapter.

On the other hand, the Internet community has also developed open standards for media streaming. This includes the Real Time Streaming Protocol (RTSP) defined in RFC 2326 [7], the Real-time Transport Protocol (RTP) and the RTP Control Protocol (RTCP), first introduced in RFC 1889, later revised in RFC 3550 [8], which became an official standard in May 2004.

6.2.1 Real-Time Streaming Protocol (RTSP)

The Real-Time Streaming Protocol (RTSP) is an application-layer protocol designed to control the delivery of media data (e.g., play, pause, and seek) with embedded timing information, such as audio and video. The protocol is independent of the lower-layer protocol. Thus, RTSP can be carried over TCP, UDP, or other transport protocols. The syntax of RTSP shares many similarities with HTTP/1.1, thus simplifying implementation and deployment. However, besides the syntax similarities, RTSP differs from HTTP in many important ways.

First, unlike HTTP, RTSP is a stateful protocol, thus requiring the host to maintain state information of a streaming session across multiple RTSP requests. Second, both the RTSP server and client can issues RTSP requests. Finally, the media data are to be delivered out-of-band, i.e., using a separate protocol such as, but not limited to, the Real-time Transport Protocol.

In a typical streaming application (see Figure 6.3), the client will first obtain a *presentation description file* using out-of-band methods (e.g., through the web using HTTP). The presentation description file describes one or more *presentations*, each composed of one or more synchronized media streams. The presentation description file also contains properties of the

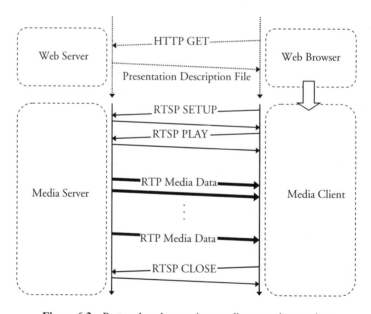

Figure 6.3 Protocol exchanges in a media streaming session

media streams, such as the encoding format, to enable the client to select and prepare for playback of the media. Each controllable media stream is identified by a separate RTSP URL, which is similar to HTTP UTL in that it identifies the server hosting the media stream and the logical path identifying the media stream. Note that the media streams in a presentation may come from multiple servers and each stream is controlled via a separate RTSP session. The interested readers are referred to RFC 2326 for the specification of the protocol.

6.2.2 Real-Time Transport Protocol (RTP)

RTP is designed for transporting data in real-time applications such as audio and video conferencing. The protocol has been designed to be independent from the lower-layer protocol which ultimately carries the RTP packets. In the Internet RTP packets are often carried over UDP datagrams, which provides multiplexing of RTP flows within the same host (using different UDP port numbers for different flows). RTP also supports data delivery over both unicast and multicast network transports (e.g., IP multicast). A control protocol – RTP Control Protocol (RTCP) – is defined as part of the standard to provide control functions such as synchronization, reception statistics reporting (e.g., loss and delay jitter), participants monitoring, etc.

It is worth noting that RTP/RTCP on their own do not provide quality-of-service control/ guarantee or perform network resource reservations. The protocols are designed to provide the necessary framework, such as header fields (sequence number, payload identification, etc.) in RTP and quality feedbacks (loss and delay jitter) in RTCP, for developers to implement their own quality-of-service mechanisms which are likely to be network- and application-specific. Thus, the RTP/RTCP protocols are often extended and integrated into the application instead of existing as standalone general purpose transport protocols like UDP and TCP.

Another point worth noting is that the standard RFC 3550 does not define how the media data is to be stored inside the RTP payload. This is specified in a profile specification in separate RFCs, such as RFC 3551 (a profile for audio and video data) [9].

Figure 6.4 depicts the header format for RTP packets. The header is divided into a 12-byte fixed header that exists in all RTP packets, and a variable part containing optional headers such

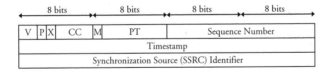

Fixed part of 12 bytes, variable optional headers.

- V is the version number (V=2 in current RTP version)
- P is padding bit, set if there are padding bytes in the payload.
- X set if there is one header extension after the fixed header.
- CC counts the number of contributing source identifiers following the fixed header.
- M used as a marker (e.g., for frame boundary), defined by a profile.
- PT is the payload type as defined in the profile.

Figure 6.4   Format of the RTP packet header

as the contributing source identifiers (typically inserted by RTP mixers), followed by the media data payload. A payload type field is included in every RTP packet to enable the sender (or a RTP mixer) to dynamically switch to a different encoding format in the middle of a session. This feature will be useful in media adaptation, e.g., switching to a lower bit-rate codec when available network bandwidth drops. Every RTP packet also includes a sequence number and a timestamp. The sequence number specifies the position of the payload within the media stream being carried by the RTP flow. Thus, the receiver can always resequence the incoming data into a proper media data stream even if out-of-order data delivery occurs in the underlying network.

If a multimedia stream contains multiple media streams, such as an MPEG system stream that includes an audio stream and a video stream, then the individual media streams should be delivered over separate RTP flows, in this case, one for audio and one for video. Note that the timestamp used in each RTP flow is not measured in real time but a sampling instant derived from a clock that increments monotonically and linearly in time. The clock frequency is application and payload format dependent. Thus, the timestamps from, say, an audio stream and a video stream may not be directly comparable. Instead, the sender will periodically send Sender Report packets using RTCP to communicate to the receivers the proper interpretation of the timestamps for synchronization purpose. There are many other features in RTP/RTCP and it is beyond the scope of this chapter to cover all the features. Interested readers are referred to RFC 3550 for more details.

The RTSP/RTP/RTCP protocols have since gained increasing support from the Internet community as well as from commercial vendors. Many commercial streaming products now also support RTSP/RTP/RTCP streaming in addition to their proprietary streaming protocols. Nevertheless vendors have kept some advanced features within their proprietary protocols, such as multi-rate-encoded media streams used in adaptive media streaming.

6.3 Summary

Media data transmission comprises two separate issues – protocol and scheduling. With the standardization of RTSP/RTP/RTCP set of protocols, more and more media streaming applications will support this set of Internet standards, thereby enhancing the inter-operability of media servers and clients from different vendors. Scheduling, on the other hand, is a more complex problem involving issues in admission control, resource allocation/reservation, and data transmission scheduling. Ideally, if the media data have a fixed playback data rate, and the network supports resource allocation (i.e., ability to allocate and guarantee a given amount of bandwidth from a source to a destination for the duration of the media streaming session), then the problem of transmission scheduling is nothing more than keeping the transmission rate the same as the playback data rate.

In practice, both assumptions may be invalid – the media playback data rate may not be constant and the available network bandwidth may also fluctuate. For example, the choice of media encoder (CBR versus VBR codec) determines if the media playback data rate will vary; and the underlying network architecture, e.g., whether it is a best-effort network or one supporting QoS guarantee, determines if the available network bandwidth will fluctuate. In Chapter 7 we investigate the scheduling problem for streaming variable bit-rate media over a mixed-traffic network that supports resource allocation/reservation. This scenario is representative of residential broadband networks operated by a service provider, who have control over the design and operation of the physical network. For best-effort networks such as the Internet, resource reservation is obviously unavailable and thus we need to approach the media streaming problem from another angle – adapting the media to the network bandwidth available, which will be covered in Chapter 8.

References

[1] J. Postel, *Transmission Control Protocol*, RFC 793, September 1981.

[2] J. Postel, *User Datagram Protocol*, RFC 768, August 1980.

[3] V. Jacobson, Congestion Avoidance and Control, *Computer Communication Review*, vol. 18, no. 4, Aug. 1988, pp. 314–329.

[4] M. Allan, V. Paxson and W. Stevens, *TCP Congestion Control*, RFC 2581, April 1999.

[5] D.M. Chiu and R. Jain, Analysis of the Increase and Decrease Algorithms for Congestion Avoidance in Computer Networks, *Computer Networks and ISDN Systems*, vol. 17, no. 1, June 1989, pp. 1–14.

[6] R. Fielding, J. Gettys, J. Mogul, H. Nielsen, and T. Berners-Lee, *Hypertext Transfer Protocol – HTTP/1.1*, RFC 2068, January 1997.

[7] H. Schulzrinne, A. Rao, and R. Lanphier, *Real Time Streaming Protocol (RTSP)*, RFC 2326, April 1998.

[8] H. Schulzrinne, S. Casner, R. Frederick, and V. Jacobson, *RTP: A Transport Protocol for Real-Time Applications*, RFC 3550, July 2003.

[9] H. Schulzrinne and S. Casner, *RTP Profile for Audio and Video Conferences with Minimal Control*, RFC 3551, July 2003.

7

Streaming Variable Bit-Rate Media Streams

Compared to streaming constant bit-rate media streams, it is substantially more complex to stream variable bit-rate media streams due to the inherent bit-rate variations and the need to maintain playback continuity at the media client. If the network supports bandwidth reservation, then the challenge is to determine how much bandwidth to reserve for streaming the VBR media stream. Obviously it will be wasteful to reserve the peak bandwidth of the VBR media stream for the entire duration of the media stream, and so it is necessary not only to reserve bandwidth, but also to adjust the amount of bandwidth from time to time to reduce bandwidth wastage. In this chapter we first introduce media bit-rate smoothing techniques to smooth out the rate variations for more efficient bandwidth utilization. Next, we investigate media streaming in mixed-traffic networks where traffics from data services as well as other continuous media streams compete for the same pool of network resources. We develop a monotonic decreasing rate scheduler that can guarantee streaming performance in mixed-traffic networks and evaluate its performance.

7.1 Introduction

Future broadband networks will support a wide variety of services with very different traffic characteristics. Among them, multimedia applications such as video-on-demand (VoD) are expected to consume a significant portion of the bandwidth. Therefore, the efficient transmission of delay-sensitive variable bit-rate (VBR) video data [1] is likely to be one of the key challenges in managing resources in such networks. Apart from the frame-by-frame bit-rate fluctuations that are also found in constant bit-rate (CBR) videos, VBR videos tend to exhibit long-range bit-rate variations in time scale of minutes. Such fluctuations complicate the admission of video streams and the scheduling of video data transmission to provide performance guarantee.

To tackle this problem, researchers have studied various ways to reduce the bit-rate variations of prerecorded VBR videos – *video bit-rate smoothing* [2–14]. Video bit-rate smoothing is a technique to reduce bit-rate variations in the retrieval and transmission of VBR encoded videos.

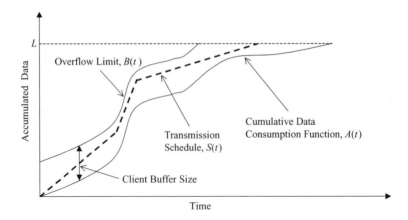

Figure 7.1   A feasible piecewise-smooth transmission schedule for VBR video delivery

The principle of smoothing is *work-ahead*, i.e., by transmitting data at a bit-rate higher than the playback bit-rate during periods of lower playback rates. Excess video data are then buffered at the client side so that the transmission bit-rate can be reduced during periods of high playback rates by consuming video data from the buffer for playback. Note that smoothing not only reduces bit-rate variations, but also reduces the peak data rate as well.

Let $A(t)$ be the cumulative data consumption function for a video (see Figure 7.1), defined as the amount of data that needs to be accumulated at the client for playback t seconds after playback starts. Let $S(t)$ be the transmission schedule for the video, defined as the amount of data transmitted to the client t seconds after playback starts. Ignoring network delay, processing delay, and interactive playback controls, it is clear that a feasible transmission schedule must not be lower than $A(t)$ for all t so that the client will not run out of video data during playback:

$$S(t) \geq A(t) \tag{7.1}$$

On the other hand, if the client buffer size is limited to, say, b bytes, then the transmission schedule cannot be too aggressive either, or else client buffer overflow will occur. This buffer constraint can be represented by a function $B(t)$ defined as

$$B(t) = A(t) + b \tag{7.2}$$

and thus to prevent buffer overflow we must ensure that

$$S(t) \leq B(t) \tag{7.3}$$

Together, the two curves $A(t)$ and $B(t)$ define the feasible region for all feasible transmission schedules:

$$B(t) \geq S(t) \geq A(t) \tag{7.4}$$

Clearly, there is an infinite number of feasible transmission schedules than can fit within the feasible region. Thus, one can pick a transmission schedule to optimize various measures of the system's performance. For example, McManus and Ross [2] suggested dividing the entire video stream into fixed size intervals and then transmitting each interval with a constant bit-rate to enable control of the rate adjustment frequencies. Feng *et al.* investigated smoothing algorithms to minimize the number of rate increases [3], and to minimize the number of rate changes [4]. In another two studies by Feng [5, 6], the author observed that in some cases, an algorithm targeted at minimizing a certain parameter might make too aggressive prefetches or allow too large buffer residency times. The author then proposed a rate-constrained smoothing algorithm [5] and a time-constrained smoothing algorithm [6] to solve these problems. Chang *et al.* suggested that transmitting at a constant rate yields lower overhead and complexity. Therefore, they proposed a smoothing algorithm that switches a single constant transmission rate on and off to adapt to the video bit-rates [7]. Salehi *et al.* investigated the optimal smoothing algorithm [8] that produces smoothing schedules with minimum peak rates and rate variations. Interested readers are referred to the excellent study by Feng and Rexford [10] for a detailed survey and comparison of various smoothing algorithms.

Besides smoothing algorithms based on finding a path inside the feasible region, there are also other related studies in this area. For example, Zhang [9] proposed smoothing using buffers located in multiple intermediate nodes in the network. Zhao and Tripathi [11] proposed an algorithm to multiplex smoothed VBR streams to further reduce bit-rate variations. Liu *et al.* [12] observed that scene changes in a video usually correlate with bit-rate variations, and thus proposed an algorithm to detect scene changes to allocate a constant bit-rate for each scene. For real-time videos, Rexford *et al.* [13] proposed an online, lossless smoothing algorithm that uses a sliding window. Liew and Tse [14] proposed using client buffer occupancy to control encoding parameters for smoother encoder output. In another study, Duffield *et al.* [15] used network status feedback to control the encoding parameters.

7.2 Streaming in Mixed-Traffic Networks

After smoothing is performed, the transmission schedule of a VBR video will be reduced to a series of constant-rate segments (see Figure 7.1). The media server can then reserve bandwidth for these segments before transmitting them over the network to the client. As long as the bandwidth reservations are successful, timely delivery of the video data to the client can be guaranteed. However, two factors in practice often affect the effectiveness of this approach.

First, although the bit-rate of each smoothed segment is constant, the system still needs to successfully complete the bandwidth reservation process before the next segment can be transmitted. The adjustments needed may contain both downward adjustments (switching from a higher bit-rate to a lower bit-rate), and upward adjustments (switching from a lower bit-rate to a higher bit-rate). The former case is straightforward as less network resource will be required, but the latter case is more complicated. In particular, if the network concurrently carries traffic from other applications (e.g., Web, FTP, other video streams, etc.), it is conceivable that the upward adjustments could fail when the additional bandwidth is not available *at that moment*. Clearly this will result in either disruption of the video stream or severe quality degradation such as playback jitter. As the *instantaneous* bandwidth consumption in a network with mixed traffics is inherently unpredictable, this problem is unavoidable unless one dedicates a portion of the network resources to a video stream. However, this clearly will result in significant over-engineering and thus defeat the whole purpose of smoothing in the first place.

Second, the processing delay of bandwidth adjustment may introduce a subtle problem. Regardless of the resource reservation protocols adopted, a sender (e.g., a video server) that desires to adjust a connection's bandwidth must first send protocol messages to one or more network controllers (e.g., routers). The network controllers may in turn need to contact other controllers along the path of the connection before the request can be granted or denied. In any case, this process will take time and the time it takes will depend on a lot of factors, such as the network topology, the reservation protocol adopted, the current utilization of the network, the number of resource-reservation requests being processed, loss of control messages, etc. The point is, not only the processing itself takes time, the time it takes also varies. This creates another problem in upward bit-rate adjustments as delay or even transmission losses may occur if an adjustment cannot be completed in time. Conceivably one can issue the upward adjustments well ahead of time to prevent delay/loss but estimating the correct lead-time is by no means trivial.

We address the two previously discussed problems in this chapter by developing a scheduler for transmitting VBR videos that can provide deterministic performance guarantee in a mixed-traffic network and is immune to random delays in processing network resource reservation requests. The principle of the scheduler, called Monotonic Decreasing Rate (MDR) scheduler, is to eliminate upward bit-rate adjustments altogether. That is, the transmission schedule is composed of a series of segments, of which each segment is assigned a bit-rate strictly lower than the previous segment. Now without the need for upward bit-rate adjustment, resource reservations are guaranteed to be successful. Moreover, the timing of the bit-rate adjustments is no longer critical as video data transmission will not be affected by a later-than-expected downward bit-rate adjustment.

Intuitively, one will expect the MDR scheduler to require more client buffer as video data are transmitted more aggressively than other smoothing algorithms. Using real-world VBR video bit-rate traces, we quantify the trade-off and show that for some video streams, the buffer requirement is indeed increased when compared to smoothing algorithms. To tackle this problem, we develop an Aggregated Monotonic Decreasing Rate (AMDR) scheduler to enable one to control the buffer requirement to the same level as smoothing algorithms. Surprisingly, simulation results show that the AMDR scheduler can achieve performance comparable to existing smoothing algorithms even when equipped with the same buffer requirement. Thus, using the AMDR scheduler, one can provide performance guarantee in streaming VBR videos over mixed-traffic networks with no trade-off in terms of admission complexity, network utilization, client waiting time, and client buffer requirement.

7.3 Monotonic Decreasing Rate Scheduler

As discussed in Section 7.2, the fundamental limitation of existing smoothing algorithms is the need for upward rate adjustments, of which correct operation depends on the successful and timely completion of network resource reservations. To remove this limitation, we can use only downward rate adjustments in the transmission schedule. In other words, the initial transmission rate will be the highest, with each subsequent rate lower than the previous one. We call this algorithm the monotonic decreasing rate (MDR) scheduler for obvious reasons.

In this chapter, we focus on prerecorded videos. The MDR schedule for a video is computed offline and is stored with the video for use during video streaming. We present an algorithm to compute the MDR schedule in the next section and derive several properties of the algorithm in Section 7.3.2 to Section 7.3.4.

7.3.1 Computing the MDR Schedules

Compared to existing smoothing algorithms, the MDR property introduces an additional constraint – only downward rate adjustments can be used. Note that although this reduces the set of possible schedules within the feasible region, it still does not uniquely determine the transmission schedule for a given video. In fact, there are still an infinite number of possible MDR schedules within the feasible region.

To select a MDR schedule, we need to consider the resultant resource requirements. The choice of the MDR schedule can affect the peak transmission rate and the client buffer requirement, both of which should be minimized. Interestingly, it turns out that we can always compute a MDR schedule that has minimum peak rate *and* minimum client buffer requirement, among all possible MDR schedules.

We define a MDR transmission schedule with the set of rate-time tuples: $\{r_i, T_i | i = 1, 2, \ldots, n\}$, where r_i and T_i are the transmission rate and commencing time for the ith segment in the transmission schedule as depicted in Figure 7.2; and n is the total number of segments in the MDR transmission schedule. For a MDR transmission schedule, the rates will be monotonic decreasing, i.e., $r_i > r_j$, for all i, j where $n \geq j > i \geq 1$.

To compute the schedule, we begin from the origin as depicted in Figure 7.2, and assign the first segment with the highest transmission rate, i.e.,

$$r_1 = \max \left\{ \left. \frac{A(t) - A(0)}{t} \right| \forall t > 0 \right\} \tag{7.5}$$

and mark the time, denoted by T_1, at which the rate is maximized. The tuple $\{r_1, T_1\}$ then represents the first segment of the MDR transmission schedule. Next, we repeat this process with T_1 as the starting point to obtain $\{r_2, T_2\}$, and so on. In general, the transmission rate for the next segment can be computed from

$$r_{i+1} = \max \left\{ \left. \frac{A(t) - A(T_i)}{t - T_i} \right| \forall t > T_i \right\} \tag{7.6}$$

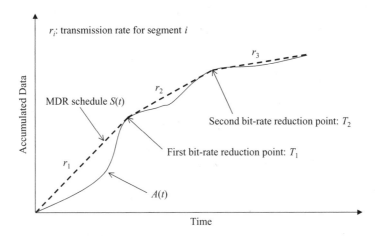

Figure 7.2   A monotonic decreasing rate schedule generated by the MDR scheduler

until it reaches the end of the video. It can be shown (see Appendix) that the above procedure guarantees that the generated transmission schedules are monotonic decreasing. The transmission schedule is then defined from the resultant rate-time tuples $\{r_i, T_i\}$:

$$S(t) = \int_0^t s(\tau)d\tau, \quad \text{where } s(\tau) = r_i, \text{ for } T_i \leq \tau < T_{i+1} \tag{7.7}$$

The rate-time tuples are then stored together with the video data. The video server will simply schedule the transmission of the video according to this MDR transmission schedule. Its monotonicity property ensures that once a stream is admitted, there will always be sufficient system bandwidth for the whole duration of the video stream, even if there are other random traffics such as web, file transfer, etc., in the system.

This MDR scheduler has several additional desirable properties, namely modest admission complexity, minimum peak rate, and minimum client buffer requirement. These are discussed in the following sections.

7.3.2 Admission Complexity

We first consider admission complexity, defined as the number of computations needed to determine if a new video stream can be admitted to a system with finite bandwidth. For existing smoothing algorithms with transmission schedules consisting of both upward and downward rate adjustments, it is necessary to check the system's bandwidth availability to determine if admitting the new stream will exceed the system capacity.

Let U be the total system capacity and $U(t)$ be the system utilization at time t. Suppose the new stream request arrives at time t_0, then the system can admit the new stream if and only if there is sufficient system capacity available for the entire duration of the new video stream, i.e.,

$$U(t) + S(t - t_0) \leq U, \quad \text{for all } t \text{ from } t_0 \text{ to } (t_0 + L) \tag{7.8}$$

In practice, we do not compute equation (7.8) in the continuous time domain as I/O schedules are likely to be organized into service rounds (e.g., disk retrieval rounds). Let δ be the round length. Then, for a video of length L, there will be $w = L/\delta$ rounds. For clarity, we refer to the system scheduler's cycles as rounds, counting from zero from the start-up of the system; and refer to a video title's data unit to be retrieved and transmitted in a round as a block, counting from zero from the beginning of the video.

Let $u_i(i = 0, 1, \ldots)$ be the system's utilization in round i, and $v_j(j = 0, 1, \ldots, w - 1)$ be the transmission rate of block j of the video, which can be computed from the transmission schedule $\{r_i, T_i\}$. Then, for a new client arriving at round A, the admission process will require w additions to compute the new aggregate bandwidth utilization for round A to round $A + w - 1$:

$$u_i = u_i + v_{i-A}, \quad i = A, A + 1, \ldots, A + w - 1 \tag{7.9}$$

and w comparisons to determine if the network capacity is exceeded in any of those w rounds. The admission complexity is then of order $O(2w)$ for a successful admission.

For an unsuccessful admission, the complexity will be lower as the admission test in equation (7.9) can be stopped once the system utilization is exceeded in any of the w rounds. In this case, the client will have to wait until the next round to repeat the admission test. This process repeats until either the client is admitted or the client leaves the system due to excessive wait. Therefore, the admission complexity is further multiplied by the waiting time.

By contrast, the MDR scheduler requires a very simple admission test with only one single computation. As MDR schedules are all monotonic decreasing, this implies that the available system bandwidth utilization u_i is a non-increasing series. It follows that if the first transmission rate can be accommodated, i.e.,

$$u_A + v_0 \leq U \tag{7.10}$$

then the rest of the schedule is guaranteed to not exceed the system capacity as well:

$$u_i + v_{i-A}, \quad i = A, A + 1, \ldots, A + w - 1$$
$$\leq u_A + v_{i-A}, \quad u_i \text{ is non-increasing}$$
$$\leq u_A + v_0, \quad v_j(j = 0, 1, \ldots, w - 1) \text{ is non-increasing}$$
$$= U, \quad (7.10) \tag{7.11}$$

If the admission is successful, then the system utilization u_i's will be updated according to equation (7.9). Otherwise, the admission test will be repeated in the next round. There is one key difference compared to general smoothing algorithms: the system utilization update needs to be performed once only, even if the client has to wait and perform multiple rounds of admission tests. This is because the admission test can be completed using equation (7.10) without computing equation (7.9) under the MDR scheduler. This leads to significantly lower complexity in the admission process.

7.3.3 Peak Transmission Rate

The first rate in a MDR schedule is the peak transmission rate. Compared to the video's data consumption function $A(t)$, this peak rate r_1 is bounded from the above by the consumption function's peak rate:

$$r_0 \leq \max \left\{ \frac{dA(t)}{dt} | \forall t \geq 0 \right\} \tag{7.12}$$

The equality will only occur when the peak rate of the cumulative data consumption function appears right at the origin. In this case, any feasible transmission schedule with zero start-up delay will have a start-up rate larger than or equal to the start-up rate of the cumulative data consumption function. This is stated in Theorem 7.1:

Theorem 7.1. *The MDR scheduler generates schedules with the minimum peak rate among all feasible schedules with zero start-up delay.*

Proof. The peak rate of a MDR schedule is the first rate r_1, so it is sufficient to prove that $S(t)$ has the smallest slope among all feasible schedules for t in $0 \leq t < T_1$. We prove this by contradiction. Let $Y(t)$ be a feasible schedule with zero start-up delay:

$$Y(t) \geq A(t) \tag{7.13}$$

and has bit-rate no larger than the MDR schedule $S(t)$ before the first bit-rate reduction point T_1, i.e.,

$$Y'(t) \leq S'(t), \quad \text{for } 0 \leq t < T_1 \tag{7.14}$$

As $S(t)$ is a straight line in the range $(0, T_1)$, equality in equation (7.14) holds only if $Y(t)$ is equivalent to $S(t)$. As we assumed they are different, that implies

$$Y'(t) < S'(t), \quad \text{for } 0 \leq t < T_1 \tag{7.15}$$

Integrating equation (7.14) on both sides with respect to t:

$$\int_0^{T_1} Y'(t)dt < \int_0^{T_1} S'(t)\,dt \tag{7.16}$$

and we obtain

$$Y(T_1) < S(T_1)$$
$$= A(T_1) \tag{7.17}$$

which implies there will be a buffer underflow at the point T_1. This contradicts our assumption that $Y(t)$ is a feasible schedule and thus the result follows.   ∎

7.3.4 Client Buffer Requirement

Similar to video smoothing, the MDR scheduler also requires the client to buffer video data ahead of their playback schedule. Given a MDR schedule $S(t)$, the buffer requirement is the maximum difference between the transmission curve $S(t)$ and the data consumption curve $A(t)$:

$$B = \max \{S(t) - A(t) | \forall t \geq 0\} \tag{7.18}$$

As discussed in Section 7.3.1, there are infinitely many feasible transmission schedules that are also monotonic decreasing. The one defined in Section 7.3.1 however, has the minimum client buffer requirement as stated in Theorem 7.2:

Theorem 7.2. *The MDR scheduler generates schedules with the minimum buffer requirement among all feasible monotonic decreasing rate schedules.*

Proof. We will prove by contradiction. Let $X(t)$ be a feasible monotonic decreasing rate schedule, i.e.,

$$X(t) \geq A(t) \tag{7.19}$$

Assume $X(t)$ has lower buffer requirement than the MDR transmission schedule $S(t)$, then

$$\exists t_0 \quad \text{such that } S(t_0) > X(t_0) \geq A(t_0) \tag{7.20}$$

We know that $S(t)$ must coincide $A(t)$ at the bit-rate reduction points, i.e.,

$$S(T_i) = A(T_i), \quad \text{for } i = 1, 2, \ldots, n \tag{7.21}$$

where n is the number of bit-rate reduction points. Now $X(t)$ cannot be lower than $S(t)$ at the bit-rate reduction points $\{T_i \mid i = 0, 1, 2 \ldots n\}$. This implies that the t_0 in equation (7.20) cannot be the bit-rate reduction points:

$$t_0 \neq T_i, \quad \text{for } i = 1, 2, \ldots, n \tag{7.22}$$

However, as $S(t)$ is constructed with straight lines connecting the bit-rate reduction points, $X(t)$ cannot be lower than $S(t)$ in between two consecutive bit-rate reduction points either:

$$t_0 \notin (T_{i-1}, T_i), \quad \text{for } i = 2, 3, \ldots, n \tag{7.23}$$

Otherwise $X(t)$ will be convex in the range (T_{i-1}, T_i), which contradicts with the assumption that $X(t)$ has monotonic decreasing rates. From equations (7.21) and (7.22), we conclude that t_0 does not exist and the result follows. ∎

7.4 Performance Evaluation

We evaluate performance of the MDR scheduler and compare it to Optimal Smoothing [1] in this section. To obtain realistic performance results, we collected the video bit-rate traces of 274 different videos from DVD movies for simulation. These are full-length (average 5,781 seconds long and 4,348 MB in size), MPEG-2 encoded videos with an average bit-rate of 6.02 Mbps. The bit-rate varies from below 0.5 Mbps to over 18 Mbps. Long-range (minutes to tens of minutes) bit-rate variations are common in these real-world MPEG-2 encoded videos.

We implemented the MDR scheduler presented in Section 7.3.1 in software and used it to compute the transmission schedule for the videos. We also implemented the Optimal Smoothing algorithm [1] for comparison purposes. The generated transmission schedules are then fed into a simulator developed using CNCL [16] to obtain simulation results.

The simulation model consists of a system with clients connecting through a 1 Gbps backbone network to a server storing the 274 VBR videos. We assume that the backbone network is the bottleneck of the system. For simplicity, we ignore delay and loss in the network. New stream requests are generated according to a Poisson process with various mean inter-arrival times to simulate different system utilization. A new stream request randomly selects a video from the 274-video collection with uniform probability. Note that we adopt the uniform popularity instead of the Zipf popularity model [17] because the video titles have varying bandwidth requirements and lengths. Consequently, using the Zipf popularity will result in large variations in the simulation results, depending on which of the video titles happened to be picked as the hot titles. To obtain more consistent results for comparison, we therefore adopt the uniform popularity model.

Table 7.1   Comparison of admission complxity between the Optimal Smoothing and the MDR schedulers

Scheduler	Unsuccessfull Admission #		Successfull Admission	
	Comparisons	Additions	Comparisons	Additions
Temporal smoothing				
complexity at $\rho = 0.6^*$	0.00 (0.00)	0.00 (0.00)	5782.52	5782.52
complexity at $\rho = 0.7^*$	0.00 (0.00)	0.00 (0.00)	5782.20	5782.20
complexity at $\rho = 0.8^*$	0.19 (0.17)	0.19 (0.17)	5782.04	5782.04
complexity at $\rho = 0.9^*$	5.32 (4.89)	5.32 (4.89)	5782.08	5782.08
MDR scheduler				
complexity at $\rho = 0.6^*$	0.00 (0.00)	0.00 (0.00)	1	5782.60
complexity at $\rho = 0.7^*$	0.00 (0.00)	0.00 (0.00)	1	5782.28
complexity at $\rho = 0.8^*$	0.17 (0.17)	0.17 (0.17)	1	5782.11
complexity at $\rho = 0.9^*$	4.83 (4.83)	4.83 (4.83)	1	5782.16

Notes: *Numerical results are measured as the average number of computations required to admit a client at a given average network utilization (ρ), averaged over requests for all 274 videos.

Numbers in parentheses are the average number of unsuccessful admission tests performed for each request.

To admit a new stream, network resource reservations are performed according to the generated transmission schedules on a per-stream basis. Admission test for schedules generated by optimal smoothing is performed with a round length of one second (cf. Section 7.3.2). Each simulation run simulates a duration of 3,000 days. We summarize the results in the following sections.

7.4.1 Admission Complexity

Table 7.1 compares the admission complexity of Optimal Smoothing and the MDR scheduler. The simulation results are obtained from counting the average number of computations required to admit a new client. We separate the computations incurred in unsuccessful and successful admissions. For unsuccessful admissions, the computation complexity is comparable for the MDR scheduler and the Optimal Smoothing scheduler. By contrast, the MDR scheduler requires significantly fewer computations than the Optimal Smoothing scheduler for successful admissions, which dominates the total admission complexity.

7.4.2 Waiting Time versus System Utilization

To evaluate the bandwidth efficiency of the MDR scheduler, we collected the mean and worst-case client waiting times for both schedulers and plot the results in Figure 7.3 a,b for three system bandwidth settings. The results show that the MDR scheduler achieves performance similar to Optimal Smoothing for all three system bandwidth settings and across system utilization from 10% to 90%. This suggests that the MDR scheduler can guarantee VBR video delivery in mixed-traffic networks with negligible trade-off in latency – the key performance metric experienced by the end users.

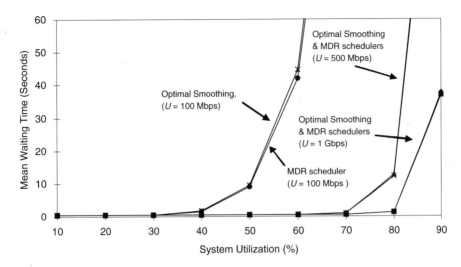

Figure 7.3a Comparison of mean waiting time versus system utilization for the Optimal Smoothing and the MDR schedulers (system bandwidth is equal to U)

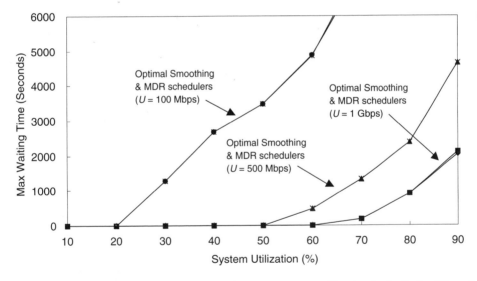

Figure 7.3b Comparison of worst-case waiting time versus system utilization for the Optimal Smoothing and the MDR schedulers (system bandwidth is equal to U)

7.4.3 Client Buffer Requirement

Although Theorem 7.2 shows that the generated MDR schedule always has the minimum buffer requirement among all monotonic decreasing rate schedules, the actual buffer requirement still depends on $A(t)$ or the bit-rate profile of the video. Figure 7.4 shows the distribution of client buffer requirements for the 274 videos tested. The average amount of client buffer required is 76.4 MB. Of the 274 tested videos, 78% require no more than 100 MB of client buffer and

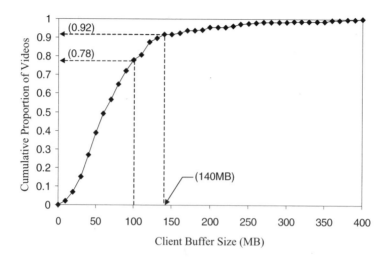

Figure 7.4   Distribution of client buffer requirement for MDR transmission schedules

92% require no more than 140 MB of client buffer. The worst-case client buffer requirement is 394.5 MB. This result is strikingly close to the 20/80 rule, also known as the Pareto's principle –78% of videos require no more than 20% (100 MB/394.5 MB) of client buffer.

With the trend towards integrating multiple information and entertainment services ranging from the Web, network gaming, to digital video recorder into a home entertainment center device, the added buffer requirement can easily be accommodated. Nevertheless, the 20/80 observation does suggest that the client buffer utilization will be low most of the time. We investigate in the next section an alternative solution that provides better control on the buffer requirement.

7.5 Aggregated Monotonic Decreasing Rate Scheduler

Results from the previous section show that the MDR scheduler can achieve performance comparable to Optimal Smoothing, and yet can provide guaranteed video delivery in a mixed-traffic network. The trade-off, as Section 7.4.3 reveals, is the increased client buffer requirement, which in a few rare cases reaches close to 400 MB. While even this amount of client buffer can easily be accommodated in future home entertainment centers with built-in hard disk, there are still two inefficiencies.

First, our results show that 78% of the video titles in our collection of 274 full-length video titles require no more than 20% of the worst-case buffer requirement. This suggests that the client buffer utilization will be low most of the time and most of the reserved buffer space will be unused.

Second, although video titles with exceedingly large client buffer requirements are rare, the MDR scheduler cannot prevent such cases as the buffer requirement depends on the individual video's bit-rate profile. In other words, the worst-case buffer requirement is, in theory, unbounded.

To tackle these two deficiencies, we introduce in this section an Aggregated Monotonic Decreasing Rate (AMDR) scheduler that applies the MDR principle to aggregated network flows so that relaxed transmission schedules can be used to accommodate those rare video titles that otherwise require very large buffer requirements.

7.5.1 Bandwidth Reservation

The large buffer requirement in those rare videos is a result of the MDR scheduler's monotonicity property. However, as we proved in Section 7.3.4, the MDR scheduler already achieves minimum buffer requirement among all monotonic decreasing rate schedules, implying that the only way to reduce the buffer requirement is to relax the monotonicity requirement.

Under the AMDR scheduler, the client buffer requirement, say B, is specified by the service provider as a design parameter. For videos with buffer requirements smaller than or equal to B, they are delivered using the original MDR schedules. By contrast, for videos that have buffer requirements larger than B, they are delivered using buffer-constrained schedules generated using temporal smoothing algorithms such as Optimal Smoothing.

Now obviously temporal smoothing in general does not guarantee monotonicity and this implies that we can no longer guarantee video delivery for these videos in a mixed-traffic network. To tackle this problem, we can over-allocate bandwidth for these video streams such that the *reserved bandwidth allocations* are kept monotonic decreasing. For example, consider the transmission schedule $\{r_i, T_i \mid i = 1, 2, \ldots, n\}$ generated by a temporal smoothing algorithm with buffer constraint B. For the rare videos, there exists i, j such that $r_i < r_j$ for $i < j$. We can maintain the monotonicity property by applying the following procedure to the transmission schedule:

$$s_i = \begin{cases} r_i, & \text{if } r_i > r_{i+1} \\ r_{i+1}, & \text{otherwise} \end{cases}, \quad i = n - 2, n - 3, \ldots, 1 \qquad (7.24)$$

to obtain the bandwidth reservation schedule $\{s_i, T_i \mid i = 1, 2, \ldots, n\}$ as shown in Figure 7.5. Note that the transmission schedule is not changed, only the bandwidth reservations are modified to maintain the monotonicity property, albeit at the expense of some unused network bandwidth.

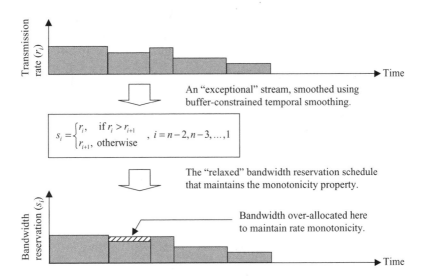

Figure 7.5 Preserving the monotonicity property by over-allocating bandwidth

To reduce the inefficiency due to bandwidth over-allocation, we observe that a video server often serves many video stream simultaneously. The data for these video streams typically go through the same backbone network before reaching the access networks. For a network link carrying more than one video stream, if we can ensure that the aggregate traffic conforms to the monotonicity property, the delivery of the individual streams is also guaranteed, even with mixed network traffics. In this case, we are applying the MDR principle to the aggregate traffic flow instead of individual video streams.

When a video stream with a transmission schedule generated by the MDR scheduler is admitted to the system, say at round A, we simply add the transmission schedule to the aggregate bandwidth utilization to obtain the new system utilization:

$$u_i = u_i + v_{i-A}, \quad i = A, A+1, \ldots, A+w-1 \tag{7.25}$$

The bandwidth reservation schedule will then be set equal to the system utilization, i.e., $s_i = u_i$.

On the other hand, when a video stream with transmission schedule generated by the Optimal Smoothing algorithm is admitted to the system at round A, we will need to perform an additional step to maintain monotonicity for the aggregated bandwidth reservations. We first compute the system utilization using equation (7.25). Then we apply a procedure similar to equation (7.24) to compute a MDR bandwidth reservation schedule by over-allocations:

$$s_i = \begin{cases} u_i, & \text{if } u_i > u_{i+1} \\ u_{i+1}, & \text{otherwise} \end{cases}, \quad i = A+w-2, A+w-3, \ldots, A \tag{7.26}$$

as shown in Figure 7.6. Again, the bandwidth over-allocations only affect the amount of network resources reserved. The individual video stream's transmission schedule is not affected.

7.5.2 Admission Complexity

The admission complexity of the AMDR scheduler depends on whether the requested video is delivered using a MDR transmission schedule or an Optimal Smoothing transmission schedule. For the MDR case, the admission complexity is the same as in the original MDR scheduler, i.e., one computation for the admission test, and $O(w)$ computations for updating the system utilization series.

For the Optimal Smoothing case, the admission complexity is higher than the MDR case but, interestingly, lower than the original Optimal Smoothing case. This is because in the AMDR scheduler, the bit-rates in the bandwidth reservation schedule $\{s_i\}$ are non-increasing. This enables the system to perform the admission test by checking only the initial rate and the rate-increasing rounds.

Again, assume the client arrives at time slot A, with a transmission schedule $\{v_i\}$. We define a round i in the transmission schedule as rate increasing if $v_i > v_{i-1}$. Let there be g such rate increasing rounds, with the round number denoted by $h_i, i = 1, 2, \ldots, g$. To simplify notations, we also define $h_0 = 0$ to represent the initial round. With these notations we can then define the admission test as:

$$s_{h_i+A} + v_{h_i} \le U, \quad \text{for } i = 0, 1, \ldots, g \tag{7.27}$$

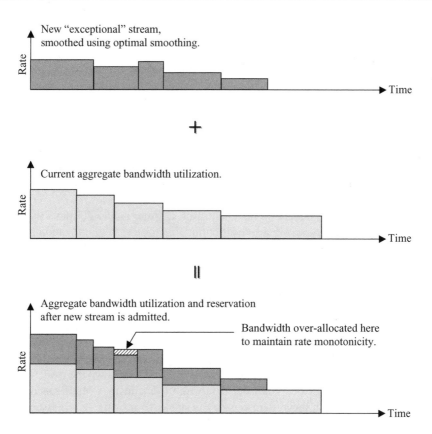

Figure 7.6   Preserving the monotonicity property in the Aggregated MDR scheduler

The monotonicity property of the bandwidth reservation schedule implies that

$$s_{i+A} + v_i \leq U \Rightarrow s_{i+A+1} + v_{i+1} \leq U \tag{7.28}$$

Starting with $i = h_0, h_1, \ldots, h_g$, and then applying equation (7.28) recursively we can show that if a new stream's transmission schedule satisfies equation (7.27), the whole transmission schedule can be added to the bandwidth reservation schedule without exceeding the system capacity.

The admission test therefore requires $(g + 1)$ additions and comparisons, instead of the w $(w >> g)$ additions and comparisons in the original temporal smoothing case. Once the admission test is successful, then the new stream's transmission schedule will be added to the aggregate system utilization using equation (7.25) and then the system can compute the new bandwidth reservation schedule according to equation (7.26).

Assume a proportion of $\alpha(0 \leq \alpha \leq 1)$ of the video collection can be admitted using MDR schedules under a given client buffer size constraint. Then, for successful admissions, the admission complexity is equal to $O(1 + (1 - \alpha)(g + w))$ comparisons and $O(w)$ additions. For unsuccessful admissions, again the complexity will be lower as the admission test is stopped as soon as the system utilization is exceeded in a time slot.

7.5.3 Performance Evaluation

We evaluate the AMDR scheduler's performance in this section using simulation. The simulation set-up is identical to the one in Section 7.4 except for two differences: (a) the MDR scheduler is replaced by the AMDR scheduler; and (b) the client buffer size is fixed.

Table 7.2 shows the admission complexity for the AMDR scheduler and the Optimal Smoothing scheduler. Comparing the results with those in Table 7.1, we observe that the AMDR scheduler requires more computations than the MDR scheduler. This is because we have set a client buffer size constraint of 32 MB for the AMDR scheduler and consequently a proportion of the videos are scheduled using the Optimal Smoothing scheduler, which requires more admission computations. Nevertheless, the resultant admission complexity is still less than the Optimal Smoothing scheduler for both successful and unsuccessful admissions.

Next we investigate the impact on the client waiting time. Figure 7.7 plots the mean and worst-case client waiting times versus client buffer size for a system utilization of 90%. We also simulated lower system utilization settings of 60% to 80% but both schedulers perform nearly identically and so the results are not shown here. From Figure 7.7, we observe that with smaller buffer sizes, both AMDR and Optimal Smoothing achieve similar waiting time. At larger buffer sizes, AMDR slightly outperforms Optimal Smoothing and ultimately converges to the MDR curve that has no buffer size constraint. The performance difference between AMDR and Optimal Smoothing is due to the fact that MDR schedules are more aggressive at the beginning of the video stream, where the transmission bit-rate is highest. This results in more work-ahead as compared to Optimal Smoothing and thus the MDR scheduler is able to utilize any unused bandwidth to reduce the bit-rate requirements down the road.

As the AMDR scheduler over-allocates bandwidth to maintain a MDR schedule, it may become less efficient when the system capacity is small. To investigate this issue, we repeat

Table 7.2 Comparison of admission complexity between the Optimal Smoothing and the AMDR schedulers (client buffer size fixed at 32MB)

Scheduler	Unsuccessfull Admission #		Successfull Admission	
	Comparisons	Additions	Comparisons	Additions
Temporal smoothing				
complexity at $\rho = 0.6^*$	0.00 (0.00)	0.00 (0.00)	5782.52	5782.52
complexity at $\rho = 0.7^*$	0.01 (0.00)	0.01 (0.00)	5782.20	5782.20
complexity at $\rho = 0.8^*$	4.15 (0.20)	4.15 (0.20)	5782.04	5782.04
complexity at $\rho = 0.9^*$	74.00 (5.15)	74.00 (5.15)	5782.08	5782.08
AMDR scheduler				
complexity at $\rho = 0.6^*$	0.00 (0.00)	0.00 (0.00)	5008.15	5781.70
complexity at $\rho = 0.7^*$	0.00 (0.00)	0.00 (0.00)	5007.47	5781.38
complexity at $\rho = 0.8^*$	0.20 (0.20)	0.20 (0.20)	5007.15	5781.22
complexity at $\rho = 0.9^*$	5.19 (5.17)	5.19 (5.17)	5007.08	5781.26

Notes: *Numerical results are measured as the average number of computations required to admit a client at a given average network utilization (ρ), averaged over requests for all 274 videos.

Numbers in parentheses are the average number of unsuccessful admission tests performed for each request.

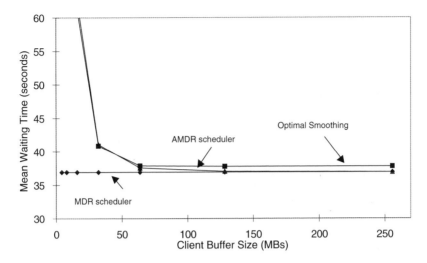

Figure 7.7a Mean client waiting time versus client buffer size
Note: The MDR scheduler does not have a buffer constraint.

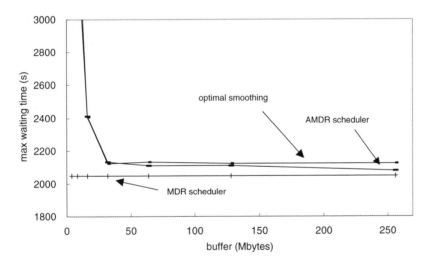

Figure 7.7b Worst-case client waiting time versus client buffer size
Note: The MDR scheduler does not have a buffer constraint.

the simulations for a range of system capacity from 100 Mbps to 1 Gbps and plot the mean and worst-case waiting times in Figure 7.8. Comparing different schedulers, we observe that AMDR and Optimal Smoothing have nearly identical performance while the MDR scheduler consistently achieves lower waiting time. This is expected, as the MDR scheduler is not subject to buffer size constraint, which in this case equals to 32 MB. Comparing the waiting time against the system bandwidth, we can see that the waiting time increases significantly at lower system bandwidth settings. Nevertheless, the differences between the AMDR scheduler and

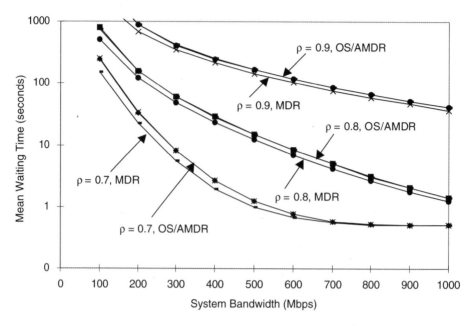

Figure 7.8a   Mean client waiting time versus system capacity

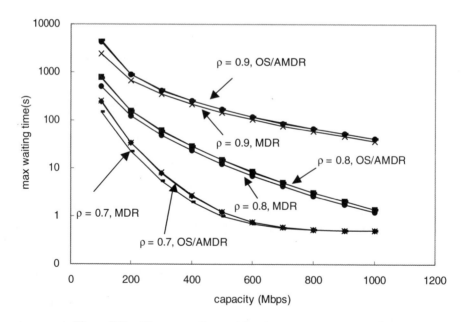

Figure 7.8b   Worst-case client waiting time versus system capacity

the optimal smoothing scheduler are negligible even for extremely small system bandwidth (e.g., 100 Mbps). This shows that the overhead incurred in maintaining a MDR schedule in the AMDR scheduler is negligible.

7.6 Summary

By scheduling the transmission of video data in a monotonic-decreasing manner, we can deliver VBR videos in a mixed-traffic network with deterministic performance guarantee. This enables the service provider to exploit the available bandwidth to support other non-delay-sensitive data services and thus improves network utilization. Extensive simulations using 274 real-world VBR video bit-rate traces showed that the MDR scheduler can achieve good performance in terms of waiting time under the same network utilization, and is comparable to that of Optimal Smoothing, while still be able to guarantee playback continuity. For applications that require a bounded client buffer requirement, the AMDR scheduler can be applied and results showed that the performance is nearly identical to Optimal Smoothing even for a buffer size as small as 32 MB. Thus, using the AMDR scheduler one can provide performance guarantee in streaming VBR videos over mixed-traffic networks with no trade-off in terms of admission complexity, network utilization, client waiting time, and client buffer requirement.

Appendix
Proof of MDR Scheduler's Monotonicity Property

Theorem 7.3. *Transmission schedules generated by the MDR scheduler are guaranteed to comprise monotonic decreasing rates.*

Proof. We prove the theorem by contradiction. Let r_i and r_{i+1} be the transmission rate of the ith and $(i + 1)$th segments of a feasible schedule $S(t)$ generated by the MDR scheduler. Graphically, let $\dot{r}$ be the slope of the line connecting $S(T_{i-1})$ and $S(T_{i+1})$, where T_{i-1} and T_{i+1} are the $(i - 1)$th and $(i + 1)$th rate reduction points.

Assume $r_i < r_{i+1}$, i.e., the rate allocated are not monotonic decreasing. Then we have:

$$\frac{S(T_i) - S(T_{i-1})}{T_i - T_{i-1}} < \frac{S(T_{i+1}) - S(T_i)}{T_{i+1} - T_i} \tag{7.29}$$

or

$$[S(T_i) - S(T_{i-1})](T_{i+1} - T_i) < [S(T_{i+1}) - S(T_i)](T_i - T_{i-1}) \tag{7.30}$$

We expand equation (7.30) to obtain.

$$S(T_i)T_{i+1} - S(T_i)T_i - S(T_{i-1})T_{i+1} + S(T_{i-1})T_i$$
$$< S(T_{i+1})T_i - S(T_{i+1})T_{i-1} - S(T_i)T_i + S(T_i)T_{i-1} \tag{7.31}$$

We cancel the $S(T_i)T_i$ term on both sides and after rearranging we obtain:

$$S(T_i)T_{i+1} - S(T_{i-1})T_{i+1} - S(T_i)T_{i-1} < S(T_{i+1})T_i - S(T_{i+1})T_{i-1} - S(T_{i-1})T_i \tag{7.32}$$

Add $S(T_{i-1})T_{i-1}$ to both sides and we obtain:

$$S(T_i)T_{i+1} - S(T_{i-1})T_{i+1} - S(T_i)T_{i-1} - S(T_{i-1})T_{i-1}$$
$$< S(T_{i+1})T_i - S(T_{i+1})T_{i-1} - S(T_{i-1})T_i - S(T_{i-1})T_{i-1} \quad (7.33)$$

Then factorize equation (7.33) to get:

$$[S(T_i) - S(T_{i-1})](T_{i+1} - T_{i-1}) < [S(T_{i+1}) - S(T_{i-1})](T_i - T_{i-1}) \quad (7.34)$$

which is equivalent to

$$\frac{S(T_i) - S(T_{i-1})}{T_i - T_{i-1}} < \frac{S(T_{i+1}) - S(T_{i-1})}{T_{i+1} - T_{i-1}} = r' \quad (7.35)$$

Since the transmission schedule coincides with the data consumption curve at bit-rate reduction points, i.e., $S(T_i) = A(T_i)$ for all i, we have

$$\frac{A(T_i) - A(T_{i-1})}{T_i - T_{i-1}} < \frac{A(T_{i+1}) - A(T_{i-1})}{T_{i+1} - T_{i-1}} \quad (7.36)$$

Now according to the MDR scheduler (cf. equation (7.6)):

$$r_i = \max\left\{ \frac{A(t) - A(T_{i-1})}{t - T_{i-1}} \middle| \forall t > T_{i-1} \right\} \quad (7.37)$$

From equation (7.36), we must have r_i equal to

$$r_i = \frac{A(T_{i+1}) - A(T_{i-1})}{T_{i+1} - T_{i-1}} \quad (7.38)$$

which violates the definition of r_i. This contradicts the assumption that the schedule $S(t)$ is generated by the MDR scheduler and therefore the schedule with increasing rates cannot be generated by the MDR scheduler and the result follows. ∎

References

[1] T.V. Lakshman, A. Ortega, and A.R. Reibman, VBR Video: Tradeoffs and Potentials, *Proceedings of the IEEE*, vol. 86, no. 5, May 1998, pp. 952–973.

[2] J. McManus and K. Ross, Video on Demand over ATM: Constant-Rate Transmission and Transport, *Proceedings of IEEE INFOCOM*, Mar. 1996, pp. 1357–1362.

[3] W. Feng and S. Sechrest, Critical Bandwidth Allocation for the Delivery of Compressed Video, *Computer Communications*, vol. 18, no. 10, Oct. 1995, pp. 709–717.

[4] W. Feng, F. Jahanian, and S. Sechrest, Optimal Buffering for the Delivery of Compressed Prerecorded Video, *ACM Multimedia Systems Journal*, Sep. 1997, pp.

[5] W. Feng, Rate-constrained Bandwidth Smoothing for the Delivery of Stored Video, *Proceedings of SPIE Multimedia Networking and Computing*, Feb. 1997, pp. 58–66.

[6] W. Feng, Time Constrained Bandwidth Smoothing for Interactive Video-on-Demand Systems, *Proceedings of International Conference on Computer Communications*, Nov. 1997, pp. 291–302.

[7] R.I. Chang, M.C. Chen, J.M. Ho, and M.T. Ko, Designing the ON-OFF CBR Transmission Schedule for Jitter-free VBR Media Playback in Real-time Networks, *Proceedings of the Fourth International Workshop on Real-Time Computing Systems and Applications*, Oct. 1997, pp. 2–9.

[8] J.D. Salehi, S.L. Zhang, J. Kurose, and D. Towsley, Supporting Stored Video: Reducing Rate Variability and End-to-end Resource Requirements through Optimal Smoothing, *IEEE/ACM Transactions on Networking*, vol. 6, no. 4, Aug. 1998, pp. 397–410.

[9] J. Zhang, Using Multiple Buffers for Smooth VBR Video Transmissions over the Network, *Proceedings of International Conference on Communication Technology*, vol. 1, Oct. 1998, pp. 419–423.

[10] W. Feng and J. Rexford, Performance Evaluation of Smoothing Algorithms for Transmitting Prerecorded Variable-bit-rate Video, *IEEE Transactions on Multimedia*, vol. 1, no. 3, Sep. 1999, pp. 302–312.

[11] W. Zhao and S.K. Tripathi, Bandwidth-efficient Continuous Media Streaming through Optimal Multiplexing, *Proceedings of International Conference on Measurement and Modeling of Computer Systems*, Apr. 1999, pp. 13–22.

[12] H. Liu, N. Ansari, and Y.Q. Shi, Dynamic Bandwidth Allocation for VBR Video Traffic Based on Scene Change Identification, *Proceedings of International Conference on Information Technology: Coding and Computing*, March 2000, pp. 284–288.

[13] J. Rexford, S. Sen, J. Dey, W. Feng, J. Kurose, J. Stankovic, and D. Towsley, Online Smoothing of Live, Variable-Bit-Rate Video, *Proceedings of International Workshop on Network and Operating Systems Support for Digital Audio and Video*, May, 1997, pp. 249–258.

[14] S.C. Liew and D.C.-Y. Tse, A Control-theoretic Approach to Adapting VBR Compressed Video for Transport over a CBR Communications Channel, *IEEE/ACM Transactions on Networking*, vol. 6, no. 1, Feb. 1998, pp. 42–55.

[15] N.G. Duffield, K.K. Ramakrishnan, and A.R. Reibman, SAVE: An Algorithm for Smoothed Adaptive Video over Explicit Rate Network, *IEEE/ACM Transactions on Networking*, vol. 6, no. 6, Dec. 1998, pp. 717–728.

[16] A. Speetzen, M. Junius, M. Steppler, M. Büter, D. Pesch, et al., *Communication Networks Class Library*, available at http://www.comnets.rwth-aachen.de/doc/cncl.html.

[17] G. Zipf, *Human Behaviour and the Principle of Least Effort*, Addison-Wesley, 1994.

8

Adaptive Media Streaming

In this chapter we consider networks where only best-effort service is supported. This type of network, including the Internet, does not guarantee the delivery of data from the media server to the media client. Moreover, even if the data are successfully transported to the client, the time it takes can vary. The media server has no control over the amount of bandwidth available and can only use whatever is available at the time. Obviously, it will be very challenging to stream high-quality media over such best-effort networks and at the same time maintain playback continuity.

As the network provides no guarantees whatsoever, the only alternative is to adapt the media stream to the network bandwidth available. This is made possible by using layered-video codec or media transcoders as discussed in Chapter 2. In this chapter, we develop an adaptation algorithm to adjust the output bit-rate of multi-layer-encoded media or transcoders according to the measured available network bandwidth and the estimated client buffer occupancy. With this adaptation algorithm, the media server can then adjust the bit-rate of the media stream dynamically (e.g., reducing bit-rate when available bandwidth drops) so that playback continuity at the client can be improved.

8.1 Introduction

The lack of end-to-end quality-of-service (QoS) support in today's Internet has caused significant difficulties for the deployment of media streaming services such as video broadcasting and video-on-demand. In particular, when the network becomes congested, significant packet losses will arise, leading to corrupted or even dropped video frames.

Given QoS support is unlikely to be widely available in the near future, researchers have resorted to another approach to tackle this problem. Specifically, a number of pioneering researchers have investigated algorithms to adapt the video bit-rate to the network bandwidth available [1–6]. For example, when the network becomes congested, the sender will reduce the bit-rate of the encoded video to alleviate the congestion. Clearly, reducing the bit-rate will also degrade the visual quality. Nevertheless, reducing the video bit-rate in a controlled manner at

Scalable Continuous Media Streaming Systems   Jack Y. B. Lee
© 2005 John Wiley & Sons, Ltd.

the sender will result in far better visual quality than attempting to recover from data loss at the receiver.

To perform video adaptation we must tackle two fundamental challenges. First, the sender must be able to dynamically control or convert the video bit-rate to the desired value. This can be accomplished by means of scalable video coding [7] and transcoding [8–10] as discussed in Chapter 2. Second, an adaptation algorithm is needed to estimate the network bandwidth available, and subsequently determine the bit-rate to be used for converting and transmitting the video stream. This chapter focuses on the second challenge, i.e., design of the rate adaptation algorithm.

This problem has been studied by a number of researchers, including the studies by Rejaie *et al.* [4] and Assunção and Ghanbari [5] who adopted UDP as the network transport; and the studies by Cuetos and Ross [1], Cuetos *et al.* [2], and Jacobs and Eleftheriadis [3] which adopted TCP as the network transport.

A common property of these adaptation algorithms is the existence of a configurable operating parameter [1–2], which is typically used in the feedback loop of the algorithms. Not surprisingly, as will be illustrated in Section 8.6, the choice of this operation parameter can significantly affect the performance of the rate adaptation algorithm. Unfortunately, to optimize this parameter for the best performance will require *a priori* knowledge of the available network bandwidth over the entire duration of the video session. This is clearly not possible in practice and thus poses significant difficulties to deploying these rate adaptation algorithms.

In this chapter, we address this issue by presenting a rate adaptation algorithm that does not have any configurable parameter. In other words, prior knowledge of the available network bandwidth is not needed to run the rate adaptation algorithm. Our results show that compared to the existing algorithms, the presented algorithm can achieve comparable or even better performance and does so without the need to tweak any operating parameters.

8.2 Related Work

In this section, we review some related work on adapting video to cope with the bandwidth fluctuations in the Internet. Assunção and Ghanbari [5] and Kanakia *et al.* [6] proposed adapting the data pumping rate and hence the bit-rate of video by adjusting the quantizer scales of the frames. The data pumping rate is controlled dynamically based on the buffer occupancy of the bottleneck switch. The idea is to maintain the buffer occupancy of the bottleneck switch at a *safe* level to avoid losses of video packets due to buffer overflow. Their approach can effectively adapt to network congestion and reduce packet losses. The only limitation is that we need to know the buffer occupancy at the bottleneck router. This may require modification to the router firmware to support this function.

Rejaie *et al.* [4] proposed an adaptive streaming scheme running on its proprietary RAP congestion control protocol [11] with the support of multi-layer-encoded videos. They proposed the criteria for adding or dropping a layer of video based on the current sending rate of its rate-based RAP congestion control protocol. This approach requires the commercial streaming player software to support playback of layered-encoded video.

Jacobs and Eleftheriadis [3] proposed a video adaptation scheme that controls the video bit-rate using transcoding. The proposed system runs on a custom-designed semi-reliable congestion control protocol. The main idea of the adaptation algorithm is to maintain the server buffer occupancy at a certain level. The algorithm assumes that video data are input into the server buffer at its playback rate and injected into the network at the rate that its congestion control algorithm allows. Therefore, when the server buffer builds up, it infers that the available network bandwidth drops and thus reduces the video bit-rate to decrease the server buffer occupancy. Its use of transcoding to control the video bit-rate and server-side information to decide the appropriate video bit-rate makes it readily deployable on existing streaming platforms.

Cuetos et al. [1, 2] proposed an adaptive rate control for streaming Fine-Grained Scalable (FGS) video. It adapts the video by controlling the video bit-rate of the enhancement layer of the FGS-coded video based on the client buffer occupancy and network bandwidth information. The heuristic rate control algorithm aims to maintain the client buffer occupancy at around the target level, while at the same time minimizing the video bit-rate variation introduced by adaptation. Although the control algorithm was originally designed to work with FGS-coded video, it can also be applied to non-FGS videos using transcoding techniques. The only limitation is the need to configure a system control parameter which can substantially affect the algorithm's performance. Our results show that the optimal value of the parameter can vary considerably over a wide range under different network traffic patterns (cf. Section 8.6) and so it may be difficult to optimize the algorithm for use in the Internet, where network traffic patterns are generally very difficult to predict.

8.3 System Model

In this chapter we consider a video streaming system that streams pre-encoded video data using TCP as the network transport to the receiver for playback. Despite the shortcomings of TCP in the context of media streaming (cf. Section 6.1), due to its aggressive congestion control and enforced error-recovery, it does possess a number of appealing features.

First, TCP is intrinsically TCP-friendly and thus fairness with other TCP traffics is automatically guaranteed. Second, using TCP the sender can stream video using, say, the standard HTTP protocol to the client. As most, if not all, video players in the market support HTTP-based video streaming and playback, compatibility is greatly enhanced. Third, for security reasons, many companies and ISP block or throttle UDP traffic at their firewalls and gateways, thus making UDP-based video streaming impossible. By contrast, TCP/HTTP streaming can pass through firewalls in the same way as web traffic. Finally, to perform bandwidth estimation, the sender will need some form of feedbacks from the client. If we use the UDP transport, then the client will need to be modified to send explicit feedbacks to the sender to enable bandwidth estimation so that rate adaptation can be performed. By contrast, TCP with its built-in flow control already can provide *implicit* feedbacks to the sender and thus no modification to the client is necessary. Again, this will greatly enhance the compatibility of the rate-adaptation algorithm to the existing video player software. Nevertheless, the rate-adaptation algorithm presented in this study can also be applied to UDP-based video streaming with appropriate support from the client's player software, such as streaming over RTP and using RTCP to report reception statistics back to the server.

Figure 8.1 shows the key components in the video streaming system. Assume the video data are encoded at a constant bit-rate of r_{max} bps. The rate controller can convert the encoded video to any bit-rate between r_{max} and r_{min} (e.g., using scalable video coding [7] or transcoding [8–10]). Note that there is a lower limit r_{min} on the achievable video bit-rate in the model, for example, the bit-rate of the base layer in FGS encoded video [7] or the lowest achievable bit-rate in transcoding [8–10] (cf. Section 2.5).

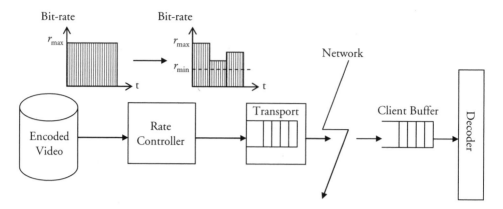

Figure 8.1   Block diagram of the system model

In practice, even with a transcoder, the video bit-rate may not be changed at any arbitrary time due to the structure of the coding algorithm (e.g., group of pictures). Thus, in the system model we assume video transcoding is performed in discrete video segments of fixed playback duration of M seconds. The rate controller will then determine the target bit-rate for the next video segment based on estimation of the client's buffer occupancy. We denote the average bit-rate for the kth video segment by r_k.

The transcoded video segments are then transmitted to the client using TCP. Note that the server does not limit the transmission rate here and simply sends the transcoded video data as fast as TCP will allow. This ensures that available network bandwidth is fully utilized. Here, we assume that the total size of server buffer in between the media server application and the network (e.g., the buffer inside the socket library and TCP) is a known constant, denoted by Z.

At the receiver, many existing video players will prefetch a certain amount of video data before starting playback to absorb the inevitable bandwidth fluctuations. We denote the playback duration of the prefetched video data by B_p seconds. Depending on the specific player software, B_p can be a fixed value known to the server, or it can be configurable by the users. If it is the latter case and the existing player software does not report this value to the server, the server will simply assume the worst case of no prefetch, i.e., $B_p = 0$ second, in performing rate adaptation.

In the case of client buffer starvation during video playback, it is assumed that playback will be paused until B_p seconds' worth of video data is again buffered at the client as depicted in Figure 8.2. Let G be the video frame rate. After the late arrival of frame i, the playback is paused and the client will have to rebuffer frames i to $(i + B_p \times G - 1)$ before resuming playback. This *rebuffering* mechanism is common among commercial video player software such as Microsoft's Media Player and RealNetworks' Video Player.

8.4 Client Buffer Occupancy and Network Bandwidth Estimation

The objective of the rate adaptation algorithm is to prevent playback starvation caused by client buffer underflow. To prevent buffer underflow, the server will need to estimate the available network bandwidth as well as the client buffer occupancy, in terms of seconds' worth of video data.

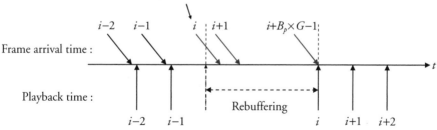

Figure 8.2   Rebuffering after buffer underflow at the client

Estimation of the client buffer occupancy is performed every time the server completes submitting a video frame to the network transport for delivery. For example, if the common socket library is used, then this is equivalent to completing all send function calls for the video frame. Let t_i be the completion time of submitting video frame i for transmission, and let f_i be the index of the oldest frame (i.e., with the smallest index number) that has not yet been completely received by the client at time t_i. Now as the server will submit data for transmission as fast as the transport allows, we can assume that the intermediate buffer at the server is always full, i.e., there are Z bytes of data accumulated awaiting transmission. Thus, we can estimate f_i from

$$f_i = \max n \quad \text{s.t.} \sum_{k=n}^{i} s_k \geq Z \tag{8.1}$$

where s_i is the size of frame i.

Similarly, after frame $i + 1$ is submitted for transmission, we can compute f_{i+1} using equation (8.1). Now if $f_{i+1} > f_i$, then we know that frame f_i to frame $f_{i+1}-1$ must have arrived at the client during the time from t_i to t_{i+1}. Assume in this short interval the frames arrive at the client at a constant rate. Then we can estimate the arrival time of frame k, denoted by T_k, from

$$T_k = t_i + \frac{k + 1 - f_i}{f_{i+1} - f_i}(t_{i+1} - t_i) \quad k \in [f_i, f_{i+1} - 1] \tag{8.2}$$

Note that in equation (8.2) we ignored network and processing delays in receiving ACKs from the client. Our simulations show that this does not have significant impact on the algorithm's performance.

Knowing the arrival time of each video frame, we can then proceed to estimate the client buffer occupancy. First, we denote K as the first frame buffered at the client during the initial prefetch period or the rebuffering period. K is initialized to 1 before streaming begins, and when rebuffering occurs, K is set to be the frame that arrives late, e.g., $K = i$ in the example shown in Figure 8.2. Let B_i (in seconds of video data) be the client buffer occupancy when frame i arrives at the client. Then we can estimate the client buffer occupancy B_i according to the following rules:

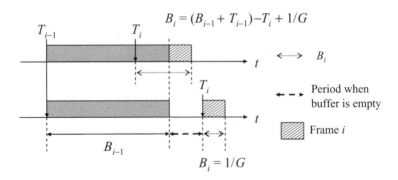

Figure 8.3 Two ways to estimate B_i when $i > K + B_p \times G - 1$

Case 1 ($i \leq K + B_p \times G - 1$): In this case, frame i belongs to the initial prefetch part of the video or the rebuffering period, i.e., the player has not yet started/resumed decoding the received video data. Thus the buffer occupancy is simply equal to:

$$B_i = (i - K + 1)/G \tag{8.3}$$

Case 2 ($i > K + B_p \times G - 1$): In this case, the way to estimate B_i depends on whether or not frame i has arrived before all the data in the client buffer is consumed as illustrated in Figure 8.3. If $(T_{i-1} + B_{i-1}) - T_i \geq 0$, that means frame i has arrived before the client buffer becomes empty, then B_i is estimated as:

$$B_i = (B_{i-1} + T_{i-1}) - T_i + 1/G \tag{8.4}$$

Otherwise, if $(T_{i-1} + B_{i-1}) - T_i \geq 0$, that means the client buffer has been empty for a period of time before frame i arrived, then B_i is simply equal to the time value of a frame, i.e.:

$$B_i = 1/G \tag{8.5}$$

where K is set to i.

The previous derivations enable us to estimate B_i at the instant frame i arrives at the client. However, a subtle complication arises due to buffering inside the server. Specifically, in order to determine the bit-rate for frame $i + 1$, we need to estimate the client buffer occupancy when frame i arrives at the client. However, due to buffering inside the server, some previous frames including frame i may not have been transmitted yet so we will need to estimate the arrival times of these frames inside the server buffer.

Let n_i be the index of the last frame of segment i, we have to predict B_{n_i} at time t_{n_i} while frame f_{n_i} to frame n_i are still in the server buffer and then use the predicted B_{n_i} to perform adaptation of segment $i + 1$. Assuming the remaining data in the server buffer at time t_{n_i} will arrive at the client at a constant rate of D_{i+1}', which is also the estimated TCP throughput for sending the segment $i + 1$, the arrival times of the remaining frames are estimated as follows:

$$T_k = t_{n_i} + \frac{1}{D_{i+1}'} \sum_{j=f_{n_i}}^{k} F_j(t_{n_i}) \forall k \in \left[f_{n_i}, n_i \right] \tag{8.6}$$

where $F_i(t)$ is the remaining amount of data of frame i at time t. With t_k, $k \in [f_{n_i}, n_i]$, we can then estimate B_{n_i}.

To estimate D_{i+1}', we simply take the rate at which segment i was submitted into the server buffer as the estimated value, i.e.,

$$D_{i+1}' = \frac{\sum\limits_{k=n_{i-1}+1}^{n_i} s_k}{t_{n_i} - t_{n_{i-1}}} \tag{8.7}$$

This is because the rate at which data are submitted into the server buffer is equal to the rate at which data leave the server buffer, and so it reflects the transmission rate.

8.5 Rate Adaptation

Armed with a mean to estimate the client buffer occupancy and network bandwidth, the next challenge is to devise an adaptation algorithm to control the video bit-rate to prevent client buffer underflow.

8.5.1 Segment-Based Rate Control

As video data are transcoded and transmitted in fixed-duration segments, the server must determine the target bit-rate before transcoding a video segment for transmission. The server determines the target bit-rate based on two factors, namely, the estimated client buffer occupancy and the estimated network bandwidth available, both can be estimated using techniques described in Section 8.4.

Suppose segment i has just been submitted to the server buffer, with the estimated D_{i+1}' and B_{n_i}, we can predict the client buffer occupancy after transmitting the segment $i + 1$ to the client, i.e., $B_{n_{i+1}}$, from:

$$B_{n_{i+1}} = B_{n_i} + M - \frac{Mr_{i+1}}{D_{i+1}'} \tag{8.8}$$

where the last term is the predicted time taken to send the whole $(i + 1)$th segment to the client. Rearranging equation (8.8), we can obtain

$$r_{i+1} = \left[1 - \frac{B_{n_{i+1}} - B_{n_i}}{M}\right] D_{i+1}' \tag{8.9}$$

From equation (8.9), we can relate the video bit-rate r_{i+1} with the estimated client buffer occupancy (represented by $B_{n_{i+1}}$). Our goal is to adjust the video bit-rate to maintain the client buffer occupancy to above a given threshold B_T so that short-term bandwidth variations can be absorbed. In practice, $B_T = B_p$ when B_p is known, otherwise it can be configured according to properties of the system/network or according to experiments. Simulations show that the system performance is not sensitive to this setting.

Now if $B_{n_i} < B_T$, then it implies that the client buffer occupancy is below the threshold. Hence the server will reduce the video bit-rate to raise the buffer occupancy to B_T by substituting $B_{n_{n+1}} = B_T$ in equation (8.9) to obtain:

$$r_{i+1} = \left[1 - \frac{B_T - B_{n_i}}{M}\right] D_{i+1}' \tag{8.10}$$

Otherwise if $B_{n_i} \geq B_T$, then it implies that the client buffer occupancy is above the threshold. In this case the server will simply maintain the current client buffer occupancy by setting $B_{n_{i+1}} = B_{n_i}$ in equation (8.9) to obtain r_{i+1}. This is a conservative strategy to reduce the likelihood of buffer underflow. Thus, we have:

$$r_{i+1} = D_{i+1}' \tag{8.11}$$

Finally, the server checks and limits the computed video bit-rate to the feasible range $[r_{min}, r_{max}]$ by

$$r_{i+1} = \min\{r_{\max}, \max\{r_{\min}, r_{i+1}\}\} \tag{8.12}$$

Note that in contrast to previous works [1–3], this adaptation algorithm has no control parameter that requires either offline or online optimization. This has practical significance as it is not easy to optimize the control parameters without knowledge of the available network bandwidth.

8.5.2 Preemptive Rate Control

In our trace-driven simulations, we find that the available network bandwidth can occasionally drop drastically to a very low value. These sudden bandwidth drops do not appear to be predictable and thus can result in client video playback starvation.

The fundamental problem is that the adaptation algorithm is executed only when a new video segment is to be transmitted. Thus, if bandwidth drops significantly, then the transmission of the current video segment will stall. The adaptation algorithm cannot react in this case as the current video segment has not yet been completely transmitted. Meanwhile the client will continue consuming video data for playback and thus may run into buffer underflow.

To tackle this problem, we can use a *preemptive scheduling* technique to shorten the delay for the adaptation algorithm to react to changing network conditions. Instead of waiting indefinitely for a video segment to be completely submitted into the server buffer, the scheduler will timeout after Mr_{i+1}/D_{i+1}' seconds, which is the expected time required to submit the $(i + 1)$th video segment into the server buffer. If not all video data can be submitted, then the remaining yet-to-be-submitted data will be discarded and the remaining video segment transcoded again according to the new estimates on the client buffer occupancy and the available network bandwidth.

Note that preemptive rate control requires the video transcoder to be able to adjust the video bit-rate in between a video segment. The implementation will be highly dependent on the video compression algorithm employed and further study is required to identify the constraints and tradeoffs of this technique.

8.6 Performance Evaluation

In this section, we use a trace-driven simulator written in ns-2 [12] to evaluate the performance of the presented adaptation algorithm (denoted by AVS) and compare it with the current state-of-the-art algorithm proposed by Cuetos and Ross [1–2] (denoted by CR).

Figure 8.4 depicts the simulated network topology. We use the common NewReno TCP [13–14] as the transport protocol to deliver the video data to the client. Cross-traffic is generated from a packet trace file obtained from Bell Labs [15–16].[1] The trace file captured 94 hours of network traffic passing through a firewall. We divide the 94-hour trace file into 94 1-hour trace files and run a simulation for each 1-hour trace file to evaluate the algorithms' performance under different cross-traffic scenarios.

Both the streaming traffic and the cross-traffic share a link of R Mbps as shown in Figure 8.4. For each simulation, we adjust R so that the network has just sufficient bandwidth to stream the video, i.e., $R = r_{max} + \bar{c}$, where $\bar{c}$ is the average data rate of the cross traffic. We summarize the system settings in Table 8.1.

We use two performance metrics, namely, rebuffering ratio and average video bit-rate, to evaluate the algorithms' performance. Rebuffering ratio is defined as the proportion of frames

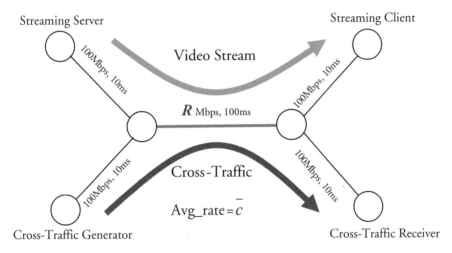

Figure 8.4   Network topology used in simulations

Table 8.1   System parameters used in simulations

Parameters	Value
Prefetch duration	5 seconds
M	1 second
r_{max}	1.1 Mbps
r_{min}	200 kbps
Video Length	3000 seconds
TCP MSS	1500 bytes

received during rebuffering periods. Playback of these frames is delayed so their amount represents the quality degradation resulting from buffer underflow. Average video bit-rate, on the other hand, represents video quality. Higher average video bit-rate generally produces better visual quality during playback.

8.6.1 Sensitivity to Prefetch Duration

The presented rate adaptation algorithm makes use of knowledge of the client's initial prefetch duration in estimating the client buffer occupancy. However, if this is not known, then it simply assumes no prefetch is performed. To investigate the performance impact of such knowledge, we run two sets of simulations for all 94 traffic traces, one set with the prefetch duration known to the server and the other set simply assuming no prefetch. In both cases the client has a prefetch duration of 5 seconds.

Figure 8.5 shows the rebuffering ratio and average video bit-rate for all 94 traces for the two cases. The result is a bit surprising as the rebuffering ratios of all traces drop by 50% on average, when the prefetch duration is not known, while the average video bit-rates for the case when the prefetch duration is not known drops only 2% on average. Not knowing the prefetch duration, the algorithm will simply assume the client does not prefetch, which means that the estimated client buffer occupancy will be lower than the actual one. As a result, the algorithm will become more conservative and the client buffer occupancy will be maintained at a higher level than expected, thus reducing the likelihood of buffer underflow.

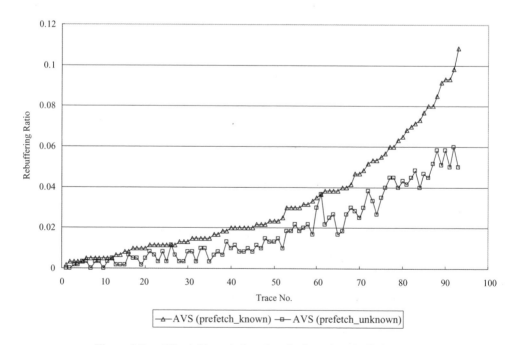

Figure 8.5a   Effect of knowledge of prefetch on the rebuffering ratio

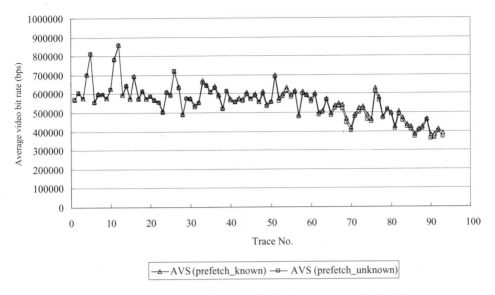

Figure 8.5b   Effect of knowledge of prefetch on the transcoded video bit-rate

8.6.2 Effectiveness of Preemptive Rate Control

To investigate how much performance gain can be obtained from preemptive rate control, we run two sets of simulations for all 94 traffic traces, one with segment-based rate control and the other with preemptive rate control and plot the results in Figure 8.6.

In all 94 traces, preemptive rate control achieves lower underflow ratios compared to segment-based rate control. On average, the rebuffering ratio is reduced by 43% when preemptive rate control is used while the average bit-rate is only 2% lower. Nevertheless, preemptive rate control does require more complex transcoders and thus further investigation is needed to quantify the gains and the tradeoffs.

8.6.3 Comparison with the CR Algorithm

In this section, we compare the presented rate adaptation algorithm (the AVS algorithm) with the current state-of-the-art algorithm proposed by Cuetos and Ross [1–2] (the CR algorithm). In the CR algorithm, there is a control parameter α ($0 \leq \alpha \leq 1$) that can affect the algorithm's performance. To find the optimal value for α, it is necessary to know the network bandwidth availability over the entire duration of the video session. This is clearly not possible in practice.

Thus, to obtain performance results for the CR algorithm, we ran 2,000 simulations with the control parameter α varied from 0 to 1 with a step size of 0.0005. We find that the optimal value for α depends heavily on the particular traffic trace chosen, and can range from 0 to 0.25 over the 94 traces. The optimal value for α is defined as the value that brings the rebuffering ratio to within 1% of the minimum and gives that maximum average video bit-rate.

As the optimal α is not known *a priori*, in comparing CR with AVS, we use the rebuffering ratio and the video bit-rate averaged over all 2,000 simulations, as well as the optimal results for comparison. The results are shown in Figure 8.7a, 8.7b. We observe that the AVS

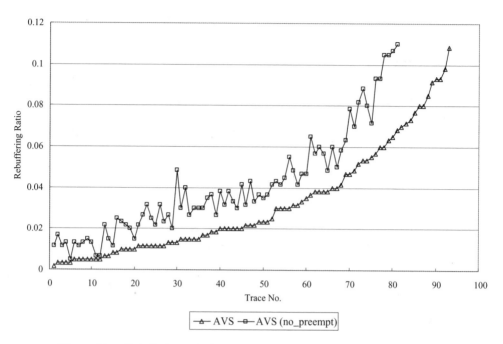

Figure 8.6a   Rebuffering ratio for AVS with and without preemptive rate control

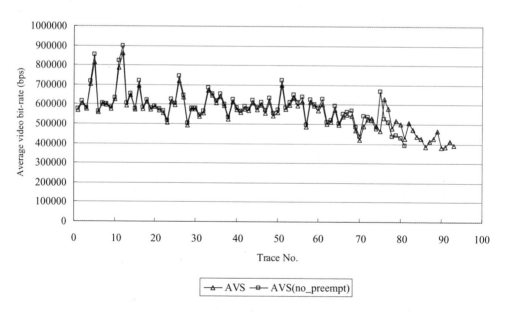

Figure 8.6b   Average video bit-rate for AVS with and without preemptive rate control

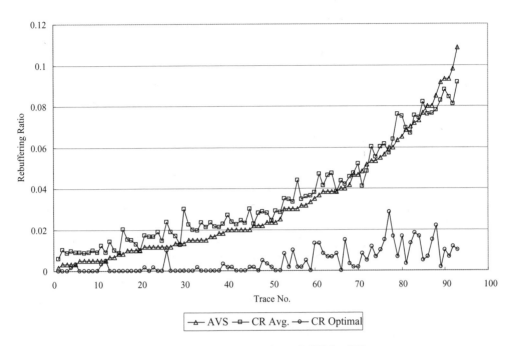

Figure 8.7a Rebuffering ratios of CR and AVS for different traces

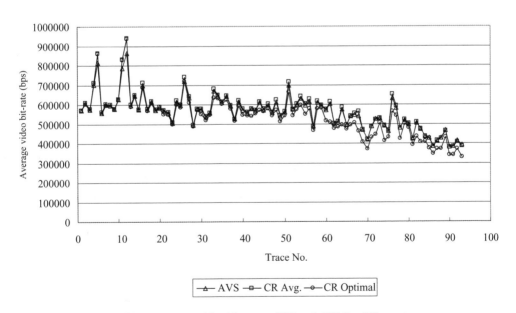

Figure 8.7b Average video bit-rates of CR and AVS for different traces

algorithm gives lower rebuffering ratio in most traces than the average-case performance of CR. Averaging over all 94 traces, the AVS algorithm can achieve 20% lower rebuffering ratio than the CR algorithm, while the average video bit-rate is only 0.7% lower than the CR algorithm, showing that both algorithms can make efficient use of the network bandwidth.

Although the optimal rebuffering ratio of the CR algorithm is much lower, it requires offline optimization which is impossible in practice. By contrast, the AVS algorithm does not require any *a priori* knowledge of the network bandwidth available or tuning of any control parameter and thus will be simpler to deploy.

8.7 Summary

In this chapter we presented a rate adaptation algorithm for streaming video over the Internet which only supports best-effort service. The algorithm has two unique features to maximize its compatibility with existing video player software. First, we showed that the rate adaptation algorithm can be applied to streaming video over TCP/HTTP, which is compatible with most of the existing video player software. Second, the rate adaptation algorithm performs network bandwidth and client buffer occupancy estimations using only local information. Thus, explicit feedbacks from the client are not needed and hence existing video player software can be supported. Moreover, the presented algorithm does not need any parameter tuning or *a priori* knowledge of the available network bandwidth to perform well, thus simplifying the deployment of the adaptation algorithm in practice.

Note

1. Network traces used in the simulations belong to the NLANR project sponsored by the National Science Foundation and its ANIR division under Cooperative Agreement No. ANI-9807479, and the National Laboratory for Applied Network Research.

References

[1] P. de Cuetos and K.W. Ross, Adaptive Rate Control for Streaming Stored Fine-Grained Scalable Video, *Proc. of NOSSDAV*, May 2002, pp. 3–12.

[2] P. de Cuetos, P. Guillotel, K.W. Ross, and D. Thoreau, Implementation of Adaptive Streaming of Stored MPEG-4 FGS Video over TCP, *Proc. of IEEE Multimedia and Expo*, 2002, pp. 405–408.

[3] S. Jacobs and A. Eleftheriadis, Streaming Video using Dynamic Rate Shaping and TCP Congestion Control, *Journal of Visual Communication and Image Representation*, vol. 9, no. 3, 1998, pp. 211–222.

[4] R. Rejaie, M. Handley, and D. Estrin, Architectural Considerations for Playback of Quality Adaptive Video over the Internet, *Technical Report 98-686, USC-CS*, Nov. 1998.

[5] P.A.A. Assunção and M. Ghanbari, Congestion Control of Video Traffic with Transcoders, *Proc. of IEEE Int. Conf. Communications*, vol. 1, June 1997, pp. 523–527.

[6] H. Kanakia, P.P. Mishra, and A.R. Reibman, An Adaptive Congestion Control Scheme for Real Time Packet Video Transport, *IEEE/ACM Tran. of Networking*, vol. 3, no. 6, Dec. 1995, pp. 671–682.

[7] W. Li, Overview of Fine Granularity Scalability in MPEG-4 Video Standard, *IEEE Tran. of Circuits and Systems for Video Tech.* vol. 11, no. 3, March 2001, pp. 301–317.

[8] P.A.A. Assunção and G. Mohammed, A Frequency-Domain Video Transcoder for Dynamic Bit-Rate Reduction of MPEG-2 Bit Streams, *IEEE Tran. of Circuits and Systems for Video Tech.*, vol. 8, no. 8, Dec. 1998, pp. 923–967.

[9] B.K. Natarajan and B. Vasudev, A Fast Approximate Algorithm for Scaling Down Digital Images in the DCT Domain, *Proc. of IEEE Int. Conf. Image Processing*, vol. 2, Oct. 1995, pp. 241–243.

[10] H. Sun, W. Kwok, and J.W. Zdepski, Architecture for MPEG Compressed Bitstream Scaling, *IEEE Tran. of Circuits and Systems for Video Technology*, vol. 6, no. 2, April 1996, pp. 191–199.

[11] R. Rejaie, M. Handley, and D. Estrin, RAP: An End-to-End Rate-based Congestion Control Mechanism for Realtime Streams in the Internet, *Proc. of IEEE INFOCOM*, vol. 3, April 1999, pp. 1337–1345.

[12] The network simulator – ns-2. [Online]. Available: http://www.isi.edu/nsnam/ns/

[13] S. Floyd and V. Paxson, Difficulties in Simulating the Internet, *IEEE/ACM Tran. of Networking*, vol. 9, no. 4, August 2001, pp. 392–403.

[14] S. Floyd and T. Henderson, *The NewReno Modification to TCP's Fast Recovery Algorithm*, RFC 2582, April 1999.

[15] NLANR Measurement and Network Analysis Group. [Online]. Available: http://pma.nlanr.net/Traces/long/bell1.html

[16] BellLabs Internet Traffic Research. [Online]. Available: http://cm.bell-labs.com/cm/ms/departments/sia/InternetTraffic/

Part Two

Parallel Server Architectures

9

Taxonomy and Architectural Alternatives

In Part I we covered the basic principles and concepts in media streaming based on the client-server model. Two issues in particular recur frequently in the discussions – capacity and reliability. Capacity dimensioning is crucial in a media streaming system as predictable performance is often needed to sustain continuous media playback at the clients. When the maximum server capacity is reached, the server will need to deny new user requests so that the performances of existing streams are not adversely affected. As discussed in Chapter 3, we can expand the server's streaming capacity by employing striped disk array and multi-disk scheduling. Nevertheless, we cannot keep adding more disks to an existing media server to expand the streaming capacity because the server will eventually run into other bottlenecks in the server host, such as the I/O bus capacity or CPU processing limit. Thus, the capacity of a single media server is still ultimately limited.

On the other hand, reliability is also an important issue in practice, especially when providing paid streaming services to a large user population. Within the media server we can improve reliability by using technologies such as error-correcting memory chips, redundant power supplies, and RAID storage. However, failure at the server level, such as those caused by software bugs, hardware failure, and so on, will not be recoverable. Current solutions such as stand-by systems can be used to restore the system but in addition to the high costs of fully-replicated hardware, the on-going streams will also likely need to be restarted in the process of switching to the stand-by systems.

The previous discussions illustrate the limitations of the single-server architecture in implementing media streaming services. In Part II we depart from this architecture to investigate an alternative: parallel server architectures, that offer promising solutions to the above-mentioned scalability and reliability limitations. In this chapter we give an overview of the issues in designing streaming systems around parallel server architectures, and present a taxonomy to classify the many possible architectural alternatives. Subsequent chapters will investigate in more detail several architectures which provide different engineering tradeoffs.

Scalable Continuous Media Streaming Systems   Jack Y. B. Lee
© 2005 John Wiley & Sons, Ltd.

9.1 Introduction

A media server is essentially a data mover, retrieving media data from the storage devices (e.g., hard disks or RAID) and then sending them over the network to the clients for playback. Thus, the primary challenges in a media server are capacity and reliability.

Despite the rapid advances in computing and network hardware, the capacity of commodity server hardware is still rather limited. Considering the emergence of high-definition television (HDTV) where a single stream can consume tens of Mbps of bandwidth, there remain considerable challenges to building high-capacity yet cost-effective media servers. Server reliability, on the other hand, has been advancing at a much slower pace, partly due to the nature of the problem and the cost of many solutions.

Many existing media servers, such as video servers used in video-on-demand (VoD) services, are built around the single-server architecture, i.e., with the streaming server running in an independent server machine equipped with storage devices and networking interfaces. This single-server architecture is well understood and widely adopted not only in streaming servers, but also in many other types of servers, such as a web server, an FTP server, and so on. In bandwidth-demanding applications such as VoD, however, the single-server architecture will quickly run into scalability and reliability limitations.

First, the capacity of a single server cannot be expanded indefinitely, even if striped disk array is employed. If we keep adding more disks to the server, we will eventually run into other system bottlenecks, such as the I/O bus capacity, the processor's capacity, memory limit, and so on. Conceivably, we can upgrade the server hardware to higher-performance hardware platforms such as massively-parallel systems [1, 2] but the costs will be substantially increased.

Another approach is to replicate all the data to a new server to increase the system's capacity. This will double the cost of the system but nonetheless provides a linear scalability path. However, in applications where the storage requirement is large, such as in movie-on-demand applications, the multiplied storage cost could become significant. To reduce storage overhead in replication, we can selectively replicate the media objects such as videos based on their viewing popularity [3-7]. The more popular videos will be replicated across more servers to increase streaming capacity while keeping fewer copies of the less popular videos to reduce storage cost. This approach will be effective if the video popularity is known. Otherwise the load of the servers may not be balanced, thus leading to unnecessary request blockings. Moreover, the popularity itself may change during the lifetime of the video [6] and so reshuffling of the video titles between the servers will be needed from time to time.

Second, in terms of reliability, a single-server streaming system simply cannot survive server-level failures. If replicated servers are available, then the users being served by the failed server can be moved to the other replicated servers. Nevertheless this switchover process is likely not transparent and will cause at least temporary playback interruptions. Moreover, with a failed server, the system streaming capacity will be reduced and so some users may not be able to resume service immediately. As the system scales up with more and more servers, this problem will only get worse.

Interestingly, we encountered similar load balancing and reliability problems when we discussed multi-disk media servers in Section 3.5. In that context the solution was to employ striped disk array and RAID, which can achieve perfect load balance irrespective of the access popularities of the media objects, and be able to sustain non-stop service even when during a disk failure. In Part II of this book we will investigate applying similar principles to multi-server media streaming systems – parallel server architectures – and study their performance advantages and the associated tradeoffs. While our focus is on video streaming, the same principles and architectures will also be applicable to other types of streaming systems.

In the rest of this chapter we will introduce a framework for the design of parallel video server architectures. We address three central architectural issues: *video distribution architectures*, *server striping policies*, and *video delivery protocols* for parallel video servers. We present possible design alternatives and review the existing designs [8-18] in the literature in the context of the framework.

9.2 Parallel Video Distribution Architectures

The essence of parallel video servers is the *striping* of data across an array of servers. Since the data consumer (such as a video decoder) expects a single stream of video data, data streams from each server must first be resequenced and merged. We use the name *proxy* to refer to the system module responsible for resequencing and merging data from multiple servers into a coherent video stream for delivery to a client. In addition, the proxy can make use of data redundancy to mask server failures and hence achieve server-level fault tolerance (see Section 9.4.3).

The proxy is a software and/or hardware module that knows the configuration of the system (such as the number and addresses of servers, locations of data, striping policy, etc.). There are three ways to implement the proxy: (1) at the server computer – *proxy-at-server*; (2) at an independent computer – *independent proxy*; and (3) at the client computer – *proxy-at-client*. Note that we use the term *computer* to refer to the hardware performing the proxy function. In practice, this hardware may or may not be a computer in the general sense.

9.2.1 Proxy-At-Server

Figure 9.1 shows the proxy-at-server architecture. There are N_S server computers and each computer performs both as storage server and proxy. As there are likely more clients than servers,

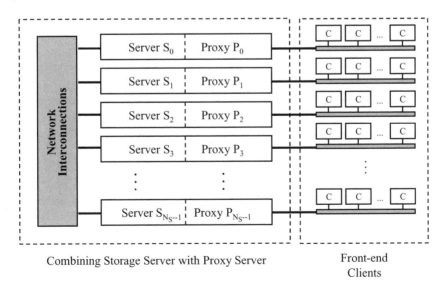

Combining Storage Server with Proxy Server   Front-end
                                                    Clients

Figure 9.1   The proxy-at-server architecture

each proxy will have to serve multiple clients simultaneously. The servers are connected locally by an interconnection network. The proxies combine data retrieved from the local storage and data received from other servers into a single video data stream for transmission to the clients.

In this architecture, the details of the system configuration can be completely hidden from the clients. One drawback of this approach is processing and communications overhead. For example, to deliver B bytes of video data from the video servers to a client, Bbytes of data must first be read from one or more servers' local storage, and then transmitted via network to the client's proxy (unless the proxy happens to share the same host as the video server, which then requires no transmission). Finally, the proxy processes the data and transmits to the client. If we assume requests are serviced evenly by all N_S servers, then on the average we need $B(2N_S-1)/N_S$ bytes of data transmission (server-to-proxy, proxy-to-client) and $B(2N_S-1)/N_S$ bytes of data reception (proxy and client) for every B bytes of data delivered from the storage servers to a client.

9.2.2 Independent Proxy

Alternatively, separate computers can be used to run the proxies. Figure 9.2 shows the independent proxy architecture using this approach. The back-end storage servers and the proxy computers are connected locally by an interconnection network. Each proxy connects to multiple clients via another external network. Similar to the proxy-at-server architecture, this independent proxy architecture also hides the server complexity from the clients. Moreover, separating the proxy from the server eliminates interference between the two processes and hence may simplify the server and proxy implementations.

Under the independent proxy architecture, data are first retrieved from the back-end server's local storage and then transmitted to the proxy. The receiving proxy then processes and

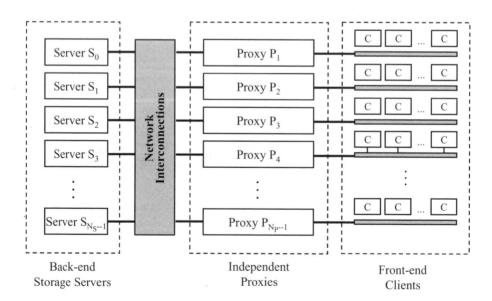

Figure 9.2 The independent proxy architecture

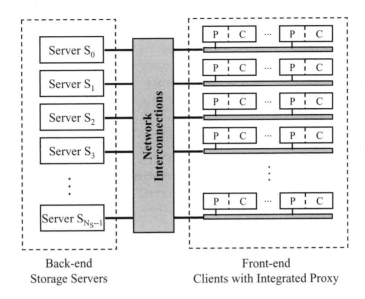

Back-end
Storage Servers

Front-end
Clients with Integrated Proxy

Figure 9.3 The proxy-at-client architecture

transmits the data to the clients. Therefore, to request B bytes of data, we need $2B$ bytes of transmission and $2B$ bytes of reception. Hence, this approach requires even more processing and network bandwidth than the proxy-at-server architecture. Moreover, additional hardware and network links are required to host and connect the proxies.

9.2.3 Proxy-At-Client

The third approach is to integrate the proxy into the client, as shown in Figure 9.3. This can be done by adding a proxy software module into the operating system or within the application. Under this architecture, a proxy requests the servers to send data directly to the client computer. After processing by the proxy, video data are then passed directly to the client application without further network communications. Hence, to retrieve B bytes from the servers, we only need B bytes of transmission from the servers and B bytes of reception at the client.

Compared to the proxy-at-server and independent proxy architectures, the proxy-at-client architecture requires only half the amount of data transfer and does not need separate hardware for the proxies. The primary advantage of the proxy-at-server and independent proxy architectures is client transparency, i.e., the complexity of communicating with multiple servers is hidden by the proxy. However, some experimental studies [13, 14] have shown that the extra complexity involved is negligible, even when running the client in low-end PC machines.

On the other hand, if the computer running a proxy fails (hardware failure, network failure, etc.) under the proxy-at-server and the independent proxy architecture, the service of all clients serving by the proxy will be disrupted. Conversely, the same situation will only affect a single client in the proxy-at-client architecture because each proxy serves only one client and the proxy runs at the client host.

9.2.4 Architecture of Existing Works

The proxy-at-server architecture has been investigated in the studies by Reddy [16], Tewari *et al.* [17], and Wu and Shu [18]. In Tewari's paper, they called this the *flat* architecture. They also considered a *two-tier* architecture that is equivalent to the independent proxy architecture. They called the proxy a delivery node, which retrieves video data from back-end storage nodes, and delivers a single video stream to the client. The independent-proxy architecture has also been investigated in another study by Buddhikot and Parulkar [11]. Unlike the previous works, they implemented the proxy functionality in their custom *ATM Port Interconnect Controller* (APIC), which also functions as the interconnection network linking the storage nodes and the external network. Lougher *et al.* [15] also investigated a *striping server* in a hierarchical network topology to perform the proxy functions. Finally, the proxy-at-client architecture has been investigated in the studies by Bernhardt and Biersack [9], Bolosky *et al.* [10], Freedman and DeWitt [12], and Lee and Wong [13, 14].

All three architectures are scalable in the sense that more servers and proxies can be added to the system to support more concurrent video sessions. However, the proxy-at-server and independent-proxy architectures suffer from the problem that a proxy failure will affect the clients connected to it. Conversely, systems based on the proxy-at-client architecture do not have the proxy reliability problem and hence only back-end storage server failures need to be considered. In the subsequent chapters we will primarily focus on the proxy-at-client architecture.

9.3 Server Striping Policies

Striping is a general technique for distributing data over multiple devices to improve *capacity* (or throughput) and potentially *reliability*. Disk array and the Redundant Array of Inexpensive Disks (RAID) [19] are among the most successful applications of the striping principle. There are also other applications, including network striping [20], and tape striping [21]. In a parallel video server, we stripe video data over multiple servers to increase the system's capacity and potentially improve the system's reliability using data redundancy. We call this *server striping* in accordance with previous work in other areas. Striping a video stream across all N_S servers is commonly called *wide striping*, and striping over a subset of the N_S servers is called *short striping* [17]. Unless stated otherwise, we will assume wide striping in the following sections and discuss the key design alternatives.

9.3.1 Time Striping

A video stream can be viewed as a series of video frames. Therefore, we can stripe a video stream in units of frames across multiple servers – *time striping*. Figure 9.4 depicts one example of how video units are striped using time striping. Assume that a stripe unit contains L frames, and the video plays at a constant frame rate of F frames per second. Then in each round of $N_s L/F$ seconds, L frames will be retrieved from each server and delivered to a client. In general, the striping size L neither need to be an integer, nor equal to or larger than 1. In particular, if $L < 1$, then it is called sub-frame striping [9]. Conversely, we simply call it frame striping for $L \geq 1$.

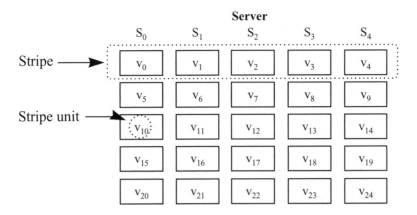

$$v_i \text{ is stripe unit } i, \text{ containing frames } ki \text{ to } k(i+1)-1$$

Figure 9.4   Striping a video stream over five servers using time striping

Existing studies using the time striping approach include those works by Biersack *et al.* [8, 9], and Buddhikot and Parulkar [11]. The study by Bernhardt and Biersack [9] adopted time striping with a granularity of one frame and also one segment of a frame – *sub-frame striping*. For sub-frame striping, a frame is sub-divided into k equal-size units and then distributed across the servers in a round-robin fashion. The key advantage of sub-frame striping is that load balance is guaranteed for both CBR and VBR video streams as each frame is striped equally across all servers. Conversely, the study by Buddhikot and Parulkar [11] used a stripe unit of k ($k \geq 1$) frames. They suggested solving the load balance problem by grouping more frames into a stripe unit to obtain a more uniform stripe unit size (see Section 9.3.4).

9.3.2 Space Striping

Time striping divides a video stream into fixed-length (in time) stripe units. Another approach is to divide a video stream into fixed-size (in bytes) stripe units – *space striping*. Space striping simplifies storage and buffer management in the servers because all stripe units are of the same size. Moreover, the amount of data sent by each server in a service round is also the same. Unlike time striping, we do not need to know the frame structure in order to perform striping, thus decoupling the striping algorithm from the encoding format.

This space striping approach has been employed in many studies [10, 12–18]. Depending on the system design, the stripe unit size can range from tens of kilobytes to hundreds of kilobytes. In most of the studies, a stripe unit is assumed to play back in a constant length of time. However, in most video compression algorithms such as MPEG, a fixed-size stripe unit will likely contain a variable number of frames and/or partial frames. Moreover, if the video is compressed using constant-quality compression algorithms (cf. Section 2.4), then the video bit-rate will become variable as well. Consequently, the decoding time for a stripe unit at the client will be variable and this may cause playback starvation unless sufficient buffering is done at the client before playback begins.

9.3.3 Placement Policies

In the previous discussions, we have assumed a round-robin placement of the stripe units across the servers in the system. If we denote the stripe units of a video stream using $v_0, v_1, \ldots$, etc., then stripe unit v_i will be stored in server ($i \mod N_S$). However, a minor problem with this policy is that server i will likely store more stripe units than server j, for $i < j$. This is because the length of a video is not always an integral multiple of the size of a stripe. Therefore, the last stripe will likely contain less than N_S stripe units, filling from server zero. To balance the storage, we can modify the round-robin policy to start striping a new video stream from different servers.

Apart from round-robin placement, Tewari *et al.* [17] also investigated a random-placement policy where the order within a stripe is permuted pseudo-randomly. They pointed out that the round-robin placement policy can introduce a *convoy effect* when one server becomes overloaded. That is, the overloading condition will shift from one server to the next due to the round-robin placement. By permuting the order of the stripe units in each stripe, this convoy effect can be avoided.

9.3.4 Redundancy

As discussed in Section 9.1, one of the principal advantages of parallel server architectures is the potential to achieve server-level fault tolerance analogous to the RAID architecture in disk arrays. Ideally, we want the system to be able to maintain continuous video playback for all active streams even when one or more servers fail.

To achieve this, we will need to introduce data and capacity redundancies into the parallel servers. In the simplest form, we can extend the striping algorithm to include a parity block for row of data blocks as depicted in Figure 9.5. The parity units are computed using the rest of the stripe units in the same stripe. When a server fails, the lost stripe unit stored in the failed

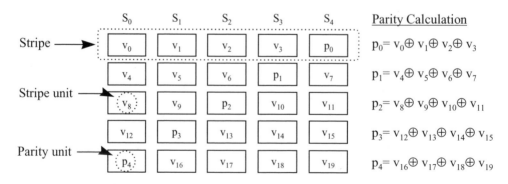

Figure 9.5   Adding redundant data to support server-level fault tolerance

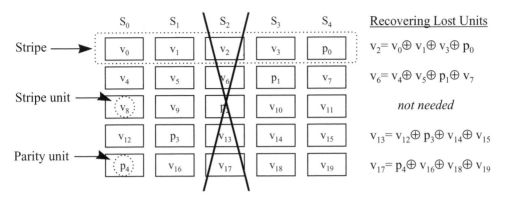

Figure 9.6   Recovering stripe units lost due to failure of server 2

server can be computed from the parity unit together with the remaining stripe units as shown in Figure 9.6.

For single-failure protection, simple parity computed from exclusive-or between the data stripe units can be used. Higher level of redundancies can be achieved by the use of more sophisticated erasure-correction codes such as the Reed-Solomon Erasure Correction code [22, 23]. Note that to perform erasure-correction computation, stripe units of the same stripe must be of the same size. Therefore, time striping algorithm which results in variable stripe unit sizes cannot easily be supported. Conversely, space striping has fixed stripe size and hence can easily be extended to incorporate redundancy.

9.4 Parallel Video Delivery Protocols

In this section, we focus on ways to deliver video data from multiple servers to a video client. The parallel video delivery requirement poses challenges in designing the application protocol's flow control, error control, and synchronization. In the following, we first consider the client pull versus the server push service model, and then discuss synchronization and fault-tolerance issues.

9.4.1 Client Pull versus Server Push

In VoD systems, there are generally two ways to request and deliver video data from a server to a client. Most studies on VoD systems let the video server send video data to the client at a controlled data rate. The client receives and buffers the incoming video data for playback through a video decoder. As shown in Figure 9.7a, once the video session has started, the video server will continue the data transmissions until the client specifically sends a request to stop it. Since the server pushes video data to the client at a controlled rate, this approach is called *server push*.

A second approach is the traditional request-response model in which the video client sends a request to the server for a particular piece of video data. As shown in Figure 9.7b, upon

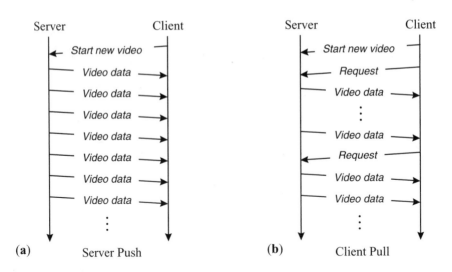

Figure 9.7 Two approaches to data delivery: server push and client pull

receiving the request, the server will retrieve the data from the disk and send it back to the client. This approach is called *client pull* for obvious reasons.

9.4.2 Inter-Server Synchronization

Most of the studies on single-server VoD systems employ the server-push delivery model. This model allows the system designer to devise periodic schedules at the server for reading data off the disk and then transmitting to the client. If we extend the server-push model to a parallel video server, a new problem arises due to the parallel transmissions from multiple independently running servers.

For example, let us consider a system with N_S servers using fixed-size space striping. To start a new video session, a client will send a request to the proxy, which in turn will send requests to all N_S servers to start a new video session. Due to variations in processing delay, network delay, and scheduling delay, the servers will start transmitting data at different time instances. It is possible that the first stripe unit will arrive at the proxy later than the subsequent stripe units. Consequently, the proxy has to buffer the later stripe units to wait for the first stripe unit to arrive for playback, thereby increasing the client buffer requirement and start-up delay.

This synchronization problem has been investigated by Biersack *et al.* in [8]. For scenarios where network delays between servers and a client are different, they proposed adding different delays to the starting times of each server to compensate for delay differences. They also extended this model to include bounded network delay jitters. In another study by Buddhikot and Parulkar [11], they designed a closely-coupled system in which each storage node is connected by a custom high-speed interconnection network (APIC). The proximity of the storage nodes enables them to be accurately synchronized through the common APIC. We will return to this synchronization issue in subsequent chapters.

9.4.3 Detecting and Masking Server Failures

In Section 9.3.4 we discussed how data redundancy can be introduced among the servers to support server-level fault tolerance. The redundant data enable the receiver to mask a server failure by computing lost stripe units stored in the failed server from the parity units together with the remaining stripe units. Using concepts similar to *Forward Error Correction* (FEC), the servers can send redundant data along with normal data to the receiver at all times. In this way, the receiver can recover lost packets using the received data together with the redundant data. This approach can be extended to parallel video servers to recover stripe units lost in failed servers.

FEC has the distinct advantage that the receiver does not need to detect a server failure. As redundant data are always transmitted and received by the receiver, lost stripe units can readily be recovered if a server fails. However, like the case in network communications, FEC incurs constant transmission overhead even when no server fails. According to coding theory, we need one redundant symbol for every lost symbol we want to recover. Therefore, if we use K to denote the number of lost symbols we want to recover per parity group (or stripe), the transmission overhead will be $K/(N_S - K)$. Note that this overhead could become significant for systems having a small number of servers, or a high level of redundancies.

Alternatively, for redundancy levels larger than one (i.e., $K > 1$), we can adopt a Progressive Redundancy Transmission (PRT) algorithm to reduce the transmission overhead. Specifically, PRT initially does not transmit all available redundant data but only a portion of them. When a server failure is detected, PRT dynamically requests the servers to begin transmitting an additional redundant unit per stripe. For example, let $K = 3$, then the system can be configured to initially transmit only one redundant unit. When a server fails, the system will be able to mask the failure immediately using the available redundant unit. At the same time the remaining servers will begin transmitting one more redundant unit per stripe to prepare for a second server failure, and so on. As servers seldom fail simultaneously (unless hit by natural disasters), this PRT algorithm can keep the transmission overhead low while still allowing the system to survive multiple server failures.

The challenge in PRT is to devise a way to detect server failures quickly and reliably. The detection method must be quick enough to ensure that video playback continuity can be sustained while the system request redundant data for recovery. On the other hand, the detection method must not generate too many false alarms to avoid sending unnecessary many redundant data to the clients. In Chapter 14 we take a closer look of FEC and PRT by modeling their availability to quantify the tradeoffs.

9.5 Summary

In this chapter, we have introduced a framework for the design of parallel video server architectures. We presented design alternatives, and reviewed existing literatures on three central architectural issues, namely video distribution architectures, server striping policies, and video delivery protocols. Table 9.1 summarizes the design choices adopted in some of the previous studies. In the next three chapters, we investigate in detail two specific parallel architectures – the concurrent-push architecture and the staggered-push architecture.

Table 9.1 Summary of design choices studied by various researchers

Researchers	Video distribution architecture	Server striping policy	Video delivery protocol	Server fault tolerance
Biersack et al. (Video Server Array)	proxy-at-client	time striping	server push	striping w/ parity; FEC
Bolosky et al. (Tiger Video Fileserver)	proxy-at-client	space striping	server push	mirroring with declustering
Buddhikot et al. (MARS)	independent proxy	time striping	server push	–
Freedman et al. (SPIFFI)	proxy-at-client	space striping	–	–
Lee et al. (Server Array and RAIS)	proxy-at-client	space striping	client pull	striping w/ parity; FEC and PRT
Lougher et al.	independent proxy	space striping	–	–
Reddy et al.	proxy-at-server, independent proxy	space striping	server push	–
Tewari et al. (clustered video server)	proxy-at-server, independent proxy	space striping	server push	–
Wu and Shu	proxy-at-server, independent proxy	space striping and time striping	server push	–

References

[1] H. Taylor, D. Chin, and S. Knight, The Magic Video-on-Demand Server and Real-Time Simulation System, *IEEE Parallel and Distributed Technology: Systems and Applications*, vol. 3, no. 2, 1995, pp. 40–51.

[2] R. Buck, The Oracle Media Server for nCube Massively Parallel Systems, *Proceedings of 8th International Parallel Processing Symposium*, 1994, pp. 670–73.

[3] N. Venkatasubramanian and S. Ramanthan, Load Management in Distributed Video Servers, *Proceedings of 17th International Conference on Distributed Computing Systems*, Baltimore, MD, USA, May 1997, pp. 528–535.

[4] C.C. Bisdikian and B.V. Patel, Issues on Movie Allocation in Distributed Video-on-Demand Systems, *Proceedings of ICC'95*, Seattle, WA, USA, June 1995, pp. 250–255.

[5] A.L. Cherenal, Tertiary Storage: An Evaluation of New Applications, PhD dissertation, University of California, Berkeley, Dec. 1994.

[6] C. Griwodz, M. Bar, and L.C. Wolf, Long-term Movie Popularity Models in Video-on-Demand Systems or The Life of an On-Demand Movie, *Proceedings of Multimedia '97*, Nov. 1997, pp. 349–357.

[7] Y.S. Chen, Mathematical Modeling of Empirical Laws in Computer Application: A Case Study, *Comput. Math. Applicat.*, Oct. 1992, pp. 77–87.

[8] E. Biersack, W. Geyer, and C. Bernhardt, Intra- and Inter-Stream Synchronization for Stored Multimedia Streams, *Proceedings of IEEE International Conference on Multimedia Computing & Systems*, 17–23, June 1996, Hiroshima, Japan.

[9] C. Bernhardt and E. Biersack, The Server Array: A Scalable Video Server Architecture, in *High-Speed Networks for Multimedia Applications*, Kluwer Press, 1996.

[10] W.J. Bolosky, J.S. Barrera, III, R.P. Draves, R.P. Fitzgerald, G.A. Gibson, M.B. Jones, S.P. Levi, N.P. Myhrvold, and R.F. Rashid, The Tiger Video Fileserver, *Proceedings of of the Sixth International Workshop on Network and Operating System Support for Digital Audio and Video*. IEEE Computer Society, Zushi, Japan, April 1996.

[11] M.M. Buddhikot and G.M. Parulkar, Efficient Data Layout, Scheduling and Playout Control in MARS, *Proceedings of NOSSDAV'95*, 1995.

[12] C.S. Freedman and D.J. DeWitt, The SPIFFI Scalable Video-on-Demand System, *Proceedings of ACM SIGMOD'95*, June 1995.

[13] Y.B. Lee and P.C. Wong, A Server Array Approach for Video-on-demand Service on Local Area Networks, *IEEE INFOCOM '96*, California, USA, March 1996.

[14] P.C. Wong and Y.B. Lee, Redundant Array of Inexpensive Servers (RAIS) for On-Demand Multimedia Services, *Proceedings of ICC'97*, Montreal, Canada, June 8–12, 1997.

[15] P. Lougher, D. Pegler, and D. Shepherd, Scalable Storage Servers for Digital Audio and Video, *Proceedings of IEE International Conference on Storage and Recording Systems 1994*, University of Keele, Keele, UK, 5–7 April 1994, pp. 140–143.

[16] A. Reddy, Scheduling and Data Distribution in a Multiprocessor Video Server, *Proceedings of Second IEEE International Conference on Multimedia Computing and Systems*, 1995.

[17] R. Tewari, R. Mukherjee, and D.M. Dias, Real-Time Issues for Clustered Multimedia Servers, *IBM Research Report RC20020*, June 1995.

[18] M. Wu and W. Shu, Scheduling for Large-Scale Parallel Video Servers, *Proceedings of Sixth Symposium on the Frontiers of Massively Parallel Computation*, Oct. 1996, pp. 126–133.

[19] D.A. Patterson, P. Chen, G. Gibson, and R.H. Katz, Introduction to Redundant Array of Inexpensive Disks (RAID), *COMPCON Spring'89*, 1989, pp. 112–117.

[20] C. Brendan, S. Traw, and J.M. Smith, Striping within the Network Subsystem, *IEEE Network*, July/August 1995, pp. 22–32.

[21] A.L. Drapeau and R.H. Katz, Striped Tape Arrays, *Proceedings of 12th IEEE Symposium on Mass Storage Systems*, IEEE Press, 1993, pp. 257–265.

[22] L. Rizzo, Effective Erasure Codes for Reliable Computer Communication Protocols, *ACM Computer Communication Review*, vol. 27, no. 2, Apr. 1997, pp. 24–36.

[23] J.S. Plank, A Tutorial on Reed-Solomon Coding for Fault-Tolerance in RAID-like Systems, *Software Practice and Experience*, vol. 27, no. 9, Sep. 1997, pp. 995–1012.

10

A Concurrent-Push Parallel Server Architecture

This chapter presents and analyzes the performance of a concurrent-push parallel server architecture for building high-capacity streaming servers from many low-cost commodity servers. The architecture adopts server striping for load balancing and extends the server-push service model for use in the parallel-server architecture. This architecture is particularly suitable for deployment in networks that support resource allocation and QoS control as the data flows generated from the servers have a constant data rate. This architecture can potentially be scaled up to more than ten thousand concurrent streams.

10.1 Introduction

In this chapter, we investigate a concurrent-push parallel server architecture for designing scalable media streaming systems such as video-on-demand (VoD) systems. We employ server striping to achieve load sharing across multiple servers without requiring the additional storage overhead incurred in replication. Furthermore, by striping using a small unit size, the system is insensitive to skewness in the videos' popularities. This architecture allows one to incrementally scale up the system capacity to more concurrent users by adding (rather than replacing) more servers and redistributing (rather than duplicating) video data among them.

In the following, we present and analyze quantitatively the concurrent-push scheduling algorithm for scheduling disk retrieval and network transmission in parallel video servers. We show that a simple extension of the server-push service model for parallel video servers does not perform well and introduce an Asynchronous Grouped Sweeping Scheme (AGSS) to improve the system performance. Next, we present a Sub-Schedule Striping (SSS) scheme to further increase the scalability of the architecture. Using numerical results with realistic assumptions, we show that the resultant architecture can be scaled up to more than ten thousand concurrent users.

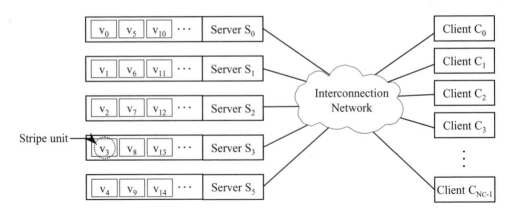

Figure 10.1 Architecture of a (5-servers) parallel video server

10.2 System Architecture

A parallel video server is composed of multiple independent servers connected by an inter-connection network (Figure 10.1). Each server has separate CPU, memory, disk storage, and network interface. This approach ensures that the scalability of the system will not be limited by resource contention. The interconnection network can be implemented using off-the-shelf packet switches like Ethernet switches or ATM switches. We denote the number of servers in the system by N_S and the number of clients by N_C. Hence the client–server ratio, denoted by Λ, is N_C/N_S. The following sections summarize the server striping algorithm, the service model, and the scheduling algorithm employed.

10.2.1 Server Striping

The principle behind the parallel video server architecture is the striping of a video stream across all servers in the system. A server's storage space is divided into fixed-size stripe units of Q bytes each. Each video title is then striped into blocks of Q bytes and stored into the servers in a round-robin manner as shown in Figure 10.1. This fixed-size block striping algorithm is called space striping, as opposed to striping in units of video frames, called time striping (cf. Section 9.3.1).

Space striping significantly simplifies the process of striping video streams encoded using inter-frame compression algorithms (e.g., MPEG), where frame size varies considerably for different frame types. Since a stripe unit in space striping is significantly smaller than a video title (kilobytes versus megabytes), this enables fine-grain load sharing (as opposed to coarse-grain load sharing in data partition) among servers. Moreover, the loads are evenly distributed over all servers independent of the skewness in video retrievals.

10.2.2 Service Model

Service model refers to the way video data are scheduled and delivered to the client. There are two service models in common use: *client pull* and *server push*. In the client-pull model, a

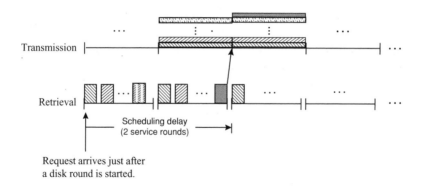

Figure 10.2   Scheduling disk retrieval and network transmission at server

client periodically sends a request to a server to retrieve video data. In this model, the data flow is driven by the client. In the server-push model, the server schedules the periodic retrieval and transmission of video data once a video session has started.

The server-push model is common among studies on single-server VoD systems [1–5]. This model allows one to design periodic schedulers [5] to optimize disk and network utilization. In the next section, we present an extension of this service model for use in parallel video servers.

10.2.3 Scheduling Algorithm

The parallel server architecture employs a *concurrent-push* algorithm to schedule disk retrievals and network transmissions at the servers. The principle behind the concurrent-push algorithm is to let all servers continuously transmit data to a client concurrently. We assume that the average video rate is homogenous for all clients, and is denoted by R_V. Since there are a total of N_S servers, each server only needs to transmit at a reduced rate of R_V/N_S to maintain an aggregate data rate of R_V.

Figure 10.2 depicts the scheduling algorithm for disk retrievals and network transmissions at each server in the system. For each video session, one block of Q bytes video data is retrieved into a disk buffer in each disk service round. To reduce seek overhead, requests within a service round can be served using the SCAN or the C-SCAN disk-arm scheduling algorithms (cf. Chapter 3). The retrieved video block is then passed to a network buffer for transmission in the next round. Therefore, if the disk service round is shorter than one transmission round, a video block will always be ready for transmission. We analyze the performance of the system under this scheduling algorithm in the next section.

10.3 Analysis of the Concurrent-Push Algorithm

In general, the internal clock of each autonomous server in the system is not precisely synchronized. Therefore, the scheduling algorithm must take this server asynchrony into account and compensate accordingly. We define *clock jitter* as the difference between the internal real-time clocks of two servers. Many algorithms for controlling clock jitter between distributed computers have been studied [6, 7] and hence will not be pursued further here. We simply assume that the maximum clock jitter between any two servers in the system is bounded and is denoted by τ. For simplicity, we ignore network delay jitter in this study. Assuming that network delay jitter is bounded (which is true in ATM networks with QoS guarantees), it is easy to see that the effect of network delay jitter can be incorporated into our performance model in the same way as clock jitter, and the same derivations are still valid.

In the following sections, we derive three key performance metrics for evaluating the parallel video server architecture, namely, server buffer requirement, client buffer requirement, and system response time.

10.3.1 Server Scheduling

Under concurrent push, the client will be receiving N_S video blocks simultaneously at an aggregate rate of R_V. The average filling time, defined as the time to completely transmit a video block of Q bytes, is given by

$$T_F = \frac{N_S Q}{R_V} \tag{10.1}$$

On the other hand, each server will be serving at most ΛN_S concurrent video sessions. Under the SCAN disk scheduler, ΛN_S video blocks will be retrieved in each service round for transmission at a rate of R_V/N_S per video stream. Hence the duration of a service round is equal to T_F in equation (10.1) and two buffers are needed for each video stream for a total of $2\Lambda N_S Q$ bytes buffers at each server.

As server clocks are not synchronous, the service round of the servers may not be aligned (see Figure 10.3). Without loss of generality, we assume a video title is striped with block zero storing at server zero. Let $T_{i,j}$ be the time server i $(0 \le i < N_S)$ starts transmitting the $(jN_S + i)$th block of a video stream. Then we can formally define *transmission jitter* as:

$$\delta = \max \left\{ \left| T_{i,j} - T_{k,j} \right| \, | \forall i, k, j \right\} \tag{10.2}$$

It may appear that the maximum clock jitter τ also bounds the transmission jitter. However, it turns out that the transmission jitter not only depends on the clock jitter, but also depends on the arrival time of a new video session request as depicted in Figure 10-4. We derive the upper bound for the transmission jitter in Theorem 10.1:

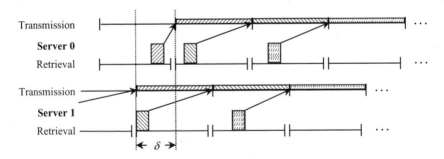

Figure 10.3 Service round misalignment between different servers

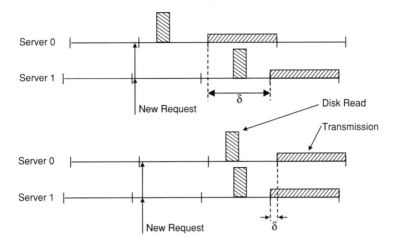

Figure 10.4 Transmission jitter depends on both clock jitter and request arrival time

Theorem 10.1. *Assume that new-session requests arrive at all servers at the same time, then the transmission jitter is bounded by*

$$\delta \leq T_F \tag{10.3}$$

Proof. Please refer to the Appendix. ∎

This bound on transmission jitter will be used to derive the amount of buffer required at the client to prevent buffer underflow and overflow respectively.

10.3.2 Video Block Consumption Model

Many studies on VoD systems assume that video data are consumed periodically by the video decoder. However, our experience in programming some off-the-shelf hardware and software video decoders reveals that the decoder consumes fixed-size data blocks only quasi-periodically.

Given the average video data rate, R_V, and block size, Q, the average time for a video decoder to consume a single block is

$$T_{avg} = \frac{Q}{R_V} \tag{10.4}$$

To quantify the randomness of video block consumption time, we first define a few notations.

Definition 10.1. *(a) Let T_i be the time the video decoder start decoding the ith video block, then the decoding-time deviation of video block i is defined as*

$$T_{DV}(i) = T_i - i T_{avg} - T_0 \tag{10.5}$$

and decoding is late if $T_{DV}(i) > 0$ and early if $T_{DV}(i) < 0$.

(b) The maximum lag in decoding, T_L, and maximum advance in decoding, T_E, are defined as:

$$T_L = \max\{T_{DV}(i)|\forall i \geq 0\} \tag{10.6}$$

$$T_E = \min\{T_{DV}(i)|\forall i \geq 0\} \tag{10.7}$$

(c) The peak-to-peak decoding-time deviation is defined as

$$T_{DV} = T_L - T_E \tag{10.8}$$

Assume the bounds T_L and T_E are known, the time between the consumption of two video blocks i and j $(j > i)$ will be bounded by

$$\max\{((j-i)T_{avg} - T_{DV}), 0\} \leq t \leq ((j-i)T_{avg} + T_{DV}) \tag{10.9}$$

We use buffers at the client to absorb these variations to prevent buffer underflow and buffer overflow during playback. Let there be $L_C = (Y + Z)$ buffers (each Q bytes) at the client, organized as a circular buffer. The client starts video playback once the first Y buffers are completely filled with video data. We prefill buffers before playback to avoid buffer underflow, and reserve the last Z buffers for incoming data to avoid buffer overflow.

10.3.3 Buffer Needed to Prevent Underflow

Since all N_S servers transmit data to a client concurrently, the client will be receiving N_S video blocks simultaneously. Hence Y must be multiples of N_S. We let $y = Y/N_S$ and consider groups of N_S buffers in the follow derivations (i.e., group zero consists of blocks 0 to N_S-1, group one consists of blocks N_S to $2N_S-1$, and so on.).

Among the N_S servers, let the earliest transmission for the first round start at time t_0, then the last transmission for the first round must start at time $t_0 + \delta$. Therefore, the time for video block group i to be completely filled, denoted by $F(i)$, is bounded by

$$((i+1)T_F + t_0 + f^-) \leq F(i) \leq ((i+1)T_F + t_0 + \delta + f^+) \tag{10.10}$$

where f^+ $(f^+ \geq 0)$ and f^- $(f^- \leq 0)$ are used to model the maximum transmission time deviation due to randomness in the system, including transmission rate deviation, CPU scheduling, bus contention, etc.

Since the client starts playing the video after filling the first y groups of buffers, the playback time for video block group 0 is simply $F(y-1)$. From Section 10.3.2, setting $T_0 = F(y-1)$ then the playback time for video block group i, denoted by $P(i)$, is bounded by

$$\{iN_sT_{avg} + F(y-1) + T_E\} \leq P(i) \leq \{iN_sT_{avg} + F(y-1) + T_L\} \tag{10.11}$$

To guarantee video playback continuity, we must ensure that a video block group arrives before playback deadline. In the worst-case scenario, the latest filling time must be smaller than the earliest playback time, i.e.

$$\max\{F(i)\} < \min\{P(i)\} \tag{10.12}$$

Now for the L.H.S., noting that $N_S T_{avg} = T_F$ (cf. equations (10.1) and (10.4)) we then have

$$\max\{F(i)\} = (i + 1)T_F + t_0 + \max\{\delta\} f^+ \qquad (10.13)$$

Using the upper-bound for δ from Theorem 10.1 we obtain

$$\max\{F(i)\} = (i + 1)T_F + t_0 + T_F + f^+ \qquad (10.14)$$
$$= (i + 2)T_F + t_0 + f^+$$

Similarly, the R.H.S. is

$$\min\{P(i)\} = i N_S T_{avg} + \min\{F(y - 1)\} + T_E \qquad (10.15)$$
$$= i T_F + y T_F + t_0 + f^- + T_E$$

Merging equations (10.14) and (10.15), we then have

$$(i + 2)T_F + t_0 + f^+ < (i + y)T_F + t_0 + f^- + T_E \qquad (10.16)$$

Rearranging, we can then obtain y:

$$y > 2 + \frac{f^+ - f^- - T_E}{T_F} \qquad (10.17)$$

Knowing the number of groups required, we can then obtain Y from

$$Y = \left\lceil 2 + \frac{f^+ - f^- - T_E}{T_F} \right\rceil N_S \qquad (10.18)$$

10.3.4 Buffer Needed to Prevent Overflow

On the other hand, to guarantee that the client buffer will not be overwhelmed by incoming video data, we need to ensure that the ith video block group starts playback before the $(i + l - 2)$th video block group is completely received, where $l = L_C/N_S$. This is because the client buffers are organized as a circular buffer, and we must have at least one group of N_S free buffers available for video blocks arriving simultaneously from N_S servers. Therefore, we need to ensure that the earliest filling time for group $(i + l - 2)$ must be larger than the latest playback time for group i:

$$\min\{F(i + l - 2)\} > \max\{P(i)\} \qquad (10.19)$$

Using derivations similar to the previous section, we can obtain the number of buffers needed to prevent buffer overflow as:

$$Z = \left\lceil 2 + \frac{f^+ - f^- + T_L}{T_F} \right\rceil N_s \qquad (10.20)$$

10.3.5 System Response Time

Response time is defined as the time from the user request for a new video session to the time actual video playback starts. This delay comprises two components: *scheduling delay*, and *prefill delay*. Scheduling delay is the time from a client sending a new-session request to the time transmission starts at the server. It is easy to see that the worst-case scheduling delay is two service rounds (see Figure 10.2):

$$D_S = \frac{2N_S Q}{R_V} \tag{10.21}$$

Prefill delay is the time from the server starts transmission to the time the first y groups of client buffers are fully filled with data. Using equation (10.10), the worst-case prefill delay can be obtained from

$$D_P = \max\{F(y-1)\} - t_0 \tag{10.22}$$

or

$$D_P = yT_F + \max\{\delta\} + f^+ = (y+1)T_F + f^+ \tag{10.23}$$

$$= \left(3 + \left\lceil \frac{f^+ - f^- - T_E}{T_F} \right\rceil \right) T_F + f^+$$

10.4 Asynchronous Grouped Sweeping Scheme

The results in the previous section reveal an important characteristic of the concurrent push algorithm, namely, the server buffer requirement, the client buffer requirement, and the response time all increase with the number of servers in the system. Therefore, the scalability of the system will either be limited by the economy of memory buffers or the tolerance of the system response time by the user. In this section, we propose an extension of the Grouped Sweeping Scheme (GSS) [8], called Asynchronous Group Sweeping Scheme (AGSS) to substantially reduce server buffer requirement, and scheduling delay.

10.4.1 Extending the Grouped Sweeping Scheme

The original GSS algorithm [8] is designed for scheduling retrieval requests in a magnetic disk. The traditional First-In-First-Out scheduling algorithm has poor disk utilization in continuous-media applications because in the worst case the disk arm may need to seek back and forth between the innermost track and the outermost track, thus wasting a lot of time in seeking (cf. Chapter 3). Instead, some systems use the SCAN scheduling algorithm to reduce seek-time overhead by serving requests while the disk arm scans across the disk surface. However, this approach requires two buffers per stream because requests may be served out of order and in the worst case, two requests for the same stream may be served in a back-to-back manner.

The GSS algorithm is designed to strike balance between minimizing seek-time overhead and minimizing buffer requirement by serving streams in groups. Streams within a group are served using SCAN to reduce seek-time overhead while the groups are served in a fixed order to reduce buffer requirement. By varying the number of groups, one can trade-off disk utilization against buffer requirement.

To extend GSS for use in parallel video servers, we propose dividing a service round into $G = gN_S$ groups, where g can be determined using the single-server model [8] to minimize buffer requirement while still meeting the playout requirement. Assume that a single server can serve at most Λ video sessions, then each group serves up to (Λ/g) video sessions. It is easy to see that this holds for two or more servers as well. Therefore, the number of disk buffers needed is reduced from ΛN_S to Λ, though we still need ΛN_S network buffers because a video block is transmitted at a lower data rate of R_V/N_S. Under this extended GSS algorithm, the total amount of server buffer required will be

$$B_{server} = QN_S\Lambda\left(1 + \frac{1}{G}\right) \tag{10.24}$$

10.4.2 Uneven Group Assignment and Admission Scheduling

The AGSS algorithm described in the previous section has a subtle problem when the servers in the system are not clock-synchronized. Figure 10.5 illustrates the problem using the arrivals of two new-session requests. As shown in Figure 10.5 while server zero assigns the two new sessions into different groups, server one assigns them into the same group. This can occur because each server assigns the new session to a group according to its own internal clock, which may be different from other servers due to clock jitter. Eventually, the group occupancy among servers may deviate in such a way that one server can accept a new video session immediately while others have to wait for an available group, thereby increasing the transmission jitter.

To reduce the transmission jitter (which also reduces buffer requirement at the client), we propose adding an admission scheduler to handle group assignment for new-session requests. To initiate a new video session, a client will first send a request to the admission scheduler, which maintains the same clock jitter bound with the servers. As new sessions are assigned solely according to the admission scheduler's clock, the scenario depicted in Figure 10.5 will not occur. To ensure that the assigned group has not started in any of the servers due to clock

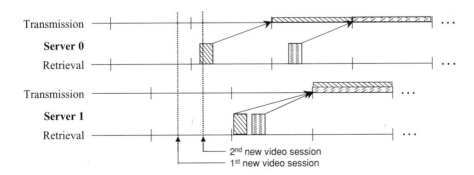

Figure 10.5   Uneven service round assignments

jitter, the admission scheduler adds an extra delay to the assignment, stated in the following theorem:

Theorem 10.2. *If the admission scheduler delays the start of a new video session by*

$$\Omega = \left\lceil \frac{\tau G}{T_F} \right\rceil + 1 \tag{10.25}$$

groups, then it guarantees that the assigned group has not started in any of the N_S servers.

Proof. Please refer to the Appendix.   ∎

Note that if the assigned group is full, the admission scheduler will sequentially check the subsequent groups until an available group is found.

10.4.3 Client Buffer Requirement

As the admission scheduler already guarantees that a new video session will be assigned to the same group in all servers, the scenario in Figure 10.5 could not occur and the transmission jitter will be the same as the clock jitter. Hence, the client buffer requirement derived in Section 10.3 becomes

$$Y = \left\lceil 1 + \frac{\tau + f^+ - f^- - T_E}{T_F} \right\rceil N_S \tag{10.26}$$

$$Z = \left\lceil 1 + \frac{\tau + f^+ - f^- + T_L}{T_F} \right\rceil N_S \tag{10.27}$$

10.4.4 System Response Time

The scheduling delay under the AGSS algorithm depends on the occupancy of the AGSS groups. Specifically, if a group as calculated from Theorem 10.2 is fully occupied, the new video session must be delayed until the next available group. In the worst case, the transmission of the first video block is delayed for $(N_S + \Omega)$ groups:

$$D_S = \left(N_S + \left\lceil \frac{\tau G}{T_F} \right\rceil + 1 \right) \frac{Q}{R_V} \tag{10.28}$$

To better evaluate the scheduling delay, we derive the average scheduling delay under a given system load. Assume that video sessions start independently and with equal likelihood at any time. Then a video session can be assigned to any one of the G groups with equal probability. Let there be n active video sessions and G groups, then the number of ways to distribute these n video sessions among G groups is a variant of the urn-occupancy distribution problem [9] and is given by

$$N(n, G, \Lambda) = \sum_{j=0}^{G} (-1)^j \binom{G}{j} \binom{G + n - j(\Lambda + 1) - 1}{G - 1} \tag{10.29}$$

To obtain the probability of having m fully-occupied groups, we first notice that there are $\binom{G}{m}$ possible combinations of picking m fully-occupied groups among G groups. Given that there are n active video sessions and m fully-occupied groups, the number of ways to distribute the remaining $(n - m\Lambda/g)$ video sessions among the remaining $(G - m)$ groups with none of those groups fully occupied can be obtained from equation (10.29) as $N(n - m\Lambda/g, G - m, (\Lambda/g) - 1)$. Hence the total number of ways for exactly m of the groups fully occupied is given by

$$N_{full}(n, m) = \binom{G}{m} N(n - m\frac{\Lambda}{g}, G - m, \frac{\Lambda}{g} - 1) \qquad (10.30)$$

The probability of having m fully-occupied groups given n active video sessions can then be obtained from

$$P_{full}(n, m) = \frac{N_{full}(n, m)}{N(n, G, \Lambda)} \qquad (10.31)$$

Knowing this, we can derive the average scheduling delay in the following way. Given m out of G groups are fully occupied, the probability for the assigned group to be available (not fully occupied) is given by

$$V_0 = \frac{G - m}{G} \qquad (10.32)$$

Hence $P_0 = (1 - V_0)$ will be the probability of the assigned group being fully occupied. It can be shown that the probability for a client to wait k additional groups provided that the first k assigned groups are all fully occupied is

$$V_k = \Pr\{(k + 1)\text{th group available} \,|\, P_k\} = \frac{G - m}{G - k}, \qquad 1 \le k \le m \qquad (10.33)$$

and the probability for the first k groups all being fully occupied is

$$P_k = \prod_{i=0}^{k-1} \left(\frac{m - k}{G - i}\right) = \frac{m!(G - k)!}{G!(m - k)!}, \qquad 1 \le k \le m. \qquad (10.34)$$

Hence, we can solve for the probability of a client having to wait k additional groups, denoted by W_k, from

$$W_k = \Pr\{(k + 1)\text{th group free} \,|\, P_k\} P_k = \frac{(G - m)m!(G - k - 1)!}{G!(m - k)!}, \qquad 1 \le k \le m. \qquad (10.35)$$

Therefore, given the number of groups that are fully occupied m, the average number of groups a client has to wait can be obtained from

$$W_{avg}(m) = \sum_{k=1}^{m} k W_k + \left\lceil \frac{\tau G}{T_F} \right\rceil + 1 \qquad (10.36)$$

Similarly, given the number of active video sessions n, the average number of groups a client has to wait can be obtained from equations (10.31) and (10.36) as follows:

$$M_{avg}(n) = \sum_{j=1}^{G-1} W_{avg}(j) P_{full}(n, j) \qquad (10.37)$$

And the corresponding average scheduling delay given a system utilization of n is

$$D_S = \frac{M_{avg}(n)Q}{R_V} \tag{10.38}$$

As the admission scheduler reduces the transmission jitter to equal to the clock jitter, the new prefill delay can be obtained by replacing δ with τ in equation (10.23):

$$D_P = \left(2 + \left\lceil \frac{\tau + f^+ - f^- - T_E}{T_F} \right\rceil \right) T_F + f^+ \tag{10.39}$$

10.5 Sub-Schedule Striping Scheme

The AGSS algorithm presented in the previous section substantially reduces the server buffer requirement as well as the scheduling delay. However, the client buffer requirement and, consequently, the prefill delay are only slightly reduced as a side effect of the admission scheduler. In this section, we consider another modification to the concurrent-push algorithm that can substantially reduce the client buffer requirement and the prefill delay.

Specifically, the analysis in Section 10.3 reveals that the main reason for the increase in client buffer requirement with the number of servers stems from the increase in the average filling time in equation (10.1). This suggests that we can reduce the buffer requirement by using smaller striping size Q. However, as the server retrieves data from the disk in units of Q bytes, reducing the striping size will adversely affect disk retrieval efficiency.

To solve this problem, we propose decoupling the transaction size for disk retrieval and transmission from the striping size – *sub-schedule striping* (SSS). In particular, we maintain the disk transaction size at Q bytes but use a striping size (denoted by U) inversely proportional to the number of servers in the system (Figure 10.6):

$$U = Q/N_S \tag{10.40}$$

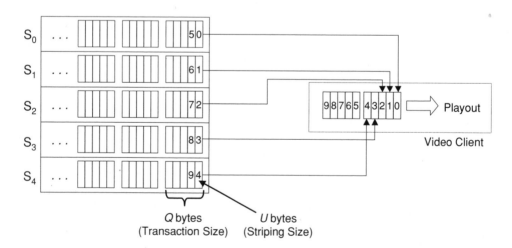

Figure 10.6 Data organization in sub-schedule striping

Hence the disk will retrieve N_S stripe units in a single transaction. Note that the client continues to consume video data in blocks of Q bytes and hence the video-block consumption model in Section 10.3.2 remains valid. However, a video block now contains stripe units transmitted from all N_S servers (Figure 10.6) rather than from a single server as in the original algorithm. Consequently, the client buffer size Y and Z no longer need to be multiples of N_S.

Sub-schedule striping requires no modification to the server as the transaction size remains the same. Therefore, the server buffer requirement, as well as scheduling delay are the same as before.

To model the effect on the client buffer requirement, we note that a Q-bytes video block comprises fragments from all N_S servers. Hence, the filling time for a video block would be affected by the transmission jitter among servers. Specifically, the filling time for video block i of a video stream started at time t_0 is bounded by

$$((i + 1) T_{avg} + t_0 + f^-) \leq f(i) \leq ((i + 1) T_{avg} + t_0 + f^+ + \tau) \tag{10.41}$$

Using similar derivations, the client buffers needed to prevent underflow and overflow can be found to be:

$$Y > 1 + \left(\frac{f^+ - f^- - T_E + \tau}{T_{avg}} \right) \tag{10.42}$$

$$Z > 1 + \left(\frac{f^+ - f^- + T_L + \tau}{T_{avg}} \right) \tag{10.43}$$

and the time to prefill the first Y client buffers is

$$D_P = Y T_{avg} + f^+ + \tau \tag{10.44}$$

Now both the client buffer requirement and prefill delay no longer depend on the number of servers in the system.

10.6 Performance Evaluation

In this section, we evaluate the performance of the parallel video server architecture studied in this chapter using numerical results. Table 10.1 lists the values for the key system parameters used in the calculation. The parameters T_E and T_L are determined empirically by collecting the video block consumption times of a hardware MPEG-1 decoder.

10.6.1 Server Buffer Requirement

Figure 10.7a plots the per-server buffer requirement versus the number of servers in the system. We can observe that AGSS substantially reduces the buffer requirement. Sub-schedule striping has no effect on the server buffer requirement. Despite the reduction achieved by AGSS, the server buffer requirement still increases with the number of servers. This poses one limitation on the ultimate scalability of the system (to be discussed in Section 10.6.4). Depending on

Table 10.1 System parameters used in performance evaluation

System parameters	Symbol	Value
Video block size	Q	65,536 Bytes
Video data rate	R_V	150 KB/s
Maximum advance in decoding time	T_E	-130 ms
Maximum lag in decoding time	T_L	160 ms
Client–server ratio	Λ	10
Transmission time deviation	f^-, f^+	0 ms
Server clock jitter	τ	100 ms
AGSS parameter	g	1

the relative cost of memory and disk bandwidth, one can reduce system cost by trading disk efficiency for smaller server buffer requirement.

10.6.2 Client Buffer Requirement

Figure 10.7b plots the client buffer requirement versus the number of servers in the system. Figure 10.7b shows that AGSS substantially reduces the client buffer requirement but it still increases linearly with the number of servers. With the addition of sub-schedule striping, the client buffer requirement is constant, regardless of the number of servers in the system. This is a crucial property as it would be impractical to upgrade all clients whenever more servers are added to the system in practice.

From equations (10.42) and (10.43), it is easy to see that the client buffer requirement is insensitive to the server clock jitter. As an example, for a 16-server system with AGSS and sub-schedule striping, the client buffer requirement is only 384 KB for a server clock jitter as large as 1,000 ms.

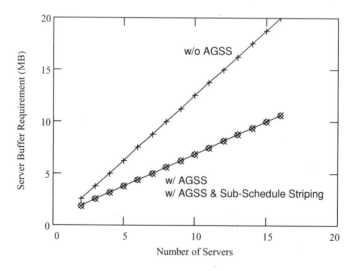

Figure 10.7a Server buffer requirement versus number of servers

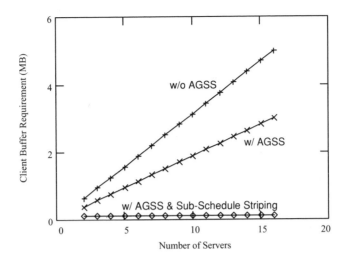

Figure 10.7b   Client buffer requirement versus number of servers

10.6.3 System Response Time

We first plot scheduling delay versus the number of servers in the system in Figure 10.8a. The case for sub-schedule striping is not plotted, as sub-schedule striping has no effect on the scheduling delay. Note that the worst-case scheduling delay is substantially reduced by AGSS, especially for large number of servers. Moreover, the average scheduling delay with AGSS is even smaller and stays relatively constant, regardless of the number of servers in the system. For example, with AGSS striping in a 16-server system, the average delay is only 1.26 seconds for system utilization as high as 90% even though the worst-case scenario is 9.18 seconds. The worst-case delay is even larger (13.98 seconds) without AGSS. Hence with AGSS, we can maintain a reasonably short scheduling delay by operating the system to within, say, 90% of the total capacity.

Figure 10.8b plots the prefill delay versus the number of servers in the system. The results show that the prefill delay is also reduced by AGSS because the worst-case transmission jitter T_F is larger than the clock jitter τ. More importantly, by using sub-schedule striping, the prefill delay becomes completely independent of the number of servers in the system.

Finally, we plot the total system response time in Figure 10.9. Clearly the proposed AGSS and sub-schedule striping can effectively maintain a small system response time (1.8 seconds for $N_S = 16$ at 90% utilization) even if the number of servers is large.

10.6.4 Scalability

The results in the previous sections have shown that both the client buffer requirement and the system response time can be kept low irrespective of the number of servers in the system. Server buffer requirement is the only factor that increases with more servers. This factor will certainly limit the ultimate scalability of the system. Nowadays, it is common to install 256MB or more memory in a PC-based server as memory price has dropped substantially. Under our system parameters and ignoring operating system overhead, a 256MB memory size will limit

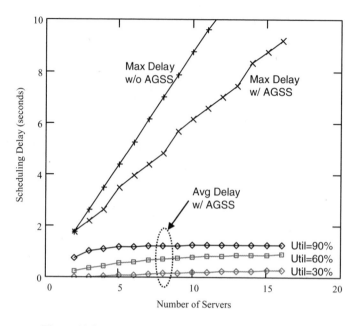

Figure 10.8a   Scheduling delay versus number of servers

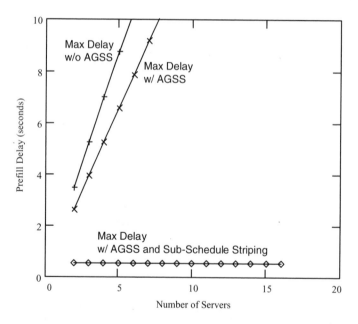

Figure 10.8b   Prefill delay versus number of servers

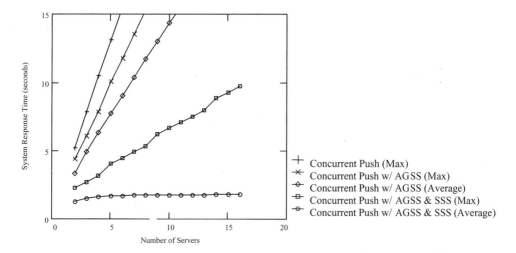

Figure 10.9   System response time versus number of servers

the scalability of the parallel video server architecture to a maximum of 408 servers serving a total of 3,672 concurrent video sessions at 90% utilization. If 1GB memory is available, the architecture can be scaled up to 14,400 concurrent video sessions using a client–server ratio of 250 at 90% utilization.

A second, more subtle limiting factor is due to the sub-schedule striping scheme. Under this scheme, the client must resequence the incoming data by copying U-bytes stripe units into the client buffer (Figure 10.6). Hence the processing overhead will likely increase with smaller striping size. Our previous experiences showed that processing overhead remains practical for software implementations running in even low-end PCs for striping size as small as 1KB. This limits N_S to 64. For larger systems, we can use more powerful server hardware with a larger client–server ratio to avoid reaching this limit. In the previous example with 1GB memory, we increase the client–server ratio to 250 to limit to a total of 64 servers. Clearly the rapid improvement in CPU speed will undoubtedly extend this limit.

10.7 Summary

In this chapter, we have presented and analyzed a concurrent-push parallel video server architecture for designing scalable video-on-demand systems. The proposed architecture employs fixed-size block striping and the server-push service model. To schedule disk retrievals and transmissions, we introduced a concurrent-push scheduling algorithm where video data are continuously transmitted from all servers to a client station. This constant-bit-rate traffic produced by the algorithm enables us to take advantage of the quality-of-service guarantees provided by the networks. To extend the scalability of the architecture, we introduced the Asynchronous Grouped Sweeping Scheme and the Sub-Schedule Striping Scheme into the architecture. Results showed that the resultant architecture can be scaled up to more than ten thousand concurrent users with acceptable buffer requirement and system response time.

Building video-on-demand systems upon parallel server architecture not only breaks through the capacity limit of a single server, but also opens the way to fault-tolerant system designs.

In the next chapter we develop fault-tolerant algorithms for the concurrent-push architecture so that even server failures can be sustained without disruption to ongoing video streams.

Appendices

A.1 Proof of Theorem 10.1

Assume server zero starts the first service round at time t_0. Since the server clocks are not precisely synchronized, we let d_i be the clock difference between server i and server zero. Hence $d_0=0$ and $\max\{|d_i - d_j| \mid \forall i, j\} = \tau$ and server i will start service round j at time $(t_0 + d_i + jT_F)$. Let t_{new} be the time a new-session request arrive at the servers. Then the request will arrive at server i during round v_i:

$$v_i = \left\lfloor \frac{t_{new} - (t_0 + d_i)}{T_F} \right\rfloor \tag{10.45}$$

and the first video block will be retrieved at round $(v_i + 1)$ and transmitted at round $(v_i + 2)$. Hence the transmission jitter between server i and server k for stripe j can be expressed as:

$$\delta_{i,k,j} = (t_0 + d_i + (v_i + 2 + j)T_F) - (t_0 + d_k + (v_k + 2 + j)T_F) \tag{10.46}$$

Substituting equation (10.45) into (10.46) we have

$$\delta_{i,k,j} = \left(d_i + \left\lfloor \frac{t_{new} - (t_0 + d_i)}{T_F} \right\rfloor T_F \right) - \left(d_k + \left\lfloor \frac{t_{new} - (t_0 + d_k)}{T_F} \right\rfloor T_F \right) \tag{10.47}$$

Without loss of generality, we can assume $d_i \geq d_k$ and let $H = \frac{t_{new} - (t_0 + d_i)}{T_F}$. Then we have

$$\delta_{i,k,j} = (d_i + \lfloor H \rfloor T_F) - \left(d_k + \left\lfloor H + \frac{(d_i - d_k)}{T_F} \right\rfloor T_F \right) \tag{10.48}$$

$$= (d_i - d_k) + T_F \left(\lfloor H \rfloor - \left\lfloor H + \frac{(d_i - d_k)}{T_F} \right\rfloor \right)$$

Noting that $\lfloor x + y \rfloor \geq \lfloor x \rfloor + \lfloor y \rfloor$ we have

$$\delta_{i,k,j} \leq (d_i - d_k) + T_F \left(\lfloor H \rfloor - \lfloor H \rfloor - \left\lfloor \frac{(d_i - d_k)}{T_F} \right\rfloor \right) \tag{10.49}$$

$$= (d_i - d_k) - T_F \left\lfloor \frac{(d_i - d_k)}{T_F} \right\rfloor$$

Finally, making use of the result that $\lfloor x/y \rfloor y \geq (x - y)$ we can then obtain

$$\delta_{i,k,j} = (d_i - d_k) - T_F \left\lfloor \frac{(d_i - d_k)}{T_F} \right\rfloor \tag{10.50}$$

$$\leq (d_i - d_k) - ((d_i - d_k) - T_F)$$

$$= T_F$$

and the result follows.

A.2 Proof of Theorem 10.2

Let the new session request arrive at the admission scheduler at time t during group $v_{new} = \lfloor tG/T_F \rfloor$. Then due to clock jitter, the current group at other servers can range from $\lfloor (t-\tau)G/T_F \rfloor$ to $\lfloor (t+\tau)G/T_F \rfloor$. To guarantee that the assigned group has not started in any of the servers implies assigning a group larger than the largest current group in any of the servers, i.e. $s_{new} = \lfloor (t+\tau)G/T_F \rfloor + 1$. Applying the inequality $\lfloor x+y \rfloor \leq \lfloor x \rfloor + \lceil y \rceil$ we have

$$s_{new} \leq \lfloor tG/T_F \rfloor + \lceil \tau G/T_F \rceil + 1 \tag{10.51}$$

Substituting into v_{new} we have $s_{new} \leq v_{new} + \lceil \tau G/T_F \rceil + 1$ and the results follows.   ■

References

[1] T.C. Chiueh, C. Venkatramani, and M. Vernick, *Design and Implementation of the Stony Brook Video Server*, Technical Report TR-16, Computer Science Department, State University of New York at Stony Brook, August 1995.

[2] F.A. Tobagi and J. Pang, StarWorks: A Video Applications Server, *IEEE COMPCON Spring'93*, 1993, pp. 4–11.

[3] R. Buck, The Oracle Media Server for nCube Massively Parallel Systems, *Proceedings of 8th International Parallel Processing Symposium*, 1994, pp. 670–673.

[4] H. Taylor, D. Chin, and S. Knight, The Magic Video-on-Demand Server and Real-Time Simulation System, *IEEE Parallel and Distributed Technology: Systems and Applications*, vol. 3, no. 2, 1995, pp. 40–51.

[5] D.J. Gemmell, H.M. Vin, D.D. Kandlur, P.V. Rangan, and L.A. Rowe, Multimedia Storage Servers: A Tutorial, *IEEE Computer*, vol. 28, no. 5, May 1995, pp. 40–49.

[6] D. Mills, Internet Time Synchronization: The Network Time Protocol, *IEEE Transactions on Comm.*, vol. 39, no. 10, Oct. 1991, pp. 1482–1493.

[7] Z. Yang and T.A. Marsland, (eds) *Global States and Time in Distributed Systems*, IEEE Computer Society Press, Los Alamitos, CA, USA, 1994.

[8] P.S. Yu, M.S. Chen, and D.D. Kandlur, Grouped Sweeping Scheduling for DASD-based Multimedia Storage Management, *ACM Multimedia Systems*, vol. 1, 1993, pp. 99–109.

[9] J.N. Lloyd and K.S. Samuel, *Urn Models and their Application*, John Wiley & Sons, 1997, pp. 125–126.

11

Fault-Tolerant Algorithms for the Concurrent-Push Architecture

One potential problem with the concurrent-push architecture, and any parallel server architectures in general, is reliability. As the system distributes video data over multiple servers, failure of a single server will cripple the entire system. Worst still, as the system is scaled up to more users, more servers will be needed and consequently the system-wide reliability will decrease accordingly. Drawing similar principles from disk array researches, we present in this chapter fault-tolerant algorithms to improve the system reliability.

In particular, we address three key problems pertaining to supporting fault tolerance in the concurrent-push architecture, namely, redundancy management, redundant data transmission protocol, and real-time fault masking. First, redundant data based on erasure codes are introduced to video data stored in the servers, which are then delivered to the clients to support fault tolerance. Despite the success of distributed redundancy striping schemes such as RAID-5 in disk array implementations, we discover that similar schemes extended to the server context do not scale well. Instead, we develop a redundant server scheme that is both scalable and consumes less server buffer. Second, two protocols are introduced to control the transmission of redundant data to the clients, namely, forward erasure correction (FEC) and progressive redundancy transmission (PRT). These two protocols achieve different tradeoffs between bandwidth overhead, implementation complexity, and client buffer requirement. Finally, we derive the amount of client buffers required so that non-stop, continuous video playback can be maintained during server failure.

11.1 Redundancy Management

To support server-level fault tolerance, we need redundant data so that a client can re-compute the unavailable video data after server failures. The problem of correcting data errors has been studied extensively in the literature. According to coding theory [1], one can encode a set of symbols with redundancies so that errors occurring within the set can be corrected later. However, server failure is slightly different in the sense that there is really no error in the coding sense. Instead, a server failure introduces erasures – the absence of data.

Scalable Continuous Media Streaming Systems   Jack Y. B. Lee
© 2005 John Wiley & Sons, Ltd.

Errors and erasures are different, because, for errors, data are still being received but the content may be corrupted. In case of erasure, the expected data are simply missing and hence no erroneous data will be received. Here we have implicitly assumed that the server is fail-stop, i.e., it stops sending out data upon failure. This type of failure could be caused by disk subsystem failure, network failure, power loss, or even software crashes. In any case, erasures are introduced into the video stream because data stored in the failed server will become unavailable.

According to coding theory, to recover an erased symbol (a unit of data) in a codeword (also called a parity group, or a stripe), one needs to encode the data with at least one redundant symbol per codeword. One well-known coding algorithm called Reed-Solomon Erasure correction (RSE) code [2, 3] can encode data with any codeword size and level of redundancies. If one needs to protect the system from only single-server failure, then an even simpler code – parity – can be used instead. For simplicity, we assume in this chapter a generic code where each additional redundant symbol can recover one erasure.

Drawing related principles from RAID [4], Figures 11.1 and 11.2 depict the proposed redundant striping policies for block striping and sub-schedule striping. The basic idea is the same – introduces one or more redundant stripe units in every stripe. The redundant units are

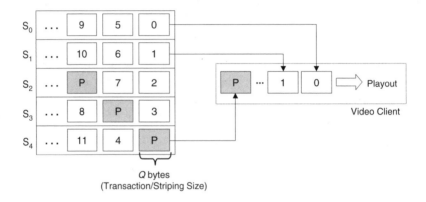

Figure 11.1   Fixed-size block striping with redundancy of one

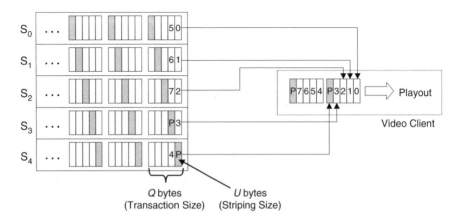

Figure 11.2   Fixed-size sub-schedule striping with redundancy of one

precomputed and distributed to the servers in a round-robin manner similar to a RAID-5 disk array. Note that a parity group spans all servers and hence the parity group size equals the number of servers in the system.

In Chapter 9, we introduced two approaches for transmitting redundant data to the clients, namely, forward erasure correction and progressive redundancy transmission. These two schemes represent different tradeoffs: FEC simplifies system implementation and has lower client buffer requirement and start-up delay in certain cases, at the expense of network bandwidth overhead during normal operation (i.e., no failure); PRT reduces this bandwidth overhead at the expense of more complicated system implementation and potentially larger buffer requirement and start-up delay. We present a FEC-based transmission scheme for concurrent push in the next section, and a PRT-based transmission scheme in Section 11.3.

11.2 Forward Erasure Correction (FEC)

As the name suggests, servers under FEC transmit redundant data regardless of server failure. As redundant data are always received, the client can re-compute unavailable data by erasure correction computation (see Figure 11.3 for the case under sub-schedule striping). Hence, one does not need to detect server failure for the sake of maintaining non-stop operation, and consequently system reconfiguration is also unnecessary. Clearly, this can greatly simplify the implementation and avoid other complications such as false alarm or undetected failure. The tradeoff is extra network bandwidth required to deliver redundant data during normal-mode operation. Specifically, with N_S servers and a redundancy level of K (i.e., up to K simultaneous server failures can be sustained), the network bandwidth overhead incurred will be given by

$$H_{FEC} = \frac{K}{N_S - K} \tag{11.1}$$

For a small-scale system (i.e., N_S small) with high level of redundancy (i.e., K large), this overhead could become prohibitive. For example, with $N_S = 3$ and $K = 1$, the overhead would become 50%. Considering that a VoD system is expected to operate mostly in normal mode,

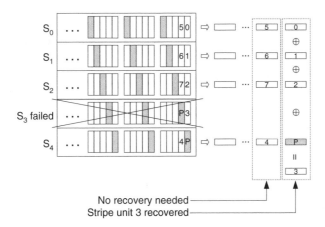

Figure 11.3   Recovery of unavailable stripe units through erasure correction code

this overhead may not be acceptable for systems with a small number of servers. The PRT scheme discussed in the next section is designed to reduce this bandwidth overhead.

11.3 Progressive Redundancy Transmission (PRT)

In PRT, the system does not transmit all the redundant data. Instead, the server transmits only a subset of the redundant data and then activates more redundant data transmissions when a server failure is detected, thus reducing the network bandwidth overhead incurred during normal-mode operation. Obviously the system must be able to detect server failures quickly so that the system can be reconfigured to send additional redundant data. By contrast, this extra step of failure detection is not needed in FEC. For simplicity, we will focus on a special type of PRT algorithm where none of the redundant data are transmitted until a server failure is detected. We will revisit the more general PRT algorithm in Chapter 13 when we compare FEC and PRT in details.

In a system with N_S servers and a video data rate of R_V, each server only needs to transmit at a rate of R_V/N_S – we call it *Min-Rate* transmission. Upon a simultaneous x-server failure ($0 < x \leq K$), the surviving servers will have to increase the transmission rate from R_V/N_S to $R_V/(N_S - x)$ to maintain the same aggregate video bit-rate. This Min-Rate transmission scheme thus requires dynamic reconfiguration of the server scheduler as well as network bandwidth allocations. Alternatively, the system can maintain the transmission rate at $R_V(N_S - K)$ – we call it *Std-Rate* transmission, even when there is no failure. The servers just skipped transmitting the redundant units. When an x-server failure occurs, the system will simply reconfigure x of the servers to start transmitting redundant data, thereby maintaining enough data for erasure correction at the clients. This approach eliminates the need to dynamically reconfigure the server scheduler and network connections.

If the network does not require per-channel resource allocation (e.g., FastEthernet), Min-Rate transmission will have no advantage over Std-Rate transmission, as the average rate is the same for both schemes. On the other hand, if the network requires per-channel resource allocation such as CBR service in Asynchronous Transfer Mode (ATM), then under Min-Rate transmission the servers will need to re-negotiate a higher bandwidth allocation from the network upon detecting a failure. However, reconfiguring hundreds or even thousands of connections simultaneously could overload the network management center, which in turn could delay the reconfiguration process significantly. Therefore, we conclude that the Min-Rate transmission scheme does not offer significant advantages over Std-Rate and is difficult to implement efficiently. By contrast, the Std-Rate transmission scheme is much simpler to implement, and so we will only consider the Std-Rate transmission scheme in the rest of the chapter.

11.3.1 Failure-Detection Protocol

As discussed in the previous section, failure detection is necessary in PRT because redundant data are not normally transmitted. The goal then is to detect a server failure quickly and accurately, so that the remaining servers can be reconfigured to begin transmitting redundant data. If the *detection delay*, defined as the time from a server fails to the time the remaining servers are notified of the failure, is too large, then video playback hiccups can occur at the clients. On the other hand, the detection algorithm should not be overly sensitive in order to avoid false alarms. We present below an admission-scheduler-based (ASB) protocol for detecting server failures.

In our previous investigations, we found that incoming control requests could be delayed for a substantial amount of time (e.g., more than one second) due to intense I/O activities at the servers. Consequently, it would be more difficult to implement server-based fault-detection protocols that can quickly detect a failure. This motivates us to implement fault-detection at the admission scheduler rather than at the servers. The admission scheduler is originally introduced to tackle the uneven group assignment problem arising from server clock jitters (cf. Section 10.4). For fault detection, we extend the admission scheduler to simulate a video client. Unlike real video clients, however, received video data are simply discarded at the admission scheduler after bookkeeping is done, and the scheduler never performs any interactive control nor will the stream ever terminate (until system shutdown). At the servers, video data destined to the admission scheduler are not retrieved from the disks, but rather generated on-the-fly. Since the generated video data will not be interpreted at the admission scheduler, the server can avoid disk overhead by sending the same buffer repeatedly after updating header information such as stream offset or sequence number.

When a server fails, it simply stops transmitting data. Hence, a server failure can be inferred by the lack of video data received at the admission scheduler. We assume that the admission scheduler is located close to the servers so that worst-case arrival deadlines are known for each and every video packet. Then the admission scheduler can declare a server to have failed if the arrival deadline is exceeded by a threshold of say, T_{ASB} seconds. This threshold is introduced to reduce the possibility of false alarms caused by unexpected data delivery delays or occasional packet losses.

Note that the admission scheduler itself could also fail. However, this type of failure will be less problematic because (a) while new streams cannot be started, the failure will not affect existing streams; and (b) compared to the video servers, the admission scheduler is much simpler and hence potentially far more reliable. For example, the admission scheduler can be diskless so that disk failure can be avoided. ECC memory can be used to protect from memory faults, etc.

11.3.2 Server Reconfiguration for Block Striping

Upon declaring that a server has failed, the admission scheduler will send messages to the surviving servers to notify them of the failure. The delay incurred will obviously be implementation-dependent. For simplicity, we assume that the failure-detection delay is bounded and the maximum is given by T_D seconds. Upon receiving the failure notification, the servers will initiate a reconfiguration process to begin transmitting redundant blocks and to *retransmit* the necessary redundant blocks.

Figure 11.4 depicts the scenario for reconfiguring a 5-server system under block striping. Note that we consider only one video stream for illustration and analysis while in practice the same process occurs for all active video streams. All algorithms and procedures still apply and no modification is needed for the multi-stream case. Note also that redundant video blocks are always retrieved, just not transmitted when there is no failure. One might notice that during normal operation, some disk bandwidth would then be wasted in retrieving redundant blocks that are not needed. It is conceivable that one can reuse this wasted bandwidth to serve extra video sessions during normal operation. However, these sessions will have to be disconnected

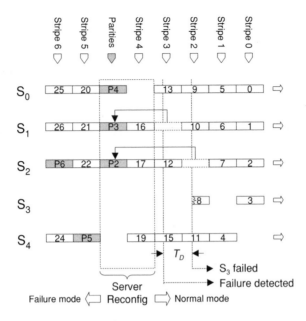

Figure 11.4   Server reconfiguration under PRT with block striping

upon server failure. More investigations are therefore needed to quantify the gains and the associated tradeoffs.

Now assuming that failure occurs during service round j, then the servers will receive the failure notification latest by round

$$k = j + \left\lceil \frac{T_D}{T_F} \right\rceil \tag{11.2}$$

where T_F is the length of a service round defined in equation (11.4) below.

Knowing the failure, the servers will transmit redundant video blocks in addition to video blocks in the next round $(k + 1)$ if there is one (e.g., P4 in Figure 11.4). However, the $(k - j + 1)$ stripes that are transmitted after the failure but before the failure is detected will have no redundant blocks transmitted (e.g., stripes 2 and 3). To enable the client to re-compute the lost stripe units, it is therefore necessary to *retransmit* the required redundant units (e.g., P2 and P3) for these stripes.

In a system with N_S servers and K redundant blocks per stripe, a maximum of $(k - j + 1)K$ redundant blocks will have to be retransmitted. Note that this is the maximum because retransmission is not needed for lost redundant blocks. Assume the failure is a simultaneous K-server failure (worst-case scenario), leaving $(N_S - K)$ working servers, the remaining servers can then retrieve and transmit $(N_S - K)$ redundant units in a service round. Hence, a maximum of

$$N_R = \left\lceil \frac{(k - j + 1)K}{(N_S - K)} \right\rceil = \left\lceil \frac{(\lceil T_D/T_F \rceil + 1)K}{(N_S - K)} \right\rceil \tag{11.3}$$

additional service rounds are required to retransmit the necessary redundant units. Consequently, transmission of subsequent stripes will be delayed by at most N_R rounds.

Now consider the recovery of stripe j (e.g., stripe 2 in Figure 11.4). At the time (round k) the failure is detected, the current disk cycle is already retrieving stripe units for the next transmission cycle $k + 1$. Hence, redundant units for stripe j can only be scheduled for retrieval in the next disk retrieval cycle, which in turn will be sent in transmission round $k + 2$. Therefore, delivery of the redundant block required to recover stripe j will be delayed by

$$N_F = (k - j + 2) = \left(\left\lceil \frac{R_V T_D}{Q(N_S - K)} \right\rceil + 2 \right) \tag{11.4}$$

service rounds. For a transmission rate of $R_V/(N_S - K)$ Bps under Std-Rate transmission, the time it takes to transmit a video block of Q bytes, i.e., length of a service round, is equal to

$$T_F = \frac{Q(N_S - K)}{R_V} \tag{11.5}$$

Therefore, using equations (11.4) and (11.5) we can compute the delay for delivering the redundant block for stripe j from

$$D_F = N_F T_F = \left(\left\lceil \frac{R_V T_D}{Q(N_S - K)} \right\rceil + 2 \right) \frac{Q(N_S - K)}{R_V} \tag{11.6}$$

Provided that $(N_S - K) \geq K$, equation (11.6) also bounds the delay for all stripes. To see why, begin with equation (11.3):

$$\begin{aligned}
N_R &= \left\lceil \frac{(k - j + 1) K}{(N_S - K)} \right\rceil \\
&\leq (k - j + 1), \quad \because ((N_S - K) \geq K) \\
&< (k - j + 2) = N_F
\end{aligned} \tag{11.7}$$

This shows that the delay experienced by stripes $\{i | i > k + 1\}$ transmitted after the failure is detected (N_R) is smaller than the delay experienced by stripe j (N_F). Therefore, the worst-case delay in equation (11.6) also bounds the delay for all stripes.

The additional delay will likely lead to video playback hiccups for the clients. If temporary service interruption can be tolerated, then the clients can simply suspend playback for D_F seconds to resynchronize with the new transmission schedule. Otherwise, we can introduce additional buffers at the client to sustain non-stop video playback during reconfiguration (Section 11.5).

11.3.3 Server Reconfiguration for Sub-Schedule Striping

Figure 11.5 depicts the server reconfiguration process for sub-schedule striping, with $N_S = 5$ and $K = 1$. Instead of considering service rounds, we consider micro-rounds – defined as the period for transmitting a stripe. Hence, a system with N_S servers will have N_S micro-rounds

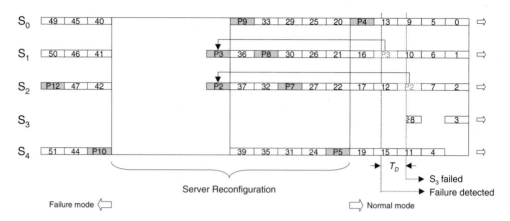

Figure 11.5   Server reconfiguration under PRT with sub-schedule striping

per service round. Note that in each service round, a server retrieves

$$Q_S = Q \frac{N_S}{(N_S - K)} \tag{11.8}$$

bytes of video data (instead of Q bytes in block striping) for every video stream and the length of a service round is

$$T_S = T_F \frac{N_S}{(N_S - K)} \tag{11.9}$$

seconds (instead of T_F seconds in block striping).

We assume that a K-server failure occurs during micro-round j and is detected in micro-round k. Similar to equation (11.2), we can obtain k from

$$k = j + \left\lceil \frac{N_S T_D}{T_S} \right\rceil \tag{11.10}$$

Once notified of the failure, the servers will begin transmitting redundant units for subsequent stripes ($> k$). As each stripe contains K redundant units, the system needs to retransmit up to $(k - j + 1)K$ redundant units. This will require up to

$$n_R = \left\lceil \frac{(k - j + 1) K}{(N_S - K)} \right\rceil = \left\lceil \frac{(\lceil (N_S - K) T_D / T_F \rceil + 1) K}{(N_S - K)} \right\rceil \tag{11.11}$$

micro-rounds for retransmitting the redundant units.

Note that this process has two subtle constraints. First, retransmission cannot start immediately in the next service round because the servers need another service round to retrieve the required redundant units. Second, even if $n_R < N_S$, the last service round for retransmission cannot be shortened because the disk requires a full service round to retrieve video blocks for transmission in the next round.

Similar to the block-striping case, the worst-case delay will be experienced by the stripe that is being transmitted when the failure occurs, provided that $(N_S - K) \geq K$. The worst-case delay can be up to

$$N_F = k - j + (N_S - 1) + N_S$$

$$= \left\lceil \frac{(N_S - K) T_D}{T_F} \right\rceil + 2N_S - 1 \tag{11.12}$$

$$= \left\lceil \frac{R_V T_D}{Q} \right\rceil + 2N_S - 1, \because T_F = \frac{Q}{R_V}(N_S - K)$$

micro-rounds, where $(k - j)$ is the worst-case delay due to failure detection, $(N_S - 1)$ is the worst-case delay to wait for the current service round to end, and N_S is the delay due to the first constraint discussed previously. Noting that the length of a micro-round is equal to T_S/N_S seconds, the delay is then given by

$$D_F = N_F \frac{T_S}{N_S} = \left(\left\lceil \frac{R_V T_D}{Q} \right\rceil + 2N_S - 1 \right) \frac{Q}{R_V} \tag{11.13}$$

seconds.

11.4 Analysis of Forward Erasure Correction

In this section, we derive the amount of client buffer needed to support fault tolerance under FEC so that non-stop playback can be sustained. Client buffers are originally introduced to absorb jitters in video-block playback times and delivery delays. To support fault tolerance using FEC, we need additional client buffers to store a complete stripe (with redundant units) for erasure-correction computation. The derivations in the following sections are based on the model introduced in Chapter 10. The overall approach is to obtain upper and lower bounds for stripe unit arrival times and stripe unit consumption times. Then, using the continuity condition, i.e., the latest arrival time for a stripe unit must not be later than the earliest consumption time, we can obtain the number of buffers required to prevent buffer underflow. We can obtain the number of buffers required to prevent buffer overflow in a similar way.

11.4.1 Buffer Requirement under Block Striping

We first consider the case for block striping. Let there be $L = (Y + Z)$ buffers (each Q bytes) at the client, organized as a circular buffer. Video playback starts once the first Y buffers are completely filled with video data. The client prefills the first Y buffers to prevent buffer underflow, and reserves the last Z buffers for incoming data to prevent buffer overflow.

Since all N_S servers transmit data to a client concurrently, the client will be receiving N_S video blocks simultaneously, of which $(N_S - K)$ blocks contain video data and the rest contain redundant data. This suggests that Y must be multiples of N_S. Therefore, we consider groups of N_S buffers (i.e., group zero consists of blocks 0 to $N_S - 1$, group one consists of blocks N_S to $2N_S - 1$, and so on.) and let $y = Y/N_S$ be the number of buffer groups prefilled.

Using techniques similar to Chapter 10, we can obtain (see Appendix A.1 for derivations):

$$Y = \left\lceil 1 + \frac{\tau + f^+ - f^- - T_E}{(N_S - K)\,T_{avg}} \right\rceil N_S \qquad (11.14)$$

for the number of buffers needed to prevent underflow, and

$$Z = \left\lceil 1 + \frac{\tau + f^+ - f^- + T_L}{(N_S - K)\,T_{avg}} \right\rceil N_S \qquad (11.15)$$

for the number of buffers needed to prevent overflow. Note that T_E, T_L are jitter bounds for video block consumption, τ is the clock jitter among servers, f^+ ($f^+ \geq 0$) and f^- ($f^- \leq 0$) are used to model the maximum transmission time deviation due to randomness in the system, including transmission rate deviation, CPU scheduling, bus contention, etc. See Table 11.1 for a summary of the symbols used in this chapter.

By setting $K = 0$ in equations (11.14) and (11.15), the equations reduce to the non-fault-tolerance version in Chapter 10. The total client buffer requirement is thus given by

$$B_{FEC}^{BS} = \left(2 + \left\lceil \frac{\tau + f^+ - f^- - T_E}{(N_S - K)\,T_{avg}} \right\rceil + \left\lceil \frac{\tau + f^+ - f^- + T_L}{(N_S - K)\,T_{avg}} \right\rceil \right) N_S Q \qquad (11.16)$$

Note the independence of equation (11.16) from T_D, as failure-detection and consequently server reconfiguration is not needed under FEC. However, we can also observe that the buffer requirement will increase when more servers are added to the system, suggesting that more buffers will be needed when scaling up the system.

11.4.2 Buffer Requirement under Sub-Schedule Striping

Under sub-schedule striping, each video block (Q_S bytes) at a server comprises multiple stripe units (U bytes each) and the size of a video block is given in equation (11.8). The client buffers now comprises $L = Y + Z$ buffer units, each of Q_S bytes. Again, we consider stripe units in groups of N_S units, i.e., group i comprises stripe units $\{i, i + 1, \ldots, (i + 1)N_S - 1\}$. Then a group of N_S stripe units will correspond to exactly one buffer unit at the client. Using similar techniques (see Appendix A.2 for derivations), the buffer requirements can be found to be

$$Y = \left\lceil \frac{f^+ - f^- - T_E + \tau}{T_{avg}} \right\rceil + 1 \qquad (11.17)$$

and

$$Z = \left\lceil \frac{f^+ - f^- + T_L + \tau}{T_{avg}} \right\rceil + 1 \qquad (11.18)$$

Surprisingly, these are the same as the non-fault-tolerant case. This counter-intuitive result is explained by the fact that each group of buffers here has the size of Q_S bytes instead of Q bytes in the non-fault-tolerant case. Hence, the system does indeed need additional buffers to support fault tolerance and the total client buffer requirement is given by

$$B_{FEC}^{SSS} = \left(2 + \left\lceil \frac{\tau + f^+ - f^- - T_E}{T_{avg}} \right\rceil + \left\lceil \frac{\tau + f^+ - f^- + T_L}{T_{avg}} \right\rceil \right) Q_S \qquad (11.19)$$

11.5 Analysis of Progressive Redundancy Transmission

Unlike FEC, PRT requires additional buffers to sustain continuous video playback during system reconfiguration. Incorporating this requirement, we derive the corresponding buffer requirement for block striping and sub-schedule striping in the following sections.

11.5.1 Buffer Requirement under Block Striping

A client operating under PRT will simultaneously receive $(N_S - K)$ instead of N_S video blocks. Therefore, a group of video blocks comprises only $(N_S - K)$ video blocks. Unlike FEC, derivations for the buffer requirements depend on whether the failure occurs before or after video playback starts. For the case where the failure occurs before video playback starts, the playback schedule will be delayed because playback cannot start until the required number of buffers are prefilled. The buffer requirements are found to be (see Appendix A.3 for derivations)

$$y_{Before} = \left\lceil 1 + \frac{\tau + f^+ - f^- - T_E}{(N_S - K) T_{avg}} \right\rceil \qquad (11.20)$$

and

$$z_{Before} = \left\lceil 1 + \frac{\tau + f^+ - f^- + T_L + D_F}{(N_S - K) T_{avg}} \right\rceil \qquad (11.21)$$

For the case where the failure occurs after video playback starts, the playback schedule will not be affected. The buffer requirements are found to be

$$y_{After} = \left\lceil 1 + \frac{\tau + f^+ - f^- - T_E + D_F}{(N_S - K) T_{avg}} \right\rceil \qquad (11.22)$$

and

$$z_{After} = \left\lceil 1 + \frac{\tau + f^+ - f^- + T_L}{(N_S - K) T_{avg}} \right\rceil \qquad (11.23)$$

Hence, the client buffer requirement is either

$$l_{Before} = y_{Before} + z_{Before} \qquad (11.24)$$

or

$$l_{After} = y_{After} + z_{After} \qquad (11.25)$$

whichever is larger. However, from equation (11.6): $D_F = N_F T_F$ and $T_F = (N_S - K)T_{avg}$. Therefore, the two equations are in fact equivalent. The total client buffer requirement is thus given by

$$B_{PRT}^{BS} = \left(2 + N_F + \left\lceil \frac{\tau + f^+ - f^- - T_E}{(N_S - K)T_{avg}} \right\rceil + \left\lceil \frac{\tau + f^+ - f^- + T_L}{(N_S - K)T_{avg}} \right\rceil\right)(N_S - K)Q$$

(11.26)

11.5.2 Buffer Requirement under Sub-Schedule Striping

To derive the client buffer requirement for sub-schedule striping, we again consider stripe units in groups of $(N_S - K)$, i.e., group i comprises stripe units $\{i, i + 1, \ldots, (i + N_S - K - 1)\}$. Now unlike FEC, each group of stripe units has the size of Q bytes, instead of Q_S bytes under FEC. Hence, the client buffer comprises $L = Y + Z$ buffer units, each of Q bytes. Proceeding the derivations in the same manner (see Appendix A.4 for details), we can obtain the total buffer requirements from

$$B_{PRT}^{SSS} = \left(2 + N_F + \left\lceil \frac{\tau + f^+ - f^- - T_E}{T_{avg}} \right\rceil + \left\lceil \frac{\tau + f^+ - f^- + T_L}{T_{avg}} \right\rceil\right)Q$$

(11.27)

From equation (11.12), we can see that N_F is proportional to N_S. This implies that the buffer requirement is also proportional to N_S. As sub-schedule striping (cf., Section 10.5) is originally introduced to maintain a constant client buffer requirement independent of system scale (i.e., N_S), the extension to PRT appears to have defeated this goal. We introduce a redundant server scheme in the next section to tackle this problem.

11.6 Redundant Server Scheme

A closer look at Figure 11.5 will reveal why buffer requirement increases with system scale in PRT. First, retransmission of redundant stripe units cannot start in the current service round. This incurs a worst-case delay equal to $(N_S - 1)T_{avg}$ seconds, which obviously is proportional to the system scale. Second, retransmissions cannot start even in the next service round due to the need to retrieve redundant stripe units, incurring another delay of $N_S T_{avg}$ seconds, which again is proportional to the system scale.

The key to the previous two observations is in the server scheduler. First, under the AGSS scheduler (cf., Section 10.4), redundant units are discarded together with the video data units once the service round ends to allow buffer reuse. Hence, if the failure-detection period spans two service rounds as shown in Figure 11.6, redundant units for the previous round will have been discarded by the time the failure is detected, rendering immediate retransmission of redundant stripe units impossible.

To tackle this problem, one can modify the AGSS scheduler so that redundant units are retained longer to cater for server failure. However, we propose a redundant server scheme (RSS) to store all redundant units centrally in one or more (K to be exact) *redundant* servers instead of distributing them over all servers. RSS has three advantages over simply increasing the buffer holding time in AGSS.

First, RSS requires only the redundant servers, instead of all servers, to have the additional memory to buffer redundant units. Therefore, the total server buffer requirement is reduced. Second, redundant units can be stored continuously on the disks in the redundant servers so that retrievals are much more efficient. By contrast, redundant units in the original distributed scheme are scattered on the disk and hence a separate disk I/O is required to retrieve each redundant unit. Third, under RSS, retransmission of the redundant units can start as soon as the failure is detected, without the need to wait for the current stripe unit to complete transmission. This is possible because the redundant servers are idle before a failure is detected.

Assume failure occurs at time t_f during the transmission of stripe j, then it will be detected latest by time $(t_f + T_D)$. Since retransmission of redundant stripe units can start immediately upon failure detection as shown in Figure 11.7, the required redundant unit will be transmitted by time $(t_f + T_D + T_{avg})$. Now let t_j be the time for which transmission of stripe j ends. Then,

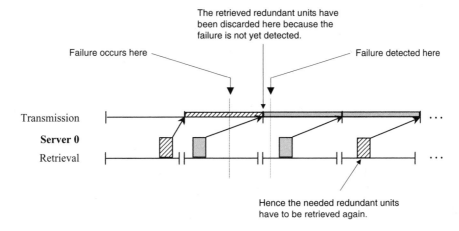

Figure 11.6   The AGSS scheduler discards retrieved video blocks once transmission is completed

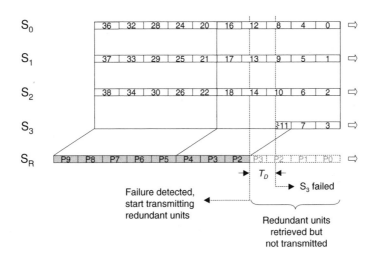

Figure 11.7   Transmission scenario for the redundant server scheme

it is easy to see that

$$t_f \leq t_j \leq t_f + T_{avg} \tag{11.28}$$

Since the client will need to wait for the redundant unit before stripe j can be re-computed, the delay incurred in receiving stripe j will be given by

$$\left(t_f + T_D + T_{avg}\right) - t_j \leq \left(t_f + T_D\right) - t_f$$
$$= T_D + T_{avg} \tag{11.29}$$

which, finally, is independent of the system scale. Using derivations similar to Section 11.5, we can obtain the client buffer requirement from

$$B_{PRT}^{RSS} = \left(3 + \left\lceil \frac{\tau + f^+ - f^- - T_E + T_D}{T_{avg}} \right\rceil + \left\lceil \frac{\tau + f^+ - f^- + T_L}{T_{avg}} \right\rceil \right) Q \tag{11.30}$$

for the case where failure occurs after playback has begun and

$$B_{PRT}^{RSS} = \left(3 + \left\lceil \frac{\tau + f^+ - f^- - T_E}{T_{avg}} \right\rceil + \left\lceil \frac{\tau + f^+ - f^- + T_L + T_D}{T_{avg}} \right\rceil \right) Q \tag{11.31}$$

for the case where failure occurs before playback begins.

To support immediate retransmission of redundant units, the redundant servers will need to retain redundant units longer than in the original AGSS scheduler. In particular, the server will need to keep retrieved redundant units (in blocks of N_S units) for

$$\left\lceil \frac{T_D}{N_S T_{avg}} \right\rceil + 1 \tag{11.32}$$

service rounds (instead of one round in AGSS). Hence, the buffer requirement for the redundant servers will be given by

$$B_{server} = Q N_S \Lambda \left(1 + \frac{1}{G} + \left\lceil \frac{T_D}{N_S T_{avg}} \right\rceil \right) \tag{11.33}$$

where Λ is the client–server ratio and G is the number of groups per service round.

11.7 Numerical Results

Based on the performance models derived in the previous sections, we compute and present numerical results in this section to illustrate the system resource requirement under various scenarios and study the sensitivity to key system parameters. Table 11.1 lists the values for the system parameters used in the calculation. The parameters T_E and T_L are determined empirically by collecting the video block consumption times of a hardware MPEG decoder.

Table 11.1   System parameters used in computing numerical results

System Parameters	Symbol	Value
Video block size	Q	65,536 Bytes
Video data rate	R_V	150 KB/s
Maximum advance in decoding time	T_E	-130 ms
Maximum lag in decoding time	T_L	160 ms
Transmission time deviation	f^-, f^+	0 ms
Server clock jitter	τ	100 ms
Failure-detection delay	T_D	2 s

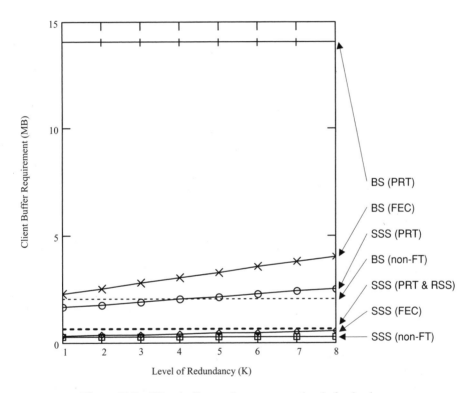

Figure 11.8   Client buffer requirement versus level of redundancy

11.7.1 Buffer Requirement versus Level of Redundancy

Figure 11.8 plots the client buffer requirement versus the level of redundancy. There are a total of $(8 + K)$ servers in the system. There are two observations. First, sub-schedule striping in general requires less client buffer than block striping. Second, sub-schedule striping with PRT and RSS is the only scheme that has constant client buffer requirement irrespective of redundancy level. Even the buffer requirement for the FEC case increases with K. This is explained by the fact that under FEC, the client must receive and process video data in

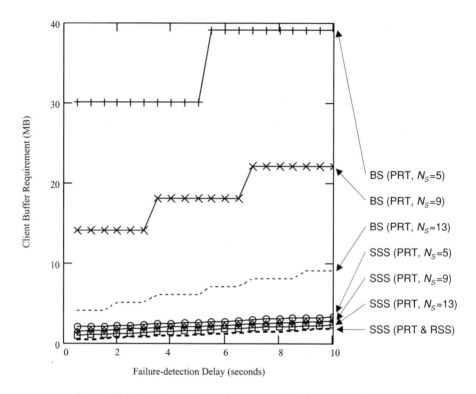

Figure 11.9   Client buffer requirement versus failure-detection delay

parity groups. Hence as K increases, so do the parity group size and, consequently, the buffer requirement. By contrast, redundant stripe units are not transmitted until failure is detected under PRT with RSS. Therefore, the buffer requirement does not depend on K at all.

11.7.2 Buffer Requirement versus Failure-detection Delay

Figure 11.9 studies the sensitivity of buffer requirement with respect to failure-detection delay for various PRT system configurations. FEC is not plotted because the buffer requirement is independent from the failure-detection delay. For all cases in Figure 11.9, the buffer requirement increases with longer failure-detection delay. The results show that sub-schedule striping again achieves lower buffer requirement in general, with PRT/RSS achieving the smallest buffer requirement.

11.7.3 Buffer Requirement versus System Scale

Figure 11.10 plots the client buffer requirement versus the number of servers in the system (i.e., system scale). The level of redundancy is one (i.e., $K = 1$) and the failure-detection delay is 2 seconds. The first result is that block striping is not scalable. This extends the results in Chapter 10 for the non-fault-tolerant case to FEC and PRT. The second result is that sub-schedule striping with PRT is also not scalable, although the slope is smaller than block striping.

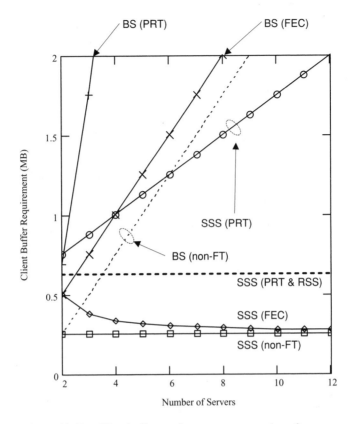

Figure 11.10   Client buffer requirement versus number of servers

Finally, we can observe that only sub-schedule striping under FEC, and under PRT with RSS are scalable, the latter being completely independent of the system scale. Interestingly, buffer requirements under FEC decrease for more servers and approach the non-fault-tolerant case. This is because the level of redundancy is fixed and hence the redundancy overhead incurred decreases when more servers are added.

11.8 Summary

In this chapter, we have investigated protocols and algorithms to support server-level fault tolerance in the concurrent push architecture. In particular, we presented and analyzed two fault-tolerant protocols: FEC and PRT, and two striping policies: block striping and sub-schedule striping. The first result is that FEC is simpler in implementation, does not require failure detection, and inherently scalable to any number of servers under sub-schedule striping. The only downside is additional network bandwidth overhead during normal operation. Surprisingly, analytical results show that PRT is not scalable if redundant data is distributed over all servers (similar to RAID-5 in disk arrays), even with sub-schedule striping. To tackle this problem, we propose storing redundant data centrally in redundant servers to avoid the reconfiguration delay. We increase the buffer holding time at the redundant servers to enable quick redundant data transmission. Analytical results prove that this redundant server scheme enables PRT to become scalable to any number of servers. Finally, we compute numerical results to show the feasibility of the proposed architecture under real-world conditions. With the proposed architecture, a concurrent-push-based parallel video server will be able to sustain multiple simultaneous-server failure and yet, can maintain non-stop, continuous video playback for all clients.

Appendices

A.1 Derivations of Buffer Requirement for Block Striping under FEC

Among the N_S servers, assume the earliest transmission for the first round starts at time t_0, then the last transmission for the first round must start at the latest by time $t_0 + \tau$, where τ is the clock jitter among servers. The time for video block group i to be completely filled, denoted by $F(i)$, is therefore bounded by

$$((i+1)T_F + t_0 + f^-) \leq F(i) \leq ((i+1)T_F + t_0 + \tau + f^+) \tag{11.34}$$

where T_F is as given in equation (11.4) and f^+ ($f^+ \geq 0$) and f^- ($f^- \leq 0$) are used to model the maximum transmission time deviation due to randomness in the system, including transmission rate deviation, CPU scheduling, bus contention, etc.

Since the client starts playing the video after filling the first y groups of buffers, the playback time for video block group 0 is simply equal to $F(y-1)$. Hence, the playback time for video block group i, denoted by $P(i)$, is bounded by

$$(i N_S T_{avg} + F(y-1) + T_E) \leq P(i) \leq (i N_S T_{avg} + F(y-1) + T_L) \tag{11.35}$$

where

$$T_{avg} = \frac{Q}{R_V} \tag{11.36}$$

is the average playback time for one video block, and T_E, T_L are the jitter bounds for video block consumption variations.

To guarantee video playback continuity, we must ensure that a video block group arrives before its playback deadline. In the worst-case scenario, the latest filling time must be smaller than the earliest playback time, i.e.,

$$\max\{F(i)\} \leq \min\{P(i)\} \tag{11.37}$$

For the L.H.S., noting that $T_F = (N_S - K)T_{avg}$ (cf. equations (11.4) and (11.16)) we then have

$$\max\{F(i)\} = (i+1)(N_S - K)T_{avg} + t_0 + \tau + f^+ \tag{11.38}$$

Similarly, the R.H.S. is

$$\min\{P(i)\} = i N_S T_{avg} + \min\{F(y-1)\} + T_E \tag{11.39}$$
$$= i N_S T_{avg} + y(N_S - K)T_{avg} + t_0 + f^- + T_E$$

Substituting equations (11.38), (11.39) into equation (11.37) gives

$$((i + 1)(N_S - K)T_{avg} + t_0 + \tau + f^+)$$
$$\leq (iN_S T_{avg} + y(N_S - K)T_{avg} + t_0 + f^- + T_E) \tag{11.40}$$

Rearranging we can then obtain y:

$$y \geq 1 + \frac{f^+ - f^- - T_E + \tau}{(N_S - K)T_{avg}} \tag{11.41}$$

Knowing the number of groups required, we can then obtain Y from

$$Y = \left\lceil 1 + \frac{\tau + f^+ - f^- - T_E}{(N_S - K)T_{avg}} \right\rceil N_S \tag{11.42}$$

On the other hand, to guarantee that the client buffer will not be overflowed by incoming video data, we need to ensure that the ith video block group starts playback before the $(i + l - 2)$th video block group is completely received, where $l = L/N_S$. This is because the client buffers are organized as a circular buffer, and we must always have at least one group of N_S free buffers to receive video blocks arriving simultaneously from N_S servers. Therefore, we need to ensure that the earliest filling time for group $(i + l - 2)$ must be greater than the latest playback time for group i:

$$\min\{F(i + l - 2)\} \geq \max\{P(i)\} \tag{11.43}$$

Using similar derivations, we can obtain the number of buffers needed to prevent buffer overflow as:

$$Z = \left\lceil 1 + \frac{\tau + f^+ - f^- + T_L}{(N_S - K)T_{avg}} \right\rceil N_S \tag{11.44}$$

A.2 Derivations of Buffer Requirement for Sub-Schedule Striping under FEC

The filling time for group i of a video stream started at time t_0 is bounded by

$$((i + 1)T_{avg} + t_0 + f^-) \leq F(i) \leq ((i + 1)T_{avg} + t_0 + f^+ + \tau) \tag{11.45}$$

Since the client starts video playback after filling the first Y groups of buffers, the playback time for video block group 0 is simply equal to $F(Y - 1)$. Hence, the playback time for video block group i, denoted by $P(i)$, is bounded by

$$(iT_{avg} + F(Y - 1) + T_E) \leq P(i) \leq (iT_{avg} + F(Y - 1) + T_L) \tag{11.46}$$

Substituting the upper bound of equation (11.45), and the lower bound of equation (11.46) into the continuity condition in equation (11.37) gives

$$((i + 1)T_{avg} + t_0 + f^+ + \tau) \leq (iT_{avg} + \min\{F(Y - 1)\} + T_E) \tag{11.47}$$

or

$$\left((i+1)\,T_{avg}+t_0+f^++\tau\right)\leq\left(i\,T_{avg}+\left(Y\,T_{avg}+t_0+f^-\right)+T_E\right) \tag{11.48}$$

Rearranging, we can obtain Y from

$$Y=\left\lceil\frac{f^+-f^--T_E+\tau}{T_{avg}}\right\rceil+1 \tag{11.49}$$

Using similar derivations, we can obtain Z from

$$Z=\left\lceil\frac{f^+-f^-+T_L+\tau}{T_{avg}}\right\rceil+1 \tag{11.50}$$

A.3 Derivations of Buffer Requirement for Block Striping under PRT

Assume that a failure occurs during the transmission of group j, then for those groups received before a failure (i.e., $0\leq i<j$), the filling time is bounded by

$$\left((i+1)\,T_F+t_0+f^-\right)\leq F_N(i)\leq\left((i+1)\,T_F+t_0+\tau+f^+\right), \quad 0\leq i<j \tag{11.51}$$

However, groups transmitted after the failure $(i\geq j)$ will be deferred due to server reconfiguration. According to Section 11.3.2, the worst-case delay due to reconfiguration is D_F seconds. Hence, the maximum filling time is bounded by

$$\left((i+1)\,T_F+t_0+f^-+D_F\right)\leq F_F(i)\leq\left((i+1)\,T_F+t_0+\tau+f^++D_F\right), \quad i\geq j \tag{11.52}$$

Merging (11.51) and (11.52) gives bounds for the general case:

$$\left((i+1)\,T_F+t_0+f^-\right)\leq F(i)\leq\left((i+1)\,T_F+t_0+\tau+f^++D_F\right), \quad \forall i \tag{11.53}$$

The bounds for $P(i)$ depend on whether a failure occurs before or after playback has begun. Specifically, if a failure occurs before playback begins, then playback will be delayed up to D_F seconds due to the need to reconfigure the servers to complete the prefill process:

$$\left(i\,(N_S-K)\,T_{avg}+F_F(y-1)+T_E\right)\leq P_{Before}(i)\leq\left(i\,(N_S-K)\,T_{avg}+F_F(y-1)+T_L\right) \tag{11.54}$$

Otherwise, if the failure occurs after playback has begun, then the playback schedule will not be affected:

$$\left(i\,(N_S-K)\,T_{avg}+F_N(y-1)+T_E\right)\leq P_{After}(i)\leq\left(i\,(N_S-K)\,T_{avg}+F_N(y-1)+T_L\right) \tag{11.55}$$

Now if the failure occurs before playback begins, then invoking the continuity condition gives

$$\max\left\{F(i)\right\}\leq\min\left\{P_{Before}(i)\right\} \tag{11.56}$$

or

$$\left((i+1)\,T_F + t_0 + \tau + f^+ + D_F\right) \le \left(i\,(N_S - K)\,T_{avg} + \min\{F_F(y-1)\} + T_E\right) \tag{11.57}$$

Substituting the lower bound of equation (11.52) into equation (11.57) and noting $T_F = (N_S - K)T_{avg}$ we get

$$\left((i+1)\,(N_S - K)\,T_{avg} + t_0 + \tau + f^+ + D_F\right) \le \left(i\,(N_S - K)\,T_{avg} + \left(y\,(N_S - K)\,T_{avg} + t_0 \right. \right.$$
$$\left. \left. + f^- + D_F\right) + T_E\right) \tag{11.58}$$

Rearranging, we can obtain y from

$$y_{Before} = \left\lceil 1 + \frac{\tau + f^+ - f^- - T_E}{(N_S - K)\,T_{avg}} \right\rceil \tag{11.59}$$

Similarly, if the failure occurs after playback has begun, then the continuity condition becomes

$$\max\{F(i)\} \le \min\left\{P_{After}(i)\right\} \tag{11.60}$$

or

$$\left((i+1)\,T_F + t_0 + \tau + f^+ + D_F\right) \le \left(i\,(N_S - K)\,T_{avg} + \min\{F_N(y-1)\} + T_E\right) \tag{11.61}$$

Solving, we can obtain y from

$$y_{After} = \left\lceil 1 + \frac{\tau + f^+ - f^- - T_E + D_F}{(N_S - K)\,T_{avg}} \right\rceil \tag{11.62}$$

Similarly for Z, we also need to consider the two cases. First, for the case where failure occurs before playback begins, we have

$$\min\{F(i+l-2)\} \ge \max\{P_{Before}(i)\} \tag{11.63}$$

or

$$\left((i+l-1)\,(N_S - K)\,T_{avg} + t_0 + f^-\right) > \left(i\,(N_S - K)\,T_{avg} + \max\{F_F(y-1)\} + T_L\right) \tag{11.64}$$

Substituting the upper bound of equation (11.52) into equation (11.64), we have

$$\left((i+l-1)\,(N_S - K)\,T_{avg} + t_0 + f^-\right) > \left(i N_S T_{avg} + \left(y\,(N_S - K)\,T_{avg} + t_0 \right. \right.$$
$$\left. \left. + \tau + f^+ + D_F\right) + T_L\right) \tag{11.65}$$

Rearranging we can obtain $z = (l - y)$ as

$$z_{Before} = \left\lceil 1 + \frac{\tau + f^+ - f^- + T_L + D_F}{(N_S - K)\,T_{avg}} \right\rceil \tag{11.66}$$

For the second case, playback is not delayed by the failure. Hence we have

$$\min\{F(i+l-2)\} \geq \max\{P_{After}(i)\} \tag{11.67}$$

or

$$\left((i+l-1)(N_S-K)T_{avg}+t_0+f^-\right) > \left(i(N_S-K)T_{avg}+\max\{F_N(y-1)\}+T_L\right) \tag{11.68}$$

Substituting the upper bound of equation (11.51) into equation (11.68), we have

$$\left((i+l-1)(N_S-K)T_{avg}+t_0+f^-\right) > \left(iN_ST_{avg}+\left(y(N_S-K)T_{avg}+t_0+\tau\right.\right.$$
$$\left.\left.+f^+\right)+T_L\right) \tag{11.69}$$

Rearranging we can obtain $z=(l-y)$ as

$$z_{After} = \left\lceil 1 + \frac{\tau+f^+-f^-+T_L}{(N_S-K)T_{avg}} \right\rceil \tag{11.70}$$

A.4 Derivations of Buffer Requirement for Sub-Schedule Striping under PRT

Assuming failure occurs during transmission of group j, then the filling time for group i of a video stream started at time t_0 is bounded by

$$\left((i+1)T_{avg}+t_0+f^-\right) \leq F_N(i) \leq \left((i+1)T_{avg}+t_0+f^++\tau\right), \quad 0 \leq i < j \tag{11.71}$$

$$\left((i+1)T_{avg}+t_0+f^-+D_F\right) \leq F_F(i) \leq \left((i+1)T_{avg}+t_0+f^++\tau+D_F\right), \quad i \geq j \tag{11.72}$$

Merging equation (11.71) and equation (11.72) gives the universal bounds for $F(i)$:

$$\left((i+1)T_{avg}+t_0+f^-\right) \leq F(i) \leq \left((i+1)T_{avg}+t_0+f^++\tau+D_F\right), \forall i \tag{11.73}$$

Similarly, the playback schedule is bounded by

$$\left(iT_{avg}+F_F(Y-1)+T_E\right) \leq P_{Before}(i) \leq \left(iT_{avg}+F_F(Y-1)+T_L\right) \tag{11.74}$$

for the case where failure occurs before playback begins, and

$$\left(iT_{avg}+F_N(Y-1)+T_E\right) \leq P_{After}(i) \leq \left(iT_{avg}+F_N(Y-1)+T_L\right) \tag{11.75}$$

for the case where failure occurs after playback has begun.

Invoking the continuity condition, we can obtain the corresponding bounds for Y as follows:

$$Y_{Before} = \left\lceil \frac{f^+-f^--T_E+\tau}{T_{avg}} \right\rceil + 1 \tag{11.76}$$

$$Y_{After} = \left\lceil \frac{f^+-f^--T_E+\tau+D_F}{T_{avg}} \right\rceil + 1 \tag{11.77}$$

Using similar derivations, we can obtain the corresponding bounds for Z as follows:

$$Z_{Before} = \left\lceil \frac{f^+ - f^- + T_L + \tau + D_F}{T_{avg}} \right\rceil + 1 \qquad (11.78)$$

$$Z_{After} = \left\lceil \frac{f^+ - f^- + T_L + \tau}{T_{avg}} \right\rceil + 1 \qquad (11.79)$$

References

[1] S.B. Wicker, *Error Control Systems for Digital Communication and Storage*, Prentice-Hall, 1995, pp. 227–34.

[2] J.S. Plank, A Tutorial on Reed-Solomon Coding for Fault-Tolerance in RAID-like Systems, *Software Practice and Experience*, vol. 27, no. 9, Sep. 1997, pp. 995–1012.

[3] L. Rizzo, Effective Erasure Codes for Reliable Computer Communication Protocols, *ACM Computer Communication Review*, vol. 27, no. 2, Apr. 1997, pp. 24–36.

[4] D.A. Patterson, P. Chen, G. Gibson, and R.H. Katz, Introduction to Redundant Array of Inexpensive Disks (RAID), *COMPCON Spring'89*, 1989, pp. 112–117.

12

A Staggered-Push Parallel Server Architecture

The concurrent-push architecture discussed in the previous two chapters is designed to take advantage of network-level QoS services and is also readily scalable to over ten thousand concurrent streams. In practice, however, not all broadband networks support resource reservation or QoS control. In this chapter we develop an alternative approach to schedule the transmissions of media data from the parallel servers – a staggered-push approach. While media servers in the concurrent-push architecture transmit media data simultaneously in proportionally reduced bit-rate, the media servers in staggered push effectively transmit data to the same client in turns, with only one of the servers sending data at any time. As a result, the data traffic will be more bursty and this is also why staggered push cannot easily take advantage of network resource reservation services.

In return, the staggered-push architecture can achieve linear scalability, i.e., the per-stream server resource requirements (e.g., buffer requirement) are invariant to the scale (number of servers) of the system. This chapter presents this staggered-push architecture, explains its admission, scheduling, and buffer management schemes, and analyzes its performance.

12.1 Introduction

In this chapter we present and analyze quantitatively a staggered-push architecture for scheduling disk retrieval and network transmission in parallel video servers. We prove a remarkable property of the staggered-push architecture – the system can be scaled up *linearly* to an arbitrary number of servers as long as the network has sufficient capacity. We discover that for loosely coupled servers like PC or workstation clusters, server–clock asynchrony could lead to inconsistent schedule assignments among different servers. To tackle this problem, we introduce an external admission scheduler to centralize admission control and perform schedule assignments. Apart from inconsistent schedule assignments, we discover that server–clock asynchrony could also lead to overlapping between data transmitted from different servers.

This could induce network congestion, leading to video packets being dropped at the network switches and routers. Worse still, the client may not be able to cope with the aggregate data rate even if the network can successfully deliver the data. To tackle this problem, we introduce an over-rate transmission scheme that can effectively prevent traffic overlapping. To evaluate the strengths and weaknesses of the proposed architecture, we use numerical results to compare and contrast staggered-push architecture with the concurrent-push architecture covered in Chapters 10 and 11 using the same system parameters and assumptions.

12.2 System Architecture

Figure 12.1 shows the architecture of a parallel video server, comprising multiple autonomous servers connected by an interconnection network. We denote the number of servers in the system by N_S and the number of clients by N_C. Hence the client–server ratio, denoted by Λ, is N_C/N_S. Each server has separate CPU, memory, disk storage, and network interfaces. A server's storage spaces are divided into fixed-size stripe units of Q bytes each. Each video title is then striped into blocks of Q bytes and stored into the servers in a round-robin manner as shown in Figure 12.1.

To schedule disk retrievals and network transmissions at the servers, we propose a *staggered-push* algorithm where the servers transmit bursts of data to a client in a round-robin manner at the average video bit-rate. Let R_V be the average video rate and assume it to be the same for all clients. Then the transmissions from the servers are staggered so that only one of the servers transmits to a receiver at any given time, depicted in Figure 12.2. In this way, there will be at most $\Lambda = N_C/N_S$ video blocks being transmitted concurrently at a server. Note that while one can potentially reduce server buffer requirement by transmitting at a rate higher than R_V, the client in turn will have to be capable of receiving at such a high data rate. This is less practical as client network connection usually has lower bandwidth and the client device (e.g., set-top box) will likely have limited processing capability.

To support staggered push, the server scheduler is divided into two scheduling levels: *micro-round* and *macro-round* as shown in Figure 12.3. Video blocks retrieved in one micro-round will be transmitted in the next micro-round. Let T_F be the average time needed to completely transmit a video block of Q bytes. Since a video block is transmitted at a rate equal to the

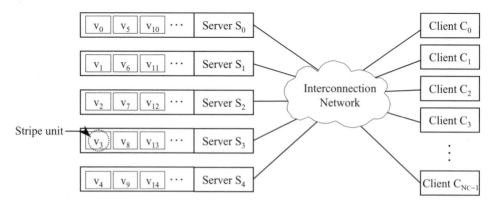

Figure 12.1   Architecture of a (5 servers) parallel video server

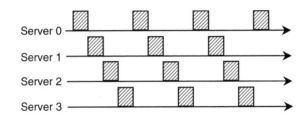

Figure 12.2 Transmission scenario for the staggered push algorithm

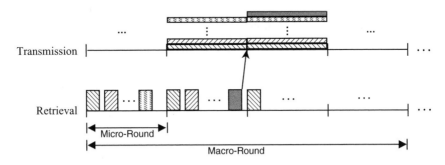

Figure 12.3 Two-level scheduler for staggered push

video data rate R_V, we can obtain T_F from

$$T_F = Q/R_V \tag{12.1}$$

In an N_S-servers system, each macro-round consists of N_S micro-rounds, and each micro-round transfers Λ video blocks. Hence, the disk will transfer up to $N_S\Lambda = N_C$ video blocks in one macro-round, with one block per video stream.

12.3 Schedule Assignment

Unexpectedly, the two-level scheduling scheme may result in inconsistent schedules among different servers if admission is performed independently at each server. Specifically, as servers are loosely coupled, the internal clock of each server in the system will not be precisely synchronized. We define *clock jitter* as the difference between the internal real-time clocks of two servers. Many algorithms for controlling clock jitter between distributed computers have been proposed [1–3] and hence will not be pursued further here. We assume that the maximum clock jitter between any two servers in the system is bounded and is denoted by τ.

With the presence of clock jitter, one server could assign two new video sessions to start with the same micro-round while another server could assign them to two different micro-rounds as shown in Figure 12.4. This can occur because each server assigns new sessions to micro-rounds according to its own internal clock, which differs from other servers due to clock jitter. As a single micro-round can serve only up to Λ video sessions, eventually one server could experience micro-round overflow although another server can admit the new video session

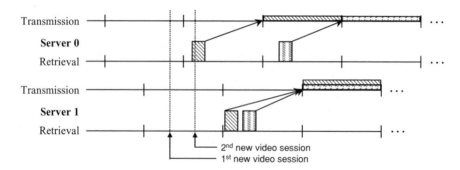

Figure 12.4   Inconsistent schedule assignment arising from server clock jitter

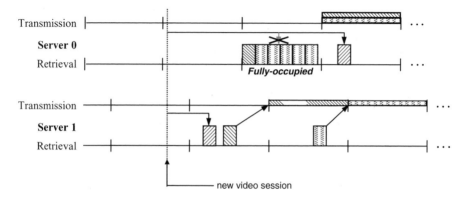

Figure 12.5   Micro-round overflow due to inconsistent schedule assignment

(Figure 12.5). While one can delay the new video session at the overflowed server until the next available micro-round, the transmission schedule will be delayed significantly and will result in severe traffic overlapping with transmissions from another server (see Section 12.4).

To solve this inconsistent schedule assignment problem, we introduce an external admission scheduler between the servers and the clients to centralize schedule assignment. To initiate a new video session, a client will first send a request to the admission scheduler. Using the same clock-synchronization protocol, the admission scheduler maintains the same clock jitter bound with the servers. As new sessions are assigned solely according to the admission scheduler's clock, the scenario depicted in Figures 12.4 and 12.5 will not occur. However, to ensure that the assigned micro-round has not started in any of the servers due to clock jitter, the admission scheduler must add an extra delay to the assignment:

Theorem 12.1. *If the admission scheduler delays the start of a new video session by*

$$\Delta = \left\lceil \frac{\tau}{T_F} \right\rceil \tag{12.2}$$

micro-rounds, then it guarantees that the assigned micro-round has not started in any of the N_S servers.

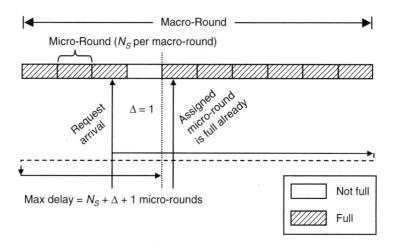

Figure 12.6   Worst-case delay in the admission process

Proof. Please refers to Appendix A.   ∎

For example, let t_A be the local time the new request arrives at the admission scheduler. Then the admission scheduler will attempt to admit the request to micro-round

$$n_A = \left\lfloor \frac{t_A}{T_F} \right\rfloor + 1 + \Delta \tag{12.3}$$

Note that we need to add one to n_A because a new request cannot join the current micro-round (it has started already). If the assigned micro-round is full, the admission scheduler will sequentially check the subsequent micro-rounds until an available micro-round is found. In the worst case shown in Figure 12.6, the transmission of the first video block will be delayed for $(N_S + \Delta + 1)$ micro-rounds:

$$D_S = T_F \left(N_S + \left\lceil \frac{\tau}{T_F} \right\rceil + 1 \right) \tag{12.4}$$

To better evaluate the delay incurred, we can derive the average scheduling delay under a given server load. Assume that there are n $(0 \le n \le \Lambda N_S)$ active video sessions, then it can be shown that (see Appendix A.2) the average scheduling delay is given by

$$D_S = \frac{Q}{R_V} \sum_{j=1}^{N_S-1} \left(\sum_{k=1}^{N_S} \frac{k(N_S - j)j!(N_S - k - 1)!}{N_S!(j-k)!} + \left\lceil \frac{\tau}{T_F} \right\rceil + 1 \right) \left(\frac{N_{full}(n, j)}{N(n, N_S, \Lambda)} \right) \tag{12.5}$$

where

$$N(n, N_S, \Lambda) = \sum_{j=0}^{N_S} (-1)^j \binom{N_S}{j} \binom{N_S + n - j(\Lambda + 1) - 1}{N_S - 1} \tag{12.6}$$

and

$$N_{full}(n, m) = \binom{N_S}{m} N(n - m\Lambda, N_S - m, \Lambda - 1). \qquad (12.7)$$

12.4 Traffic Overlapping

If server clock jitter is greater than zero, then transmissions from two or more servers destined to the same client will overlap and multiply the transmission rate in the overlapping interval (Figure 12.7). This could cause congestion at the network and the client, resulting in the packet being dropped.

To avoid traffic overlapping, we can sacrifice some server and network bandwidth, and transmit video data at a rate higher than R_V, say R_{ORT} (Figure 12.8). We call this scheme over-rate transmission (ORT) for obvious reasons. The transmission window will then be reduced to a time interval of

$$T_w = \frac{Q}{R_{ORT}} \qquad (12.8)$$

We can guarantee that there will be no transmission overlapping by ensuring that

$$T_w + \tau < T_F \qquad (12.9)$$

Rearranging, we can then obtain the minimum transmission rate needed to avoid traffic overlapping:

$$R_{ORT} > \frac{Q R_V}{Q - R_V \tau} \qquad (12.10)$$

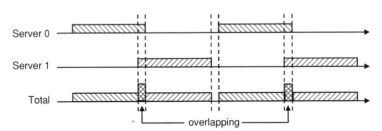

Figure 12.7   Traffic overlapping due to server clock jitter

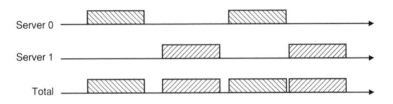

Figure 12.8   Preventing traffic overlap by over-rate transmission

Since the transmission rate must be positive and less than infinity, we have the condition that

$$\tau < \frac{Q}{R_V} = T_F \tag{12.11}$$

In other words, the server clock jitter must be smaller than a micro-round. Note that under this condition, traffic overlapping involves at most two servers and the data rate is doubled to $2R_V$ in the overlapping region. As R_{ORT} in equation (12.10) can become very large when the denominator becomes small, the useful operating range for over-rate transmission is actually limited by:

$$R_{ORT} < 2R_V \tag{12.12}$$

Substituting equation (12.10) into equation (12.12) and rearranging we can then determine the maximum clock jitter for which ORT is applicable:

$$\tau < \frac{Q}{2R_V} = 0.5T_F \tag{12.13}$$

Therefore, ORT can prevent traffic overlapping if clock jitter is less than half of a micro-round. With ORT, the maximum network bandwidth needed at each server will be increased to

$$C_{ORT} = \Lambda R_{ORT} = \frac{\Lambda Q R_V}{Q - R_V \tau} \tag{12.14}$$

12.5 Buffer Management

In this section, we present buffer management algorithms for the server and the client, and derive the respective buffer requirements. For simplicity, we ignore network delay and delay jitter. However, the effect of network delay and delay jitter can be incorporated in the same way as clock jitter and the same derivations are still valid.

12.5.1 Server Buffer Requirement

There are N_S micro-rounds in a macro-round, therefore the duration of a macro-round, denoted by T_R, is given by

$$T_R = \frac{N_S Q}{R_V} \tag{12.15}$$

As buffers are released after each micro-round, this scheduler requires only $2\Lambda Q$ buffers for each server, regardless of the number of servers and clients in the system. Therefore, existing servers do not need any upgrade when one scales up a system by adding more servers.

12.5.2 Client Buffer Requirement

Many studies on VoD system have assumed that video data are consumed periodically by the video decoder. However, as previously discussed in Section 10.3.2, hardware and software video decoders consume fixed-size data blocks only quasi-periodically. Given the average video data rate, R_V, and block size, Q, the average time for a video decoder to consume a single block is

$$T_{avg} = \frac{Q}{R_V} \qquad (12.16)$$

To quantify the randomness of video block consumption time, we employ the consumption model proposed in Section 10.3.2, reproduced below for sake of completeness.

Definition 12.1. *Let T_i be the time the video decoder starts decoding the ith video block, then the decoding-time deviation of video block i is defined as*

$$T_{DV}(i) = T_i - iT_{avg} - T_0 \qquad (12.17)$$

and decoding is late if $T_{DV}(i) > 0$ and early if $T_{DV}(i) < 20$. The maximum lag in decoding, denoted by T_L, and the maximum advance in decoding, denoted by T_E, are defined as follows:

$$T_L = \max\{T_{DV}(i)|\forall i \geq 0\} \qquad (12.18)$$

$$T_E = \min\{T_{DV}(i)|\forall i \geq 0\} \qquad (12.19)$$

The bounds T_L and T_E are implementation-dependent and can be obtained empirically. Knowing these two bounds, the playback instant for video block i, denoted by $p(i)$, is then bounded by

$$\max\{(T_0 + iT_{avg} + T_E), 0\} \leq p(i) \leq (T_0 + iT_{avg} + T_L) \qquad (12.20)$$

Buffers are used at the client to absorb these variations to prevent buffer underflow (which leads to playback hiccups) and buffer overflow (which leads to packet dropping). Let $L_C = (Y + Z)$ be the number of buffers (each of Q bytes) available at the client, organized as a circular buffer. The client prefills the first Y buffers before starting playback to prevent buffer underflow, and reserves the last Z buffers for incoming data to prevent buffer overflow.

We first determine the lower bound for Y. Let t_0 be the time (with respect to the admission scheduler's clock) when the first block of a video session begins transmission. Let d_i be the clock jitter between the admission scheduler and server i. Without loss of generality, we can assume that the video title is striped with block zero at server zero. Then the time for block i to be completely received by the client, denoted by $f(i)$, is bounded by

$$((i+1)T_F + t_0 + f^- + d_{\bmod (i,N_S)}) \leq f(i) \leq ((i+1)T_F + t_0 + f^+ + d_{\bmod (i,N_S)}) \qquad (12.21)$$

where f^+ and f^- are used to model the maximum transmission time deviation due to randomness in the system, including transmission rate deviation, CPU scheduling, bus contention, etc.

Since the client begins video playback after filling the first Y buffers, the playback time for video block 0 is simply equal to $f(Y-1)$. Setting $T_0 = f(Y-1)$ in equation (12.20) then the playback time for video block i is bounded by

$$(f(Y-1) + iT_{avg} + T_E) \leq p(i) \leq (f(Y-1) + iT_{avg} + T_L) \qquad (12.22)$$

To guarantee video playback continuity, we must ensure that all video blocks arrive before their respective playback deadlines. Therefore, we need to ensure that for all video blocks, the latest arrival time must be smaller than the earliest playback time:

$$\max\{f(i)\} < \min\{p(i)\} \quad \forall i \geq 0 \tag{12.23}$$

Using the bounds from equation (12.21) and (12.22), we can rewrite equation (12.23) as

$$(i+1)T_F + t_0 + f^+ + d_{\bmod (i,N_S)} < iT_{avg} + f(Y-1) + T_E \tag{12.24}$$

or

$$(i+1)T_F + t_0 + f^+ + d_{\bmod (i,N_S)} < iT_{avg} + YT_F + t_0 + f^- + d_{\bmod (Y-1,N_S)} + T_E \tag{12.25}$$

From equations (12.1) and (12.16) we know that $T_{avg} = T_F$, rearranging and solving for Y, we then obtain

$$Y > 1 + \left(\frac{f^+ - f^- - T_E + (d_{\bmod (i,N_S)} - d_{\bmod (Y-1,N_S)})}{T_F} \right) \tag{12.26}$$

Since $\max\{|d_i - d_j| \ \mid \ \forall i, j\} = \tau$, the worst case is

$$Y > 1 + \left(\frac{f^+ - f^- - T_E + \tau}{T_F} \right) \tag{12.27}$$

which is the number of buffers that must be prefilled before beginning video playback.

Similarly, to guarantee that the client buffer will not be overwhelmed by incoming video data, we need to ensure that the ith video block starts playback before the $(i + L_C-2)$th video block is completely received. This is because the client buffers are organized as a circular buffer. Therefore, we need to ensure that

$$\min\{f(i + L_C - 2)\} > \max\{p(i)\} \quad \forall i \geq (L_C - Z) \tag{12.28}$$

Again using the bounds from equations (12.21) and (12.22), we can rewrite equation (12.28) as

$$(i + L_C - 1)T_F + t_0 + f^- + d_{\bmod (i+L_C-2,N_S)} > iT_{avg} + f(L_C - Z - 1) + T_L \tag{12.29}$$

or

$$(i + L_C - 1)T_F + t_0 + f^- + d_{\bmod (i+L_C-2,N_S)} > iT_{avg} + (L_C - Z)T_F + t_0 + f^+$$
$$+ d_{\bmod (L_C-Z-1,N_S)} + T_L \tag{12.30}$$

Similarly, rearranging and solving for Z we obtain

$$Z > 1 + \left(\frac{f^+ - f^- + T_L + (d_{\bmod (L_C-Z-1,N_S)} - d_{\bmod (i+L_C-2,N_S)})}{T_F} \right) \tag{12.31}$$

Again noting that $\max\{|d_i - d_j| \mid \forall i, j\} = \tau$, the worst case is

$$Z > 1 + \left(\frac{f^+ - f^- + T_L + \tau}{T_F} \right) \tag{12.32}$$

which is the number of empty buffers needed to avoid client buffer overflow.

12.5.3 System Response Time

Another key performance metric of a VoD service is system response time, defined as the time from initiating a new request to the time video playback starts. Ignoring system administration and network delay issues, system response time consists of two components: scheduling delay and prefill delay. Scheduling delay is the delay incurred at the admission scheduler plus the delay incurred at the server scheduler, as derived in Section 12.3. For prefill delay, we note that the client prefills the first Y video blocks before starting video playback. Hence, the average prefill delay can be obtained from

$$D_P = \frac{YQ}{R_V} \tag{12.33}$$

and the system response time is simply the sum $D_S + D_P$.

12.6 Performance Evaluation

In this section, we evaluate the performance of the staggered-push architecture using numerical results. All results are computed using the derivations in Sections 12.3 to 12.5 with the system parameters listed in Table 12.1.

12.6.1 Design Example

To illustrate performance and resource requirements of the architecture, we first consider a design example in this section. We assume that there are 8 servers in the system, with a client–server ratio of 50 (i.e., up to 400 concurrent streams). Using the parameters in Table 12.1, the server buffer requirement is calculated to be 6.25MB. Compared to the amount of memory in today's PC, this buffer requirement is relatively small. Moreover, as conventional PCs can be expanded to 1GB or more memory, in theory a client–server ratio of over 8,000

Table 12.1 System parameters used in performance evaluation

System Parameters	Symbol	Value
Video block size	Q	65,536 Bytes
Video data rate	R_V	150 KB/s
Maximum early in decoding time	T_E	-130 ms
Maximum late in decoding time	T_L	160 ms
Client-Server ratio	Λ	10
Transmission time deviation	f^-, f^+	0 ms
Server clock jitter	τ	100 ms

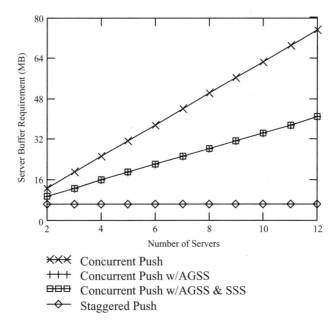

Figure 12.9   Server buffer requirement versus system scale

can be supported. Hence, server buffer requirement will not become a limiting factor to the system's scalability.

Using the same parameters, the client buffer requirement is calculated to be 256KB. This translates into an average prefill delay of 1.41 seconds. To determine the system response time, we assume that the system is at 90% utilization. Then the corresponding scheduling delay will be 0.735 seconds. Together with prefill delay, the average system response time becomes 2.146 seconds, well within acceptable limits. We perform more detailed sensitivity analysis with respect to key system parameters in the following sections.

12.6.2 Server Buffer Requirement

Figure 12.9 plots server buffer requirement versus system scale (i.e., number of servers) for both concurrent push and staggered push. This graph clearly shows the remarkable property of staggered push – constant server buffer requirement irrespective of system scale. By contrast, server buffer requirement increases with system scale under concurrent push, even with AGSS and SSS. When concurrent push is scaled up to 12 servers, server buffer requirement increases to 40.6MB compared to just 6.25MB under staggered push. Hence the ultimate scalability of the concurrent push architecture will be limited by server buffer, while the proposed staggered-push architecture can be scaled up without any upgrade to the existing servers.

12.6.3 Client Buffer Requirement

Figure 12.10 plots client buffer requirement versus system scale for both concurrent push and staggered push. We observe that concurrent push is not scalable without SSS, while staggered

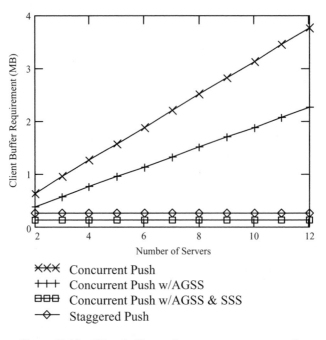

Figure 12.10 Client buffer requirement versus system scale

push has a constant client buffer requirement that will not limit scalability. Note that although client buffer requirement in concurrent push can be controlled to a constant by SSS (cf. Section 10.5), the system scalability is still limited as client processing overhead due to SSS increases with system scale. It is particularly important to maintain a constant client buffer requirement in practice as it would be very expensive (if not impossible) to upgrade every existing client devices (e.g., set-top box) whenever the system is scaled up.

In Figure 12.11, we analyze the sensitivity of client buffer requirement to server clock jitter. As the results indicate, the buffer requirement is relatively insensitive to clock jitter, even if the jitter is increased to one second. Hence one can safely employ the existing software-based, distributed clock-synchronization protocols in staggered push.

12.6.4 System Response Time

Figure 12.12 plots the system response time versus system scale. While the worst-case system response time increases linearly with more servers, the average system response time remains low (~2 seconds) for a utilization of 90%. This suggests that we can maintain a low system response time simply by limiting the system to, say, 90% utilization through admission control.

In Figure 12.13, we study the sensitivity of system response time to server clock jitter. As expected, the system response time increases for larger clock jitter values (cf. Theorem 12.1). However, given that server clock jitter can readily be controlled to within 100ms [1], the average system response time is still only 0.735 seconds for an 8-servers system at 90% utilization.

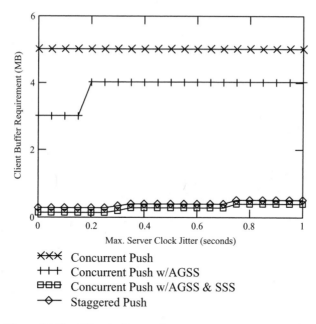

Figure 12.11　Client buffer requirement versus server clock jitter

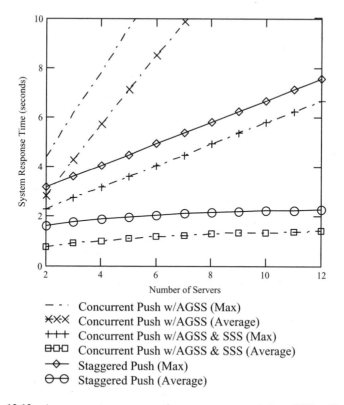

Figure 12.12　Average system response time versus system scale at (90% utilization)

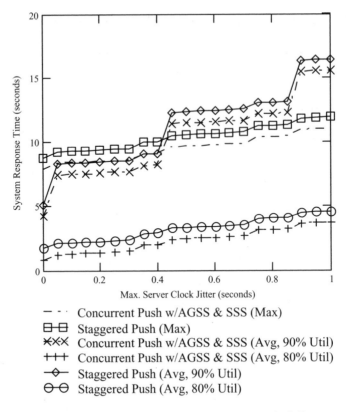

$- -$ Concurrent Push w/AGSS & SSS (Max)
⊟⊟ Staggered Push (Max)
✳✳✳ Concurrent Push w/AGSS & SSS (Avg, 90% Util)
+++ Concurrent Push w/AGSS & SSS (Avg, 80% Util)
◇◇ Staggered Push (Avg, 90% Util)
⊖⊖ Staggered Push (Avg, 80% Util)

Figure 12.13   System response time versus server clock jitter

12.6.5 Server Bandwidth Overhead

Figure 12.14 plots the ORT transmission rate versus server clock jitter for block sizes of $Q = 64$KB, 128KB, and 256KB. As clock jitter can be readily controlled to within 100ms by distributed software algorithms, the results show that over-rate transmission is applicable in all three cases. For example, with $Q = 64$KB, ORT will transmit at 1.556Mbps instead of the video bit-rate at 1.2Mbps, incurring a bandwidth overhead of 29.7%. Increasing the block size to 256KB reduces the ORT transmission rate to 1.273Mbps, or a bandwidth overhead of only 6%. Thus, the system designer can adjust the block size to balance between bandwidth cost and memory cost. In any case, compared to uncontrolled traffic overlapping which results in doubled transmission rate at 2.4Mbps, bandwidth under ORT is clearly substantially lower.

12.7 Network Resource Reservations

As the results in the previous section show, the staggered-push architecture can be scaled up to any number of servers, provided that the network has sufficient capacity. Compared with the concurrent-push architecture, staggered-push architecture achieves linear scalability at the expense of bursty network traffic (and slightly larger delay and client buffer requirement). In

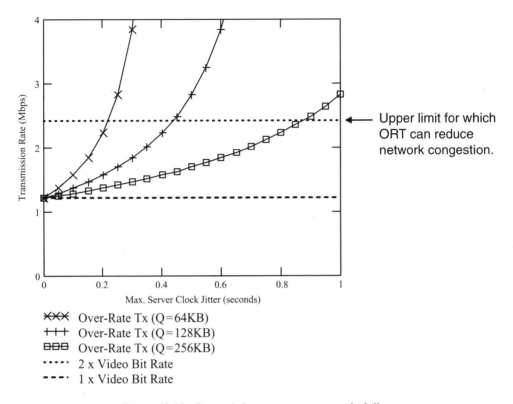

Figure 12.14   Transmission rate versus server clock jitter

particular, if we consider the network traffic between a server and a client, it is easy to see that the traffic will be in the form of bursts with an average inter-burst interval of $(N_S - 1)T_F$ seconds. By contrast, servers in concurrent push transmit to a client continuously at a constant rate, allowing easy integration with QoS offered by existing networks. Staggered push will not be able to make use of QoS available in today's QoS-enabled networks.

In practice, if the VoD system is deployed in dedicated networks with *a priori* bandwidth planning, then staggered push can still be used effectively. This is because the over-rate transmission scheme already guarantees that network congestion due to traffic overlapping will not occur, and the aggregate traffic going from the servers to a client will be close to constant bit-rate, with small gaps in between (due to over-rate transmission).

12.8 Summary

In this chapter, we have presented and analyzed a staggered-push parallel server architecture for implementing linearly scalable media streaming systems. The architecture employs fixed-size block striping for data storage, and a staggered-push scheduling algorithm for co-ordinating transmissions among multiple autonomous servers. We incorporated the effect of server clock jitter to address the inconsistent schedule assignment problem and the traffic overlapping problem. We tackled the former problem by an external admission scheduler and the latter problem by an over-rate transmission scheme. Our results showed that the over-rate transmission scheme can effectively prevent traffic overlapping with a small bandwidth overhead under clock jitter bounds achievable by existing software-based synchronization algorithms. Moreover, we showed that the server buffer requirement, the client buffer requirement, and the server bandwidth requirement are all independent of the number of servers in the system. The average system response time, though it increases slightly with more servers, remains acceptable if we limit the system to less than full utilization. These results demonstrate that the proposed architecture can be scaled up to large number of users without costly upgrade to the existing servers and clients.

Appendices

A.1 Proof of Theorem 12.1

Let t_i be the local time a new request arrives at server i ($0 \le i < N_S$), t_A be the local time the new request arrives at the admission scheduler, and Δ be the extra scheduling delay (in number of micro-rounds). Then the admission scheduler will attempt to admit the request to micro-round n_A as given in equation (12.3). For server i, the new request arrives during micro-round $n_i = \lfloor t_i / T_F \rfloor$. Hence, the problem is to find Δ so that $n_A > n_i$ for $0 \le i < N_S$, i.e., the assigned micro-round has not been started in any of the servers. Using this condition, we can then obtain the following inequality:

$$n_A > n_i \tag{12.34}$$

Expanding gives

$$\left\lfloor \frac{t_A}{T_F} \right\rfloor + 1 + \Delta > \left\lfloor \frac{t_i}{T_F} \right\rfloor \tag{12.35}$$

Rearranging gives

$$\Delta > \left\lfloor \frac{t_i}{T_F} \right\rfloor - \left\lfloor \frac{t_A}{T_F} \right\rfloor - 1 \tag{12.36}$$

Applying the inequality $\lfloor x \rfloor - \lfloor y \rfloor \le \lceil x - y \rceil$: $x, y \ge 0$, to R.H.S. of equation (12.36) we can then obtain

$$\left\lfloor \frac{t_i}{T_F} \right\rfloor - \left\lfloor \frac{t_A}{T_F} \right\rfloor - 1 \le \left\lceil \frac{t_i}{T_F} - \frac{t_A}{T_F} \right\rceil - 1 \tag{12.37}$$

Since clock jitter is bounded: $|t_i - t_A| \le \tau$, for $0 \le i < N_S$, we can rewrite equation (12.37) in terms of τ:

$$\left\lceil \frac{t_i}{T_F} - \frac{t_A}{T_F} \right\rceil - 1 \le \left\lceil \frac{\tau}{T_F} \right\rceil - 1 \tag{12.38}$$

Hence, if

$$\Delta > \left\lceil \frac{\tau}{T_F} \right\rceil - 1 \qquad (12.39)$$

or at least

$$\Delta = \left\lceil \frac{\tau}{T_F} \right\rceil \qquad (12.40)$$

then the assigned micro-round is guaranteed not to have started in any of the servers. ∎

A.2 Derivation of the Average Scheduling Delay

Assume that video sessions start independently and with equal likelihood at any time. Then a video session can be assigned to any one of the N_S micro-rounds with equal probability. Assume that there are n active video sessions, then the number of ways to distribute these n video sessions among N_S groups is a variant of the urn-occupancy distribution problem [4] and is given by

$$N(n, N_S, \Lambda) = \sum_{j=0}^{N_S} (-1)^j \binom{N_S}{j} \binom{N_S + n - j(\Lambda + 1) - 1}{N_S - 1} \qquad (12.41)$$

To obtain the probability of having m fully-occupied micro-rounds, we first notice that there are $\binom{N_S}{m}$ possible combinations of having m fully-occupied micro-rounds. Given that, the number of ways to distribute $(n - m\Lambda)$ video sessions among $(N_S - m)$ micro-rounds with none of those micro-rounds fully occupied can be obtained from equation (12.41) as $N(n - m\Lambda, N_S - m, \Lambda - 1)$. Hence, the total number of ways for exactly m of the micro-rounds fully occupied is given by

$$N_{full}(n, m) = \binom{N_S}{m} N(n - m\Lambda, N_S - m, \Lambda - 1) \qquad (12.42)$$

Hence, the probability of having m fully-occupied micro-rounds given n active video sessions can be obtained from

$$P_{full}(n, m) = \frac{N_{full}(n, m)}{N(n, N_S, \Lambda)} \qquad (12.43)$$

Assume that m out of N_S micro-rounds are fully occupied, then, the probability for the assigned micro-round to be available (not fully occupied) is given by

$$V_0 = \frac{N_S - m}{N_S} \qquad (12.44)$$

Hence, $P_0 = (1 - V_0)$ will be the probability of the assigned micro-round being fully occupied. Now provided that the assigned micro-round is fully occupied, the probability that the next micro-round is available is

$$V_1 = \Pr\{\text{next round available}| P_0\} = \frac{N_S - m}{N_S - 1} \tag{12.45}$$

This is also the probability for a client to wait one additional micro-round provided the assigned micro-round is already fully occupied. It can be shown that the probability for a client to wait k additional micro-rounds provided that the first k assigned micro-rounds are all fully occupied is

$$V_k = \Pr\{(k+1)\text{th round available}| P_k\} = \frac{N_S - m}{N_S - k} \quad 1 \le k \le m \tag{12.46}$$

We already know P_0, and it can be shown that the probability for the first k micro-rounds all being fully occupied is given by

$$P_k = \prod_{i=0}^{k-1}\left(\frac{m-i}{N_S - i}\right) = \frac{m!(N_S - k)!}{N_S!(m-k)!}, \quad 1 \le k \le m \tag{12.47}$$

Hence, we can solve for the probability of a client having to wait k additional micro-rounds from

$$W_k = \Pr\{(k+1)\text{th round free}| P_k\} P_k \tag{12.48}$$
$$= \frac{(N_S - m)m!(N_S - k - 1)!}{N_S!(m-k)!}, \quad 1 \le k \le m$$

Therefore, given m – the number of micro-rounds that are fully occupied – the average number of micro-rounds a client has to wait can be obtained from

$$W_{avg}(m) = \sum_{k=1}^{N_S} k W_k + \left(\left\lceil \frac{\tau}{T_F}\right\rceil + 1\right) \tag{12.49}$$

where the second term accounts for the additional delay as described in Theorem 12.1. Similarly, given n – the number of active video sessions – the average number of micro-rounds a client has to wait can be obtained from

$$M_{avg}(n) = \sum_{j=1}^{N_S-1} W_{avg}(j) P_{full}(n, j) \tag{12.50}$$

And the corresponding average scheduling delay given a system utilization of n is

$$D_S = \frac{M_{avg}(n)Q}{R_V} \tag{12.51}$$

Substituting equations (12.43), (12.48), (12.49), and (12.50) into equation (12.51) gives the desired result. ∎

References

[1] R. Gusella and S. Zatti, The Accuracy of the Clock Synchronization Achieved by TEMPO in Berkeley UNIX 4.3BSD, *IEEE Transactions on Software Engineering*, vol. 15, no. 7, July 1989, pp. 847–853.

[2] D. Mills, Internet Time Synchronization: The Network Time Protocol, *IEEE Transactions on Comm.*, vol. 39, no. 10, Oct. 1991, pp. 1482–1493.

[3] Z. Yang and T.A. Marsland, (eds) *Global States and Time in Distributed Systems*, IEEE Computer Society Press, Los Alamitos, CA, USA, 1994.

[4] J.N. Lloyd and K.S. Samuel, *Urn Models and their Application*, John Wiley & Sons, 1997, pp. 125–126.

13

FEC versus PRT

With data and capacity redundancy, a parallel-server streaming system can sustain server-level failures using either the Forward Erasure Correction (FEC) protocol or the Progressive Redundancy Transmission (PRT) protocol. Except for the need for failure detection, PRT is superior to FEC as it consumes significantly less bandwidth overhead for redundant data transmission. Nevertheless PRT may also reduce the reliability of the system if multiple servers fail within a short time. This chapter investigates this issue, and more generally, compares the reliability of FEC and PRT under the same conditions so that fair and meaningful comparisons can be made. Surprisingly, we discover that by allowing a small trade-off in storage overhead, PRT not only can maintain similar or even better system-level reliability, but also reduces the bandwidth overhead in sending the redundant data by more than 50%.

13.1 Introduction

One challenge inherent in all parallel server architectures is fault tolerance. In particular, server failure, while uncommon, can cripple the entire system if redundancies are not incorporated. To tackle this problem, we can employ erasure correction code to enable the client to recover data lost in failed servers (cf. Chapter 11). If the recovery is done in real-time, then the process can even be made transparent to the end user – non-stop streaming, which is highly desirable from a service-provisioning point of view.

Note that to enable the client to perform erasure correction computations, the servers need to send the redundant data units in addition to the normal data units to the clients. We introduced two such redundant data transmission protocols – Forward Erasure Correction (FEC) and Progressive Redundancy Transmission (PRT) in Chapter 9 and subsequently applied them to the concurrent-push architecture in Chapter 11.

Qualitatively, the PRT protocol is more complex as it requires the detection of server failure and the dynamic reconfiguration of the system to transmit more redundant data. Moreover, as fewer redundant data are transmitted, one would expect PRT to be less reliable than FEC. In this chapter we investigate this reliability issue quantitatively by modeling the system as a continuous-time Markov chain to derive its mean-time-to-failure (MTTF), incorporating the effects of server failure rate, server repair rate, failure detection and system reconfiguration

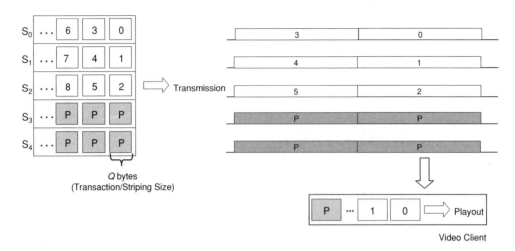

Figure 13.1   Data placement and transmission in a parallel streaming server

time. Surprisingly, instead of having a lower reliability than FEC, we discover that we can configure PRT to outperform FEC (achieving longer MTTF) *and* consume significantly less transmission bandwidth overhead (half that of FEC). We first introduce the system model in the next section and then analyze the two redundant data transmission protocols.

13.2 System Model

Figure 13.1 depicts the generic system model for a parallel streaming server. Let N_S be the number of servers in the system. Each server (denoted by S_i, $i = 0, 1, \ldots, N_S - 1$) is autonomous, and equipped with its own CPU, memory, disk storage, and network interface. This ensures that any server failure will not spread to other servers. Specifically, we assume the servers to fail independently and, when they fail, they simply stop all data transmissions.

A media object such as a video stream is first divided into fixed-size data blocks of Q bytes each, denoted by $\{b_i | i = 0, 1, \ldots\}$. To generate redundant data we use a (N_S, K) erasure correction code and compute K redundant data blocks for every $(N_S - K)$ data blocks. For example, K redundant blocks (i.e., blocks marked with 'P' in Figure 13.1) are computed from data blocks $\{b_i | i = 0, 1, \ldots, (N_S - K - 1)\}$ and together they form a parity group. Each of the data/redundant blocks in a parity group will be stored in a separate server in the system. The client can recover all $(N_S - K)$ data blocks as long as any $(N_S - K)$ data/redundant blocks of a parity group are available, thus enabling it to tolerate up to K simultaneous server failures.

13.3 Forward Erasure Correction

In the FEC protocol, all N_S data blocks in a parity group are transmitted to the clients at all time, irrespective of server failure. This protocol is simple to implement as the server schedules are fixed irrespective of individual server failures. The system does not even need to detect server failures or to reconfigure the system for degraded-mode operation. The client simply performs erasure correction computation to recover the data blocks whenever $N_S - K$ video blocks of a parity group are received.

13.3.1 System Reliability

To quantify the amount of redundancy needed to achieve a given target system MTTF, we can model the system using a continuous-time Markov chain. We assume server failures are independent and exponentially distributed with a MTTF of $1/\lambda$. Failed servers are repaired immediately and independently with a mean-time-to-repair (MTTR) of $1/\mu$. Thus, the system forms a Markov chain with state h representing the state with h failed servers (see Figure 13.2, for an example). Assume the system is configured with K redundant blocks per parity group, then the system fails when more than K servers fail, i.e., when the Markov chain enters the absorbing state $h = K + 1$. Otherwise, servers in the system in state h will have an aggregate failure rate λ_h, given by

$$\lambda_h = \lambda(N_S - h) \tag{13.1}$$

and an aggregate repair rate μ_h, given by

$$\mu_h = h\mu. \tag{13.2}$$

Thus, the MTTF of the system is equivalent to the first passage time for the system to reach state $h = K + 1$ from the initial state $h = 0$. It can be shown that the MTTF for a system with N_S servers and K redundancies using FEC is given by

$$MTTF_{FEC} = \sum_{i=0}^{K} \left(\sum_{j=0}^{i} \frac{\prod_{l=0}^{j-1} \mu_{i-l}}{\prod_{l=0}^{j} \lambda_{i-l}} \right). \tag{13.3}$$

Therefore, using equation (13.3) we can determine the amount of redundancy needed for a given target system MTTF.

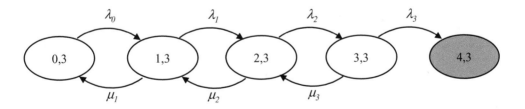

$\boxed{h,k}$: state with h servers failed and k level of redundancy
λ_h : aggregate server failure rate with h server failed
μ_h : aggregate repair rate with h server failed

Figure 13.2 A Markov chain mode for FEC with $K = 3$

13.3.2 Bandwidth Overhead

The price to pay for the low-complexity FEC protocol is transmission overheads. In particular, with N_S servers and a redundancy level of K, the bandwidth overhead incurred – defined as the ratio of extra bandwidth needed over the media bit-rate, is given by

$$H_{FEC} = \frac{K}{N_S - K} \qquad (13.4)$$

For a small-scale system (i.e., N_S small) with a high level of redundancy (i.e., K large), this overhead could become substantial. For example, with $N_S = 6$ and $K = 2$, the bandwidth overhead becomes 50%. Considering that most systems are expected to operate in normal mode most of the time with no server failure, this bandwidth overhead is clearly significant.

13.4 Progressive Redundancy Transmission

Generally speaking, media servers are usually high-end PCs with good host-level reliability (e.g., equipped with ECC memory, redundant power supply, RAID disk storage, etc.). Thus, typical server MTTF is likely to be in the range of tens of thousands of hours. Therefore, although over a long time span the system may run into multiple server failures, it is highly unlikely that more than one server will fail simultaneously within a short time interval (e.g., tens of seconds) unless catastrophic failure occurs (e.g., fire, earthquake, etc.).

Therefore, instead of sending all K redundancies at all times as in FEC, the system can initially transmit only k ($k \leq K$) of the K redundancies, thus reducing the bandwidth overhead to $k/(N_S - K)$ – Progressive Redundancy Transmission (PRT). Clearly, in this case the client can only recover from up to k simultaneous server failures. However, in PRT the system will activate transmission of more redundant data as server failures are detected. Thus, after, say, w server failures, the number of redundant data transmission will be increased from k to $(k + w)$. The key is to complete the reconfiguration quickly so that additional redundant data transmissions are activated before too many server fail.

On the other hand, whenever a server is repaired and becomes operational again, the system will deactivate an existing redundant data transmission, until the number of *excess redundancies* (i.e., number of transmitted redundant data minus number of failed servers) transmitted reduces back to k.

To implement PRT we need to address two issues. First, the system must be able to detect server failure so that it can reconfigure the system to activate transmissions of additional redundant data. The detection can be done using a number of existing protocols such as heartbeat protocol [1], or by monitoring the streaming traffic in the network (cf. Section 11.3.1). Once a failure is detected, the system can be reconfigured through some control protocols. For simplicity, we lump together the system reconfiguration time and the failure detection time and henceforth refer to the latter term only.

Intuitively, the detection time should be short so as to reduce the risk of encountering additional server failures before reconfiguration is completed, but not too short to prevent generating too many false alarms. In particular, if more than k servers fail before reconfiguration is completed, then the system will still fail, even if the number of redundancies available is larger ($K > k$). This leads to the second problem: quantifying the impact of detection delay to the system MTTF. In other words, we need to derive the system MTTF in terms of the failure rate, repair rate, total number of redundancies, initial number of redundancies transmitted, and the detection delay in order to choose the appropriate system parameters to meet a given system MTTF requirement.

13.4.1 System Reliability

Again we model the system as a continuous-time Markov chain with state denoted by (h, k), where h is the number of failed servers and k is the number of redundancies activated for transmission. Let K_{max} be the total number of redundancies in the system and K_{min} be the excess number of redundancies to be transmitted. Figure 13.3 illustrates an example with $K_{min} = 2$ and $K_{max} = 4$. In general, there are three types of state transition. Specifically, the system will transit from state (h, k) to state $(h + 1, k)$ at a rate of λ_h when an additional server fails. After the failure is detected, with a mean detection time of $1/\omega$, the system will activate a new redundancy transmission and transit to state $(h + 1, k + 1)$, thus bringing the number of excess redundancies back to K_{min}. When a failed server is repaired at a rate of μ_h in state (h, k), then the system will transit from state (h, k) to state $(h - 1, k)$ if $(k - h) \leq K_{min}$. Otherwise, i.e., $(k - h) = K_{min}$, it will transit from state (h, k) to state $(h - 1, k - 1)$ by deactivating one redundancy currently being transmitted. This again brings the number of excess redundancies back to K_{min}.

Note that there is a subtle problem associated with this model. In FEC, since all the servers in the system are operating, server failure will only occur in servers actively transmitting video

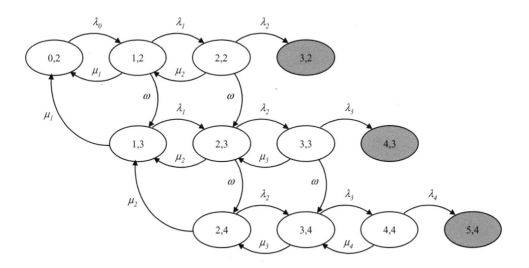

(h,k) : state with h servers failed and k level of redundancy
λ_h : aggregate server failure rate with h server failed
μ_h : aggregate repair rate with h server failed
ω : detection rate

Figure 13.3 A Markov chain for PRT with $K_{min} = 2$ and $K_{max} = 4$

data. However, in PRT, not all servers are actively transmitting data. When an idle server fails, it will undergo repair and then return to the normal state after some time. The difference is that when an idle server fails, the operation of the system is not affected since none of the data transmissions are affected by the failure. By contrast, failure of an active server will require the system to increase the level of redundancy transmitted. Thus, the failure of an idle server is equivalent to the failure of an active server with an additional redundancy activated immediately, i.e., the failure detection time is equal to zero. The previous Markov chain model, which assumes a failure-detection time of $1/\omega$, is therefore a conservative estimation of the system MTTF. Nevertheless, our numerical results show that such differences are negligible and so we will ignore this subtle complexity.

Using the Markov chain model, we can then derive the system MTTF from the first passage time of the Markov chain to reach any of the absorbing states (e.g., states (3, 2), (4, 3), and (5, 4) in Figure 13.3). For systems with small K_{min} and K_{max}, we can solve the Markov chain analytically to obtain the equation for the system MTTF. For example, when $K_{min} = 1$ and $K_{max} = 2$, the system MTTF is given by

$$MTTF = \frac{\begin{array}{c}\lambda_1\mu_1\mu_2 + 2\mu_1\lambda_1\lambda_2 + \lambda_1{}^2\lambda_2 + \lambda_1\lambda_2\omega + \lambda_2\mu_1{}^2 + \mu_1\mu_2\omega + \lambda_0\mu_1\mu_2 \\ +\mu_1{}^2\mu_2 + \omega\lambda_0\mu_2 + \omega\lambda_2\mu_1 + \mu_1\lambda_0\lambda_2 + \omega\lambda_0\lambda_1 + \omega\lambda_0\lambda_2 + \lambda_0\lambda_1\lambda_2\end{array}}{\lambda_0\lambda_1(\mu_1\mu_2 + \mu_1\lambda_2 + \lambda_1\lambda_2 + \omega\lambda_2)} \tag{13.5}$$

However, for a system with larger values of K_{min} and K_{max}, the number of equations involved increases drastically. Although still solvable in principle, the process soon becomes too tedious to perform manually. Therefore we resort to using mathematical software packages (e.g., Maple [2]) to compute the symbolic solutions. For still larger values of K_{min} and K_{max}, even this computation time can become too long. Also the resultant analytic solutions are often too complicated to be useful. For example, the solution for system MTTF of the configuration in Figure 13.3 with $K_{min} = 2$ and $K_{max} = 4$ has 474 terms in the numerator and 66 terms in the denominator. In these cases we need to resort to computing the results numerically instead of symbolically.

13.4.2 Modeling the Failure-Detection Time

So far we have assumed that the failure-detection time is exponentially distributed. In practice, the exact distribution of the failure-detection time will depend on the failure detection protocol employed and thus may not conform to the exponential distribution. To investigate the impact of the type of distribution for the failure-detection time, we extend the model by representing the failure-detection time as an Erlang-k random variable. Thus, by varying the parameter k we can obtain a series of different probability distributions, ranging from exponential distribution (for $k = 1$) to normal distribution (for large k).

To extend the system model to use Erlang-k distributed failure-detection time, we decompose the Erlang-k distribution into a series of k independent exponential random processes, each having a rate equal to $k\omega$ as shown in Figure 13.4 for $k = 2$. Thus, the mean detection time for the Erlang-k process is exactly the same as the exponential process in the original model.

Similarly, the system MTTF is the first passage time to reach any of the absorbing states, i.e., states $\{(K_{max} + i, K_{min} + i - 1, j) \mid i = 1, 2, \ldots, (K_{max} - K_{min} + 1), j = 0, 1, \ldots, k - 1\}$ in this generalized Markov chain model. For small values, of N_S, K_{min}, and K_{max}, we can solve

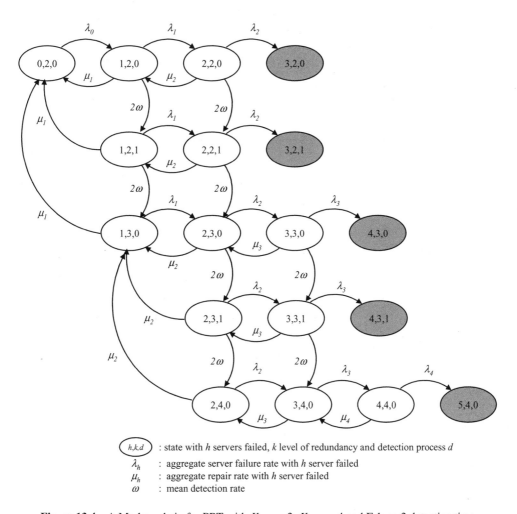

Figure 13.4 A Markov chain for PRT with $K_{\min} = 2$, $K_{\max} = 4$ and Erlang-2 detection time

it analytically to obtain the first passage time directly. For larger values, the resulting solutions are very complex and again we make use of Maple to obtain numerical solutions.

For the sake of verification, we have also developed a simulation program to measure the system MTTF. For large values of $K_{\min}$ and $K_{\max}$, the simulation time required is extraordinary long. But for smaller values of $K_{\min}$ and $K_{\max}$ the simulation time is manageable and the simulation results do confirm the correctness of the numerical results obtained from Maple.

13.5 Performance Evaluation

Using the reliability models in Sections 13.4 and 13.5, we answer in this section the question of how much bandwidth overhead can be saved by PRT under the constraint that the system reliability is at least as good as FEC. Table 13.1 summarizes the system parameters used in computing the following numerical results.

Table 13.1   System parameters used in computing numerical results

System parameter	Symbol	Value
Average Failure-Detection Time	$1/\omega$	6 sec
Node MTTF	$1/\lambda$	50,000 hrs
Node MTTR	$1/\varepsilon$	48 hrs
Number of Servers	N_S	64

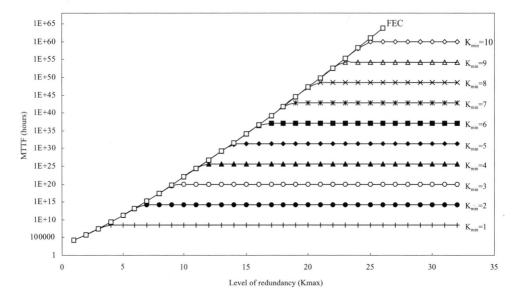

Figure 13.5   System MTTF of PRT and FEC with average detection time equal to 6 seconds

13.5.1 Effect of Detection Time Distribution

In Section 13.4, we use the Erlang-k distribution to model different types of distributions for the failure detection time. We have computed the MTTF of a system with $K_{min} = 10$ and $K_{max} = 32$ for different values of k and find that differences in the shape of the detection time distribution have negligible effect on the system MTTF. For example, with a mean detection time of $1/\omega = 60$ seconds, the system MTTF is increased by only $8.7*10^{-45}$ % when k is increased from 1 to 10. This observation reflects the fact that the detection time is many orders of magnitude smaller than the server MTTF (e.g., tens of seconds versus tens of thousands of hours) and thus changes in its distribution have little effect on the system MTTF. Therefore, we will simply use the Erlang-1 distribution, i.e., exponential distribution, to compute the numerical results in the following sections.

13.5.2 Bandwidth Overhead Reduction

Figure 13.5 compares the system MTTF of FEC and PRT versus the level of redundancy in the system (i.e., K for FEC and K_{max} for PRT). For FEC, the system MTTF increases exponentially with more redundancies as expected. For PRT, a total of 10 curves are plotted, each with a different setting of K_{min}, ranging from 1 to 10, representing the number of redundancies that are actively transmitted. There are two observations. First, the system MTTF of PRT for small values of K_{max} is similar to that of FEC with $K = K_{max}$. However, beyond a certain number of redundancies, the system MTTF levels off. This implies that for PRT with a given K_{min}, there is an *upper limit* on the achievable system MTTF even if we increase K_{max} indefinitely. Beyond that limit any additional redundancies are simply wasted.

Second, the plateau of the curves for PRT is determined by the number of redundancies actively transmitted – K_{min}, with larger K_{min} resulting in higher achievable system MTTF. This implies that to achieve a target system MTTF, there is a *lower limit* on K_{min} below which the required MTTF can never be achieved, regardless of the total number of redundancies in the system. The key is that the minimum K_{min} required will still be substantially smaller than the corresponding number of redundancies K required to achieve the same system MTTF in FEC.

For example, FEC with $K = 7$ achieves a system MTTF of 10^{15} hours. Now consider the PRT curve in Figure 13.6 with $K_{min} = 2$ and $K_{max} = 7$ the system MTTF, is only 10^{14} hours which is lower than FEC. To increase the system MTTF, we can either increase K_{max} or increase K_{min}. In this case, increasing K_{max} does not work because the system MTTF levels off to a plateau below the required system MTTF. However, increasing K_{min} to 3 does not work either because the resultant system MTTF is still below 10^{15} hours. This is expected as failure detection in PRT incurs a delay in responding to a server failure. Thus, with the same number of total redundancies $K = K_{max}$, it must have a lower system MTTF compared to

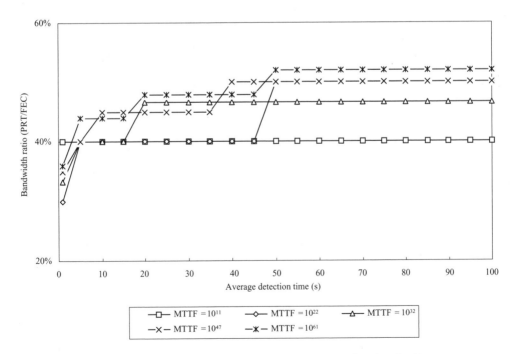

Figure 13.6   Bandwidth ratio (K_{min}/K) versus average detection time for 64 servers

FEC if $K_{min} < K_{max}$. To compensate, we can increase K_{max} from 7 to 8 and the system MTTF will become 10^{17} hours with $K_{min} = 3$, exceeding the system MTTF of FEC with $K = 7$. Therefore, in this example we can reduce the bandwidth overhead from $K = 7$ redundancies in the case of FEC to $K_{min} = 3$ redundancies in the case of PRT.

We further illustrate the bandwidth overhead savings in Figure 13.6 by plotting the bandwidth ratio, defined as the ratio K_{min}/K, versus the average failure-detection time. In all cases PRT achieves the same or a better system MTTF compare to FEC. We can observe that the reduction in bandwidth overhead is very significant, at least half of the bandwidth overhead is saved and the savings increase with shorter failure-detection time. Considering failure detection in practice is likely to be in the order of seconds, this result clearly shows the feasibility and superiority of the PRT protocol.

13.5.3 Storage Overhead

Compared to FEC, the PRT protocol has one trade-off. Reconsidering the previous example with $K = 7$ in FEC, we need to configure PRT with $K_{min} = 3$ and $K_{max} = 8$ to achieve the same or a better system MTTF than FEC. While the bandwidth overhead is reduced by $(7 - 3)/7 = 57\%$, the fact that K_{max} is larger than K means more redundancies will need to be stored, though not all are actively transmitted. In this case the additional storage overhead in PRT is equal to $(K_{max} - K)/K = 14\%$.

Figure 13.7 plots the percentage increase in storage overhead due to PRT versus the target system MTTF requirements. From the results we can see that the additional storage requirement decreases rapidly with higher target system MTTF. With the rapid decrease in storage cost, most hard disks are I/O bound rather than storage bound. Thus, the added storage requirement can be easily absorbed with little to no impact on the total cost of the system.

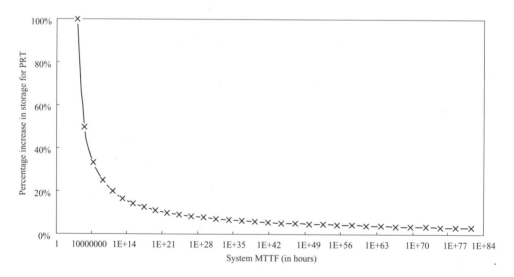

Figure 13.7   Percentage increase in storage due to PRT for different system MTTF requirement

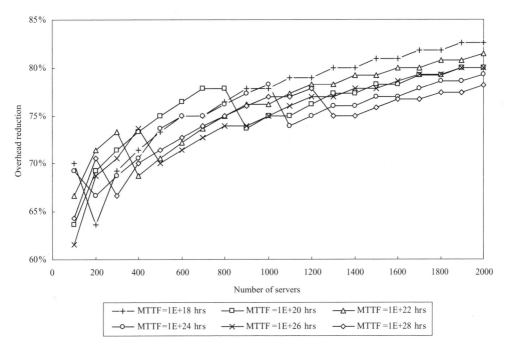

Figure 13.8 Reduction in bandwidth overhead versus number of servers with six different MTTF requirements

13.5.4 Scalability

Finally, we investigate the scalability of the PRT protocol. Specifically, we want to know if PRT can maintain the reduction in bandwidth overhead when we scale up the system to more servers. We plot in Figure 13.8 the bandwidth overhead reduction achieved by PRT versus the number of servers in the system for six different MTTF requirements. The results clearly show that the bandwidth overhead reduction achieved by PRT in fact increases with the system scale, suggesting that the PRT protocol is indeed scalable.

13.6 Summary

In this chapter, we modeled and compared the reliability of two redundant data transmission protocols, namely Forward Erasure Correction (FEC) and Progressive Redundancy Transmission (PRT). Surprisingly, the PRT protocol can achieve over 50% reduction in transmission bandwidth overhead while maintaining the same or better system reliability. The key is to encode the data with more redundant data (compared to FEC) but only transmit some of them initially. As the failure-detection time is far shorter than the server MTTF, the chance of experiencing multiple server failures within a short time, which could lead to system failure in PRT, is very small. The only trade-off in PRT is increased storage requirement, which is relatively small (e.g., storage overhead $\leq 20\%$ for $K \geq 5$) and thus can easily be accommodated in practice. The reduced bandwidth overhead will translate into lower requirement on disk throughput, server processing requirement, network bandwidth requirement, as well as client access bandwidth requirement. This enables substantial cost savings across the whole system.

References

[1] M.K. Aguilera, W. Chen, and S. Toueg, Heartbeat: A Timeout-Free Failure Detector for Quiescent Reliable Communication, *Proceedings of the 11th International Workshop on Distributed Algorithms (WDAG '97)*, Saarbrucken, Germany, Sept. 1997, pp. 126–140.

[2] Official website of Maple http://www.maplesoft.com.

14

Algorithms for Server Rebuild

In the previous chapters we have introduced the many desirable features of parallel streaming servers such as scalability and fault tolerance. In this and the next chapter we will address two practical issues resulting from the use of striped server storage. First, in this chapter we investigate the issue of server data rebuild and in Chapter 15 the issues in expanding a parallel server system.

Armed with redundant data and streaming capacity, a parallel streaming server can sustain server-level failures and maintain non-stop media playback. However, the failed server will eventually need to be repaired, or if repair is not feasible or desirable, replaced by a new server. In the latter case we will need to load the appropriate media data into the new server so that it can share the streaming workload – the server data rebuild problem. This chapter investigates this and analyzes algorithms for rebuilding data in a failed server into a new server transparently so that existing streaming sessions are not adversely affected.

14.1 Introduction

Armed with data and capacity redundancy, a parallel streaming server can operate in degraded mode without causing any interruption to the existing streaming sessions. However, the failed server will still need to be repaired or replaced and in the latter case we will also need to load appropriate media data into the new server so that it can share some of the streaming workload. This is referred to as the server data rebuild problem.

In the context of disk arrays and RAID [1], a similar data rebuild problem also exists, and in Chapter 5 we have investigated rebuild algorithms for disk arrays. Despite the similarities, parallel server differs from RAID in that bandwidth of the communications links is far more limited. For example, when considering RAID in Chapter 5 we assumed that the data bus connecting the hard disks and the main system was not the bottleneck and so the disks could retrieve and stored data into the system memory buffers as fast as the disk would allow. While this is a reasonable assumption in RAID, in parallel server the network linking up the servers will likely have more limited bandwidth, e.g., 1 Gbp using Gigabit Ethernet [2] switches. Thus, the network may become the bottleneck in the data rebuild process.

14.1.1 Sparing Schemes

Figure 14.1 shows the use of a spare server to store rebuilt data – *hot sparing*. Note that the spare server is not used under normal mode operation. When a server fails, the data in the failed server are then rebuilt and stored in the spare server. When the rebuild process finishes, the spare server simply replaces the failed server as shown in Figure 14.2.

Figure 14.3 shows another approach of allocating spare units – *distributed sparing*, in which the spare stripe units are distributed over all the servers. During normal mode operation, all servers participate in serving client requests. After a server has failed, data in the failed server are then rebuilt and stored in the spare stripe units. Unlike hot sparing, the rebuilt system has a different configuration from the original configuration (Figure 14.4). Therefore, an additional phase called *system restoration* is required to copy all rebuilt data onto another spare server. That spare server can then replace the old server to resume normal system operation.

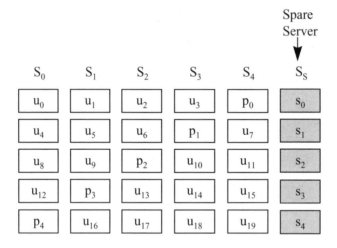

Figure 14.1   Storage configuration for hot sparing

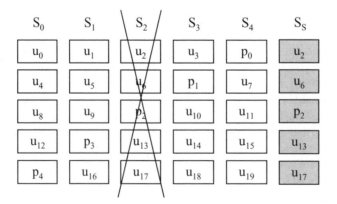

Figure 14.2   Storage configuration after lost-data rebuild in hot sparing

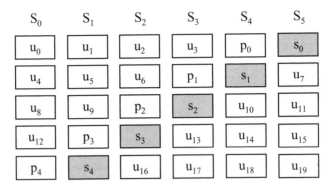

Figure 14.3 Storage configuration for distributed sparing

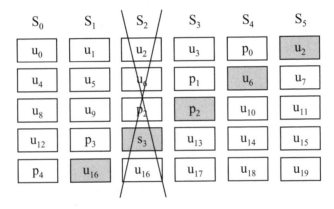

Figure 14.4 Storage configuration after lost-data rebuild in distributed sparing

Distributed sparing has the advantage that all servers are utilized under normal mode. However, the extra restoration phase increases system repair time considerably because the network and server throughout are limited. Therefore, we consider only hot-sparing in this chapter.

14.1.2 Data Rebuild Computation

To rebuild the data in the failed server, we can perform erasure correction computation using the remaining data and redundant data units. To do this, we will need to send the remaining data and redundant units to a host to perform the erasure correction computation. These data transmissions obviously will add to the streaming workload of the remaining servers. To avoid adversely affecting the on-going streaming sessions, it is therefore necessary to use only the residual capacities in the retrieval and transmission schedulers for such a purpose.

In the following we investigate five rebuild algorithms – disk migration, reloading data from back-up, baseline rebuild, distributed rebuild, and mixed distributed-baseline rebuild. The first two are simple solutions which also require extra equipment and/or human intervention in the process. The last three algorithms are automatic and transparent, automatic in the sense that no human intervention is required and transparent in the sense that existing streaming sessions are not affected in any way.

For simplicity, we consider only single-server failure in the rest of the chapter but the analysis can readily be extended to cover multiple-simultaneous server failures. Let N_S be the number of servers in the system, U be the storage capacity of each server, and S_S be the effective server transfer capacity. The effective server transfer capacity S_S is the maximum data rate at which a server can transfer data to/from the network. That is, if a server has an effective transfer capacity of $S_S = 600$Mbps, and the server is sending data at a rate of 200Mbps, then the server will only be able to receive data up to a rate of 400Mbps.

14.2 Disk Migration

It is possible that a server failure may not be caused by (unrecoverable) disk failures. In this case, we may simply disconnect the disk subsystem from the failed server and connect it to the spare server. This process is simple but requires human intervention to first identify the source of the failure, and then to migrate the disk subsystem to the spare server. It is also possible to use an electronic wiring network to connect a disk subsystem to two or more servers, including a spare server (e.g., using twin-tailed disks). The migration of disk units can then be done electronically.

14.3 Reloading Data from Back-up

If the media data in the system are also stored in a back-up storage, then we can simply load the required media data from the back-up media to the spare server. This can be done automatically without affecting the remaining active servers. The rebuild rate will then be equal to the transfer capacity of the back-up device or the spare server, whichever is smaller.

While this scheme is simple and efficient, the mechanisms needed to automate the process (selection and loading of media data) are not inexpensive, such as large juke-boxes or robotic tape libraries. Moreover, management of the back-up data will be more complex as old media data are removed and new ones uploaded to the servers from time to time. Finally, if the media data streams are backed up sequentially, then considerable searching will be needed when loading the media data into a spare server due to the server stripping scheme used.

For comparison with other rebuild algorithms, we assume the back-up device has infinite throughput and capacity. Therefore, the rebuild rate is bounded by the server throughput, or

$$R_{reload} = S_S \tag{14.1}$$

which is also the maximum rate that can be achieved by any data rebuild algorithm.

14.4 Baseline Rebuild

The two methods discussed earlier both require extra hardware/software and/or human operator support. This section presents the first automatic algorithm – baseline rebuild, that requires neither extra hardware/software support nor human intervention. The principle is to utilize the idle capacities in the remaining servers to send data/redundant units to the spare server, which

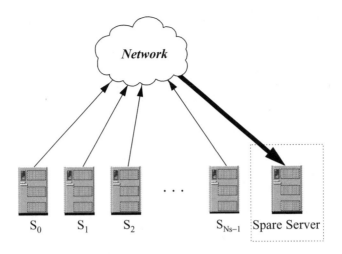

Figure 14.5   The baseline rebuild algorithm

then rebuilds the data lost in the failed server (Figure 14.5). This algorithm is similar to the baseline rebuild scheme in disk arrays [3].

To recover an unavailable data unit we need to perform erasure correction using the remaining $(N_S - 1)$ data/redundant units. Since we want to keep the process transparent to normal streaming sessions, we can only utilize the servers' idle capacities. Assuming the system is running at a utilization of $\rho \in [0,1]$, the average available transfer rate from each server will be equal to $S_S(1 - \rho)$. The aggregate data transfer rate of the remaining $(N_S - 1)$ servers is therefore equal to

$$r = S_S(1 - \rho)(N_S - 1) \tag{14.2}$$

Now as the spare server has a transfer capacity of S_S, the aggregate data rate r will exceed the spare server's transfer capacity if ρ is less than $(1 - /(N_S - 1))$. This is stated in the Theorem 14.1 which computes the upper limit on the baseline rebuild rate.

Theorem 14.1. *The data rebuild rate of baseline rebuild, denoted by* $R_{baseline}$ *is bounded by the capacity* S_S *of the spare server and is given by*

$$R_{baseline} = \begin{cases} S_S/(N_S - 1) & \text{for } \rho < (1 - 1/(N_S - 1)) \\ S_S(1 - \rho) & \text{for } \rho \geq (1 - 1/(N_S - 1)) \end{cases} \tag{14.3}$$

Proof. Case 1: $\rho < (1 - 1/(N_S - 1))$
From equation (14.2):

$$r = S_S(1 - \rho)(N_S - 1)$$
$$\geq S_S(1 - (1 - 1/(N_S - 1)))(N_S - 1)$$
$$= S_S$$

Although the aggregate transfer rate from the remaining servers is greater than S_S, the spare server can only accept data at a rate of S_S. Using erasure correction the spare server will rebuild one data unit for every $(N_S - 1)$ data/redundant units received. We have therefore a rebuild rate $R_{baseline} = S_S/(N_S - 1)$ at the spare server.

Case 2: $\rho \geq (1 - 1/(N_S - 1))$

From equation (14.2):

$$r = S_S(1 - \rho)(N_S - 1)$$
$$\leq S_S(1 - (1 - 1/(N_S - 1)))(N_S - 1)$$
$$= S_S$$

As the transfer rate from the remaining servers is less than S_S, the corresponding rebuild rate at the spare server is $R_{baseline} = r/(N_S - 1) = S_S(1 - \rho)$.   ∎

The rebuild time for a server with storage U is then given by

$$T_{baseline} = \frac{(N_S - 1)U}{\min\{S_S, S_S(1 - \rho)(N_S - 1)\}} \tag{14.4}$$

14.5 Distributed Rebuild

In baseline rebuild, the transfer capacity of the spare server can become the bottleneck even if the remaining servers have abundant idle capacities available. An alternative approach is to rebuild the unavailable data units before transferring them to the spare server. In this way, only the rebuilt data are sent to the spare server and hence the limited transfer capacity of the spare server can be better utilized.

To achieve this, we can employ a distributed rebuild scheme to distribute the rebuild computations over all the remaining servers. We first divide the unavailable data into $(N_S - 1)$ equal-size subsets, with each subset then rebuilt by one of the remaining $(N_S - 1)$ servers, as shown in Figure 14.6. The server responsible for a subset will receive the required data/redundant units from the other $(N_S - 2)$ servers, rebuild the unavailable units, and then send the rebuilt units to the spare server for storage.

To derive the rebuild rate of the distributed rebuild algorithm, we first note that the sum of transfer rates in and out of the remaining $(N_S - 1)$ servers is equal to $S_S(1 - \rho)(N_S - 1)$. Second, to rebuild each data unit we need $2(N_S - 2)$ transfers (half for transmission and the other half for reception) of data/redundant units from the other $(N_S - 2)$ servers to the rebuild server – the server responsible for rebuilding the unavailable data unit. Note that we need only $(N_S - 2)$ transmissions instead of $(N_S - 1)$ because the rebuild server already has one of the data/redundant units stored locally, and so no transfer over the network is needed. Therefore, we can compute the rebuild rate $R_{distributed}$ from

$$R_{distributed} = \frac{S_S(1 - \rho)(N_S - 1)}{2(N_S - 2) + 1} = \frac{S_S(1 - \rho)(N_S - 1)}{2N_S - 3} \tag{14.5}$$

which is also the data rate at which rebuilt data are sent to the spare server.

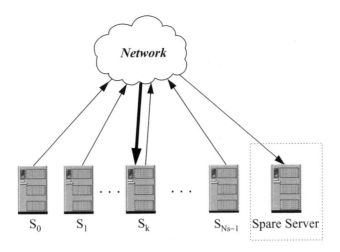

Figure 14.6   The distributed rebuild algorithm

Interestingly, being the bottleneck in baseline rebuild the spare server is no longer the limiting factor in distributed rebuild. The following theorem shows that the rebuild data rate $R_{distributed}$ is always smaller than the transfer capacity of the spare server, regardless of the system utilization ρ.

Theorem 14.2. *In distributed rebuild, the rate of data transfer from the active servers to the spare server will never exceed the capacity of the spare server.*

Proof. We note that $(N_S - 1) \leq 2(N_S - 3)$ for all $N_S \geq 2$ in (14.5) thus $R_{distributed} \leq S_S(1-\rho)$ S_S.   ∎

The corresponding rebuild time in distributed rebuild is thus given by

$$T_{distributed} = \frac{U}{R_{distributed}} = \frac{U(2N_S - 3)}{S_S(1 - \rho)(N_S - 1)} \tag{14.6}$$

14.6 Mixed Distributed Baseline Rebuild

In distributed rebuild, the spare server is never fully utilized. This is due to the fact that for every data unit rebuilt, a total of $(2N_S - 3)$ data/redundant units will need to be transferred among the remaining active servers. By contrast, the ratio is only $(N_S - 1)$ in baseline rebuild, albeit at the cost of $(N_S - 1)$ times more capacity required at the spare server. In other words, in baseline rebuild the remaining servers are underutilized while in distributed rebuild the spare server is underutilized. This suggests a mixed algorithm to rebuild part of the data using distributed rebuild and the rest by baseline rebuild to maximize the server utilizations to reduce the rebuild time.

14.6.1 Rebuild Rate Analysis

To analyze the mixed distributed-baseline rebuild, we allocate a fraction $\mu (0 \leq \mu \leq 1)$ of the server capacity to distributed rebuild, and the remaining fraction $(1 - \mu)$ to baseline rebuild. Assuming all idle capacities in the remaining servers are employed for data rebuild, then the aggregate rate at which data are transferred to the spare server will be equal to

$$R_{mixed} = \frac{S_S(1 - \rho)(N_S - 1)\mu}{2N_S - 3} + S_S(1 - \rho)(1 - \mu) \tag{14.7}$$

where the first and second terms are the rebuild data rates generated from distributed rebuild and baseline rebuild respectively.

Differentiating equation (14.7) with respect to μ, we obtain

$$\frac{d R_{mixed}}{d\mu} = S_S(1 - \rho)\left(\frac{N_S - 1}{2N_S - 3} - 1\right) \tag{14.8}$$

$$< 0 \quad \forall N_S \geq 3$$

Therefore, for systems with three or more servers reducing μ (i.e., allocating more idle capacity to baseline rebuild) always increases the aggregate data rebuild rate. When $N_S = 2$, the rebuild rate is independent of μ and simply equal to $S_S(1 - \rho)$. To find the lower bound for μ, we invoke the constraint that the aggregate rate at which data are transmitted to the spare server cannot exceed S_S:

$$\frac{S_S(1 - \rho)(N_S - 1)\mu}{2N_S - 3} + S_S(1 - \rho)(N_S - 1)(1 - \mu) \leq S_S \tag{14.9}$$

Solving for μ, we can then obtain its lower bound subject to the spare server's capacity constraint and the constraint that $\mu \geq 0$:

$$\mu = \begin{cases} \dfrac{(1 - (1 - \rho)(N_S - 1))}{(1 - \rho)(N_S - 1)\left(\frac{1}{2N_S - 3} - 1\right)}, & \text{for } \rho \leq (1 - 1/(N_S - 1)) \\ 0, & \text{otherwise} \end{cases} \tag{14.10}$$

Using this allocation ratio all idle capacities in the system will be fully utilized. Substituting equation (14.10) into equation (14.7) gives the maximum rebuild rate:

$$R_{mixed} = \begin{cases} \dfrac{S_S\left[1 + (1 - \rho)(N_S - 1)\right]}{2(N_S - 1)} & \text{for } \rho \leq (1 - 1/(N_S - 1)) \\ S_S(1 - \rho) & \text{otherwise} \end{cases} \tag{14.11}$$

14.6.2 Optimality

In this section, we derive the optimal rebuild rate for a system and show that the mixed distributed-baseline rebuild scheme can achieve this optimal rate. We assume that (1) unavailable data are rebuilt automatically without human intervention using erasure-correcting code computation; (2) all processing is performed by the remaining active servers and the spare server; and (3) the transfer capacity of the spare server is all used in rebuild. The first assumption excludes manual and hardware replacement schemes. The second assumption excludes the use of back-up devices to reload data into the spare server. The third assumption guarantees the transfer capacity available at the spare server. Note that the rebuild of unavailable units is performed either at the spare server, the remaining $(N_S - 1)$ servers, or partially done at both. Therefore, we have the next lemma.

Lemma 14.1. *For any unavailable data unit rebuilt and stored into the spare server, we need at least $(N_S - 1)$ transmissions from the remaining active servers.*

Proof. If the unavailable data unit is rebuilt by the spare server, then the $(N_S - 1)$ data/redundant units of the same parity group will need to be sent to the spare server, resulting in $(N_S - 1)$ transmissions.

If the unavailable data unit is rebuilt by one of the remaining servers, then $(N_S - 2)$ data/redundant units from servers other than the rebuild server and the spare server will need to be sent to the rebuild server, resulting in $(N_S - 2)$ transmissions. After that, the rebuild server will send the rebuilt data unit to the spare server for storage, incurring one more transmission. Together the whole process thus generates $(N_S - 1)$ transmissions.   ∎

Each server contributes its idle transfer capacity to the rebuild process either for transmission, reception, or both. Let φ $\{0 \leq \varphi \leq 1\}$ be the proportion of transfer capacity each server has used for transmission in the rebuild process. The sum of the transmission capacities of the remaining servers is then equal to

$$\varphi S_S(1 - \rho)(N_S - 1) \tag{14.12}$$

Note that the transmitted data will have to be received. Therefore, the total reception capacities must be at least as large as the total transmission capacities, i.e.,

$$(1 - \varphi)S_S(1 - \rho)(N_S - 1) + S_S \geq \varphi S_S(1 - \rho)(N_S - 1) \tag{14.13}$$

where on the L.H.S. the first term is the reception capacities of the remaining servers and the second term is the reception capacity of the spare server.

Now Lemma 14.1 shows that the rebuild rate is proportional to the sum of transmission rates by the remaining servers. Therefore, to maximize the rebuild rate we need to maximize the transmission capacity of the system, subject to the constraint in equation (14.13). Noting the constraint $0 \leq \varphi \leq 1$ we can solve for φ by rearranging equation (14.13) to give

$$\varphi = \begin{cases} \dfrac{1 + (1 - \rho)(N_S - 1)}{2(1 - \rho)(N_S - 1)}, & \text{for } \rho \leq (1 - 1/(N_S - 1)) \\[2mm] 1, & \text{otherwise} \end{cases} \tag{14.14}$$

Next we substitute equation (14.14) into equation (14.12) and invoke Lemma 14.1 to obtain the maximum achievable data rebuild rate:

$$R_{\max} = \begin{cases} \dfrac{S_S[1 + (1 - \rho)(N_S - 1)]}{2(N_S - 1)} & \text{for } \rho \leq (1 - 1/(N_S - 1)) \\[2mm] S_S(1 - \rho) & \text{otherwise} \end{cases} \tag{14.15}$$

which is exactly the same as equation (14.11). We state the result in the following theorem:

Theorem 14.2. *The mixed distributed-baseline rebuild scheme achieves the optimal rebuild rate and hence requires the minimum amount of rebuild time.*

Proof. The result follows directly from equations (14.11) and (14.15).   ∎

14.6.3 Controlling the Rebuild Time

The previous sections focus on deriving the rebuild rate and time. It is clear that the rebuild time increases with server utilization ρ. To control the rebuild time, the server could limit its utilization ρ to reserve transfer capacities for the rebuild process.

First, by setting $\rho = 0$ in equation (14.15) we can obtain the minimum achievable rebuild time:

$$T_{min} = \frac{U}{R_{max}|_{\rho=0}} = \frac{2(N_S - 1)U}{N_S S_S} \tag{14.16}$$

Second, if we want to complete the rebuild process by time $t (t \geq T_{min})$, we will need to limit the server utilization ρ to

$$\rho \leq \begin{cases} 1 + \dfrac{1}{N_S - 1} - \dfrac{2U}{t S_S} & \text{for } T_{min} \leq t \leq \dfrac{U(N_S - 1)}{S_S} \\ 1 - \dfrac{U}{t S_S} & t > \dfrac{U(N_S - 1)}{S_S} \end{cases} \tag{14.17}$$

by means of admission control.

14.7 Numerical Results

To illustrate and compare performances of the rebuild algorithms we consider a system of $N_S = 5$ active servers and one spare server. Each server has 200GB storage, so the system has a total of 1TB storage, including the redundant units. We assume a server transfer capacity of 600Mbps, e.g., using Gigabit Ethernet links.

Figure 14.7 plots the data rebuild rate versus server utilization for all the rebuild algorithms. We include the data rebuild rate for reloading data from back-up for the sake of comparison. Note that this data rebuild rate is also the upper bound. We observe that for baseline rebuild, the data rebuild rate is constant at $S_S/(N_S - 1)$ for $\rho \leq (1 - 1/(N_S - 1))$, even if the remaining active servers are lightly loaded and have idle capacities available. As the system utilization approaches one, the rebuild rate drops quickly. Distributed rebuild performs better than baseline rebuild when the server utilization is low (e.g., $\rho \leq 0.55$), but it deteriorates earlier when the system utilization increases. This is because in distributed rebuild the active servers need to receive data transmissions from other servers in addition to sending data to other servers, and thus consume considerably more transfer capacity than baseline rebuild. Finally, as expected, the mixed distributed baseline rebuild gives the best performance in all cases.

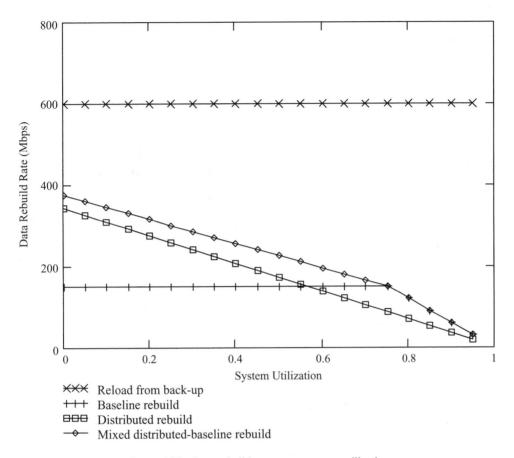

Figure 14.7   Data rebuild rate versus system utilization

Figure 14.8 shows the time required to completely rebuild a server's worth of data (200GB). Interestingly, for system utilization $\rho \geq 0.75$ performances of the baseline rebuild algorithm and the mixed distributed baseline algorithm converge. This is because at high system utilization the remaining active servers become the bottleneck, and so it is better to switch to the baseline rebuild algorithm which consumes less transfer capacities of the active servers than distributed rebuild. In this case the mixed distributed baseline algorithm simply allocates all transfer capacity to baseline rebuild.

Finally, we plot in Figure 14.9 the rebuild time versus the number of servers in the system under a system utilization of $\rho = 0.5$. The key observation is that the baseline rebuild algorithm is not scalable – the rebuild time increases with the number of servers in the system. This is because in baseline rebuild the maximum rebuild rate is limited by the transfer capacity of the spare server. Thus, as the number of servers increases, so will the number of data/redundant units that need to be sent to the spare server to rebuild an unavailable data unit, thereby resulting in longer rebuild time.

By contrast, the rebuild times of distributed rebuild and mixed distributed baseline rebuild do not increase significantly with increases in the number of servers and so are much more

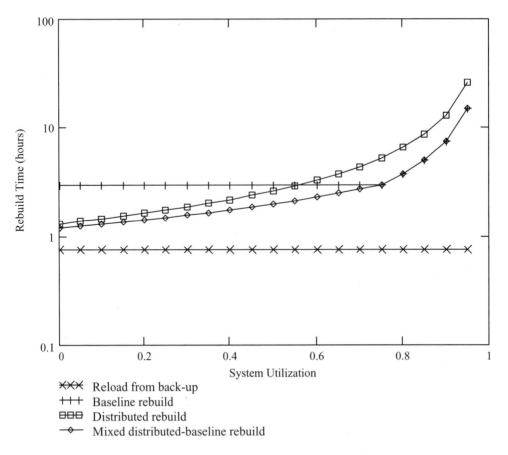

Figure 14.8   Rebuild time versus system utilization

scalable. For example, under a system utilization of $\rho = 0.5$, a mixed distributed baseline rebuild can complete the rebuild process in 2.61 hours and 2.87 hours for a 16-server and 64-server system, respectively. Thus, if no new server failure occurs within this period, the system can then resume normal operation and be ready to sustain a new server failure in the future.

14.8 Summary

In this chapter we addressed the issue of rebuilding the data lost in a failed server to a spare server so that the system can resume normal operation. Using a model incorporating the transfer capacity of the servers, the system utilization, and other system parameters, we showed that the mixed distributed-baseline rebuild algorithm can achieve the maximum rebuild rate. Moreover, the algorithm is completely automatic, and transparent to existing streaming sessions. The numerical results showed that such a data rebuild process can be completed in a reasonably short time (a few hours) and thus can enhance the reliability of the system.

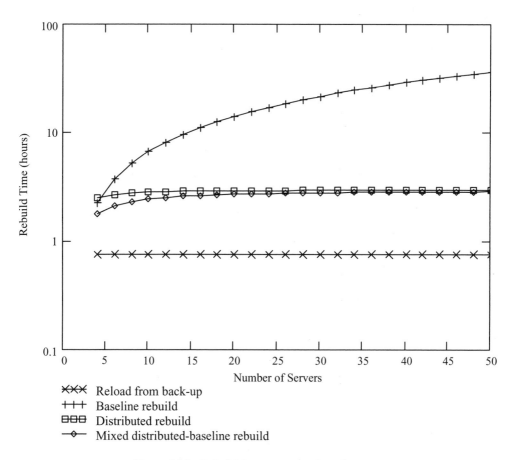

Figure 14.9   Rebuild time versus number of servers

References

[1] D. Patterson, G. Gibson and R. Katz, A Case for Redundant Arrays of Inexpensive Disks (RAID), *Proceeding of the ACM SIGMOD Conference*, June, 1988, pp. 109–116.

[2] H. Frazier, The 802.3z Gigabit Ethernet Standard, *IEEE Network*, vol. 12., no. 3, May–June 1998, pp. 6–7.

[3] J. Chandy and A.L. Narasimha Reddy, Failure Evaluation of Disk Array Organizations, *Proceedings of the 13th International Conference on Distributed Computing Systems*, 1993, pp. 319–326.

15

Algorithms for System Expansion

One of the desirable features of parallel server architectures is the incremental scalability, i.e., we can progressively add more servers to increase the system streaming capacity. In practice, a service operator will likely begin with a smaller system, and then gradually add more servers when the user population grows. This system expansion, however, creates two new challenges. First, as the media data are already stored among the existing media servers, we will need to redistribute some of the media data to the newly added servers so that they can share the streaming workload. Second, for fault tolerance, in addition to the media data, there will also be redundant data encoded from the media data using erasure correction codes. After adding more servers and redistributing the media data, however, the original redundant data unfortunately will become invalid as the size of the parity group has increased. Hence, we will need to update the redundant data so that the system's fault-tolerant capability can be maintained. This chapter presents new algorithms to solve these two challenges.

15.1 Introduction

There are two challenges in expanding a parallel streaming server with more servers. First, the newly added servers cannot share the streaming workload until a portion of the media data have been redistributed to them from the existing media servers. We call this process *data reorganization*. While simple algorithms can easily be devised, our results show that they are very inefficient and incur significant overheads in transferring data between the existing and the new servers. To tackle this problem we present in Section 15.3 a Row-Permutated Data Reorganization (RPDR) algorithm that can efficiently reorganize data in a parallel streaming server. Compared to the trivial algorithm, RPDR can reduce the reorganization overhead by over 70%. More importantly, RPDR can guarantee streaming load balance after the data have been reorganized.

Second, if the parallel streaming server employs redundant data to support fault tolerance, then the redundant data will also need to be updated after the media data are reorganized. We call this the *redundant data update* problem. In Section 15.4 we present a Sequential

Redundant Data Update (SRDU) algorithm for updating redundant data in erasure-coded distributed storage. This algorithm exploits the structure of Reed-Solomon erasure correction codes to enable the reuse of old redundant data in generating the new redundant data. This enables the algorithm to cut down the amount of data transfer by as much as 70% for the SRDU algorithm. It is worth noting that the algorithms presented in this chapter are not limited to parallel server systems, but are also applicable to any striped data storage, such as disk arrays, RAID [1], or even the emerging peer-to-peer systems. For this reason we will use the term "node" to refer to a device in the striped storage, e.g., a server in a parallel streaming server, a disk in a RAID, or a user host in a peer-to-peer streaming system.

15.2 Related Work

The problem of data reorganization has been studied previously in the context of disk arrays [2–3]. The study by Ghandeharizadeh and Kim [2] is the earliest study on data reorganization known to the author. They investigated the data reorganization problem in the context of adding disks to a continuous media server. They studied the round-robin data striping commonly found in disk arrays and proposed techniques to perform data reorganization *online*, i.e., without disrupting on-going video streams. Due to the round-robin placement requirement, a large portion of the data blocks will need to be redistributed to maintain the data placement order when a new disk is added, thus incurring significant data reorganization overhead. The advantage is that this approach can maintain perfect streaming load balance when data reorganization is completed.

In a more recent study by Goel *et al.* [3], an algorithm called SCADDAR for data placement and data reorganization is proposed for use in disk arrays. In this algorithm, each data block is initially randomly distributed to the disks with equal probabilities. When a new disk is added to the disk array, each block will obtain a new sequence number according to their randomized SCADDAR algorithm. If the remainder of this number is equal to the disk number of the newly added disk, the corresponding block will be moved to this new disk. Otherwise, the block will stay on the original disk. As SCADDAR no longer needs to maintain a strict round-robin placement order, it can reduce the reorganization overhead to levels approaching the theoretical lower bound.

However, the SCADDAR algorithm did not consider streaming load balance. If we apply SCADDAR to a streaming server, then the pseudo-random placement policy can result in significant streaming load imbalance, especially after a large number of disks have been added to the system. This load imbalance complicates data transmission scheduling and may also reduce the usable streaming capacity and/or increase the response time of the system.

Interestingly, the previous two studies can be considered as two extremes of the trade-off between data reorganization overhead and streaming load balance. In particular, Ghandeharizadeh and Kim's algorithm achieves perfect load balance at the expense of substantial data reorganization overhead, while the SCADDAR algorithm achieves near-minimal data reorganization overhead at the expense of load imbalance. The Row-Permuted Data Reorganization described in Section 15.3, by comparison, can achieve perfect streaming load balance and yet incurs significantly lower reorganization overhead than the round-robin algorithm.

15.3 Row-Permutated Data Reorganization

We present in this section the Row-Permutated Data Reorganization algorithm to redistribute media data after a new node is added to the system. We use two performance metrics, namely, data reorganization overhead and streaming load balance to compare its performance with the round-robin data reorganization algorithm [2] and the SCADDAR algorithm [3].

15.3.1 Placement Policy

In striped storage a media object is first divided into fixed-size blocks, denoted by $v_{i,j}$, where $i \in 0, 1, \ldots$ is the group number and $j \in 0, 1, \ldots, (N - 1)$ is the media block number within the same group. To maintain streaming load balance it is necessary to ensure that every block in the same group resides in a different node of the system.

As Ghandeharizadeh and Kim's study [2] showed, the data reorganization overhead incurred in maintaining the round-robin data placement order is very high. This is illustrated in Figure 15.1 which shows the data placement before and after adding a node to a 4-node system. The shaded data blocks all have to be moved from one node to another to restore the round-robin placement in the new 5-node configuration. This obviously incurs significant overhead.

In streaming applications, the client will process and play back media data sequentially (ignoring interactive playback control). This implies that if the client always receives one group of media data before playback, then the exact order in which the data blocks arrival at the client will become irrelevant. Therefore, instead of enforcing the round-robin placement order we can relax the constraint to reduce the number of block movements – row-permutated placement policy.

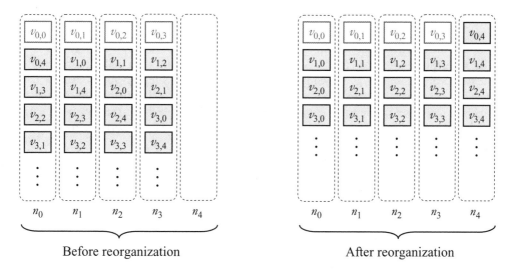

Before reorganization   After reorganization

Figure 15.1   Reorganization under round-robin placement (from 4 to 5 nodes). The shaded blocks all need to be relocated to form the new configuration

Divides media data stream into fixed-size blocks

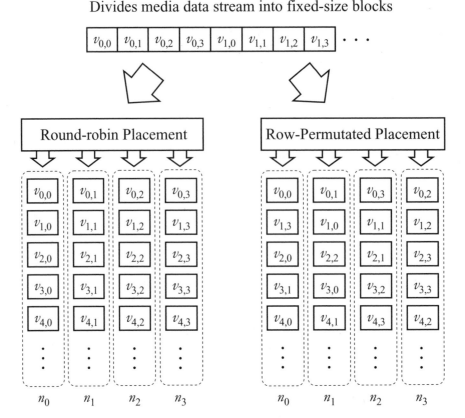

Figure 15.2   The row-permutated placement policy

Specifically, in a N-node system the first N media data blocks $\{v_{0,j} \mid j = 0, 1, \ldots, N-1\}$ will be distributed to all N nodes in pseudo-random order, with each node storing exactly one of the N data blocks as shown in Figure 15.2. This process repeats for the next N data blocks $\{v_{1,j} \mid j = 0, 1, \ldots, N-1\}$, and so on until all data blocks are distributed. As long as the client receives a whole parity group before decoding it for playback, this row-permutated placement policy can achieve perfect streaming load balance, same as the original round-robin placement policy.

15.3.2 Data Reorganization

To determine which data blocks will need to be moved after adding a new node, we first re-index all the media data blocks according to the new configuration. Figure 15.3 shows an example of reorganizing from a 4-node system to a 5-node system. For example, media blocks $v_{1,0}$ and $v_{1,1}$ will be re-indexed to $v_{0,4}$ and $v_{1,0}$ respectively in the 5-node configuration.

If we consider the first group of media blocks in the new configuration, we can see that node n_1 now needs to send two media blocks $v_{0,1}$ and $v_{0,4}$ while the new node is not used. This is the reason why load imbalance will occur if the data blocks are not reorganized.

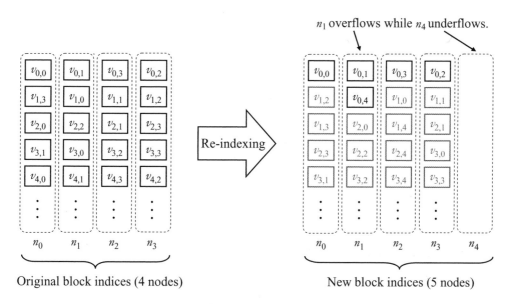

Figure 15.3 Re-indexing the first group of media data blocks

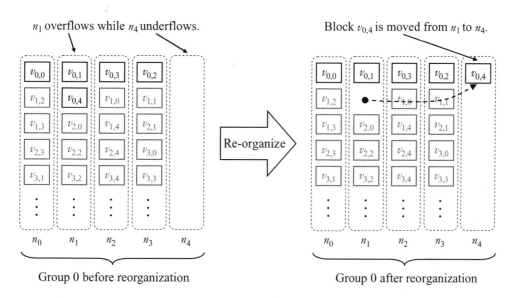

Figure 15.4 Reorganizing the first group of media data blocks

Since we do not need to maintain the strict round-robin placement order under RPDR, we can simply move the excess block from the overflow node to the underflow node. For example, we move data block $v_{0,4}$ from node n_1 to node n_4 as shown in Figure 15.4 to restore the streaming load balance of the new 5-node configuration.

As a further illustration, we consider the second group of data blocks in Figure 15.5. Now there are two overflow nodes n_0 and n_2, as well as two underflow nodes n_1 and n_4. To restore

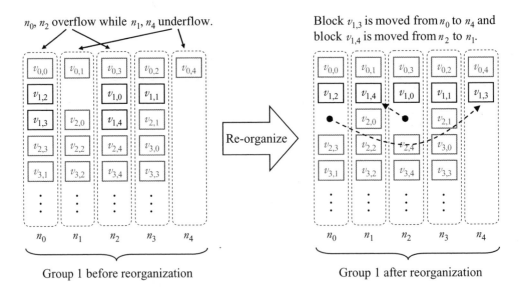

Figure 15.5 Reorganizing the second group of media data blocks

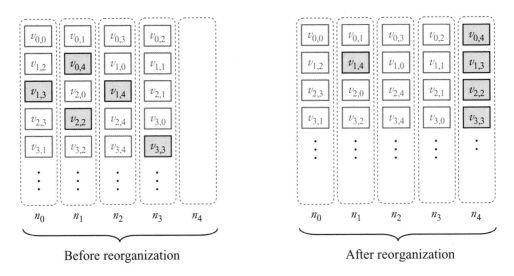

Figure 15.6 Reorganizing from 4 to 5 nodes using RPDR. Shaded media data blocks need to be relocated to form the new configuration

streaming load balance we move block $v_{1,3}$ from n_0 to n_4 and block $v_{1,4}$ from n_2 to n_1. Thus, we need two block movements to reorganize this second group.

Repeating this process we can then reorganize the whole system. Figure 15.6 shows the required block movements for the first four groups of media data blocks. Comparing it to Figure 15.1, we can clearly see the savings in block movements (e.g., from 16 down to 5). In the next section, we relax the streaming load balance constraint to further reduce the reorganization overhead.

15.3.3 Multi-Row-Permutated Data Reorganization

While perfect streaming load balance is desirable, the cost of data reorganization, which itself consumes system resources, can still be substantial. Depending on the particular system configuration (e.g., disk throughput, network bandwidth, system utilization, etc.), it may be desirable to trade off some streaming load balance to further reduce the data reorganization overhead.

By relaxing the streaming load balance constraint, we generalize the row-permutated data reorganization algorithm into a multi-row-permutated data reorganization (m-RPDR) algorithm, which also subsumes the original RPDR as the special case 1-RPDR. In m-RPDR we process the media blocks m groups at a time. Moreover, we redefine node overflow to occur only if more than m blocks from the m groups are stored in the same node. Similarly, a node underflows if it stores fewer than m blocks from the m groups under consideration.

Figure 15.7 illustrates the reorganization of the first two groups using 2-RPDR. In this example, nodes n_0 and n_2 overflow and so we move media data blocks $v_{1,3}$ and $v_{1,4}$ to n_4 to restore streaming load balance. Comparing to reorganizing the same two groups using 1-RPDR (Figure 15.4 and Figure 15.5), the number of block movements is reduced from 3 to 2. The trade-off is in streaming load balance. As shown in Figure 15.7 node n_4 stores media data blocks $v_{1,3}$ and $v_{1,4}$, and node n_1 stores blocks $v_{0,1}$ and $v_{0,4}$. Thus, if the system sends a group of N data blocks in each service round, then in the first service round node n_1 will need to send two data blocks while node n_4 will be idle, and vice versa in the second service round.

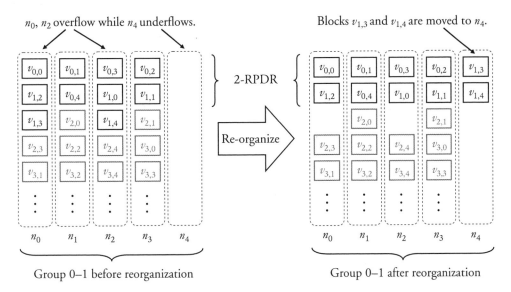

Figure 15.7 Reorganizing group 0 and 1 using the 2-RPDR algorithm

Therefore, to employ m-RPDR the system will need to be designed with this load imbalance in mind to guarantee streaming performance.

15.3.4 Performance Analysis

In this section, we evaluate and compare the multi-row-permutated data reorganization algorithm with the round-robin [2] and the SCADDAR [3] algorithms, both originally proposed for disk arrays. The performance metrics used for comparison are data reorganization overhead and streaming load balance.

The results are computed numerically for a media object of $B = 4{,}000$ blocks. Unless stated otherwise the system begins with a single node and then incrementally grows to 200 nodes by adding new nodes one by one. Each data point is averaged over 50 results obtained from using different random seeds for the random number generator (used in SCADDAR and the m-RPDR algorithms).

15.3.4.1 Data Reorganization Overhead

Data reorganization overhead is defined as the number of data blocks that need to be redistributed to bring the system back to streaming load balance. This metric can reflect the time, memory and bandwidth requirement incurred by the reorganization process.

For a system with B blocks already evenly distributed to n nodes, the minimum number of blocks that need to be redistributed when a new node is added is equal to $B/(n + 1)$, provided that perfect storage balance is to be maintained. Note that this lower bound does not consider streaming load balance. The SCADDAR algorithm in some cases incurs less overhead than the lower bound because SCADDAR does not guarantee storage balance.

For round-robin placement, we can derive the approximate reorganization overhead analytically. Consider expanding a system from $(n - 1)$ nodes to n nodes. If we divide the media blocks into groups of $\mathrm{LCM}(n - 1, n)$ blocks, where the function $\mathrm{LCM}(n - 1, n)$ computes the least common multiple of $n - 1$ and n, then the first $(n - 1)$ blocks in each group will not need to be moved (see Figure 15.1 for an example). All other blocks in the group will need to be relocated. Therefore, the reorganization overhead in round-robin placement is given by

$$B \left(\frac{\mathrm{LCM}(n - 1, n) - (n - 1)}{\mathrm{LCM}(n - 1, n)} \right) = B \left(\frac{(n - 1)n - (n - 1)}{(n - 1)n} \right)$$

$$= B \left(1 - \frac{1}{n} \right) \tag{15.1}$$

For the SCADDAR algorithm the reorganization overhead has been shown to approach the theoretical lower bound of B/n and the exact values can be obtained from simulated calculations. Similarly, we also obtain the reorganization overhead of m-RPDR using simulated calculations. Figure 15.8 compares the reorganization overhead of round-robin, SCADDAR, and m-RPDR for $m = 1, 2, 5$, and 10.

There are three observations. First, the round-robin algorithm and the SCADDAR algorithm have the highest and lowest reorganization overhead respectively. Moreover, the differences

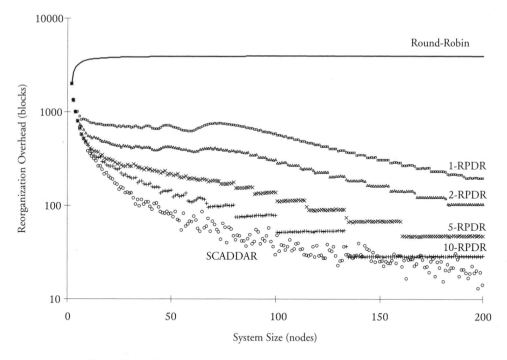

Figure 15.8 Comparison of reorganization overhead versus system size

increase when the system grows larger. This is because for the SCADDAR algorithm the overheads approximately equal to B/n and so decreases as the system grows. By contrast, overheads of the round-robin algorithm approach B as n increases (cf. equation (15.1)).

Second, the reorganization overheads of the m-RPDR algorithms are lower than the round-robin algorithm but higher than the SCADDAR algorithm. In particular, a larger value of m will result in lower reorganization overhead, thus providing a tool to trade off streaming load balance for lower reorganization overhead.

Third, the 1-RPDR algorithm can achieve lower overhead than the round-robin algorithm and yet can still achieve perfect streaming load balance. Thus, the 1-RPDR algorithm can be used in place of the round-robin algorithm whenever perfect streaming load balance is required.

So far we have assumed that the system is reorganized whenever a new node is added to the system. In practice, it may be desirable to add multiple nodes to the system at the same time to amortize the administrative overheads. Interestingly, performing data reorganization simultaneously for multiple nodes can also result in lower per-node reorganization overhead compared to adding them one by one.

Figure 15.9 illustrates the potential savings from performing data reorganization for $k = 1, 2, 3$, and 5 new nodes. The results clearly show that the per-node reorganization overhead decreases significantly (note that the vertical axis is in logarithm scale) for larger values of k. Moreover, the savings are most significant when switching from $k = 1$ to $k = 2$, suggesting that adding nodes to the system two-at-a-time can be an efficient strategy to expand the system incrementally.

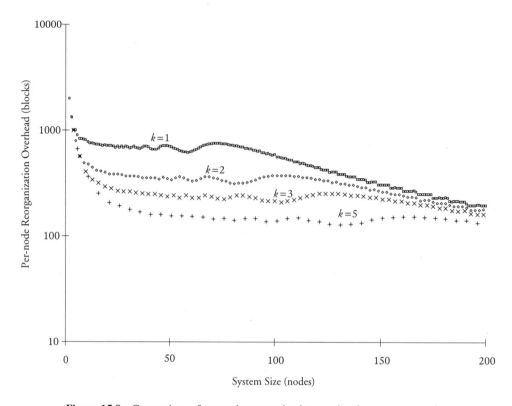

Figure 15.9 Comparison of per-node reorganization overhead versus system size

15.3.4.2 Streaming Load Balance

Most media streaming systems retrieve and transmit media data blocks in fixed-duration service rounds (cf. Chapter 3), thus if the media blocks are not distributed evenly across all nodes some nodes will be overloaded while others will be underutilized. Among the data reorganization algorithms, only the round-robin and the 1-RPDR algorithms can achieve perfect streaming load balance. All other algorithms, including SCADDR and m-RPDR with $m > 1$, will result in some degree of load imbalance.

To quantify streaming load balance, we count the number of overflow media blocks for each algorithm after data reorganization is completed, and then plot the proportion of overflow media blocks in Figure 15.10. As expected, both round-robin and 1-RPDR achieve zero overflow. The SCADDAR algorithm results in over 35% overflow blocks. The m-RPDR algorithm, on the other hand, results in fewer overflows depending on the choice of m. Finally, another desirable feature of m-RPDR is that it can guarantee that the maximum number of overflow blocks in each service round will not exceed m. This enables one to incorporate the overflow effect either by design or through admission control.

15.4 Sequential Redundant Data Update

In data reorganization we have ignored the redundant data, which will become invalid once data reorganization is performed. Obviously, to maintain the system's fault tolerance capability we

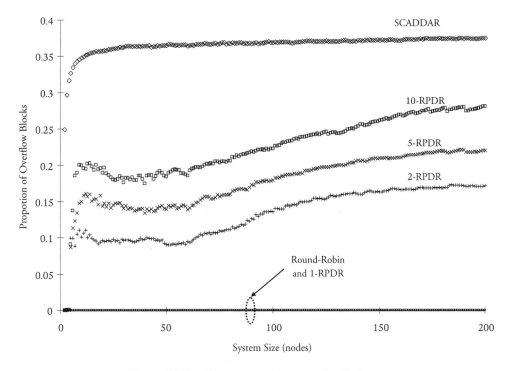

Figure 15.10   Comparison of streaming load balance

will need to update the redundant data blocks according to the newly reorganized data placements. In this section we develop a Sequential Redundant Data Update (SRDU) algorithm that exploits structure of the Reed-Solomon erasure correction (RSE) code [4–6] to substantially reduce the number of block movements required in redundant data update.

15.4.1 Redundant Data Regeneration

Figure 15.11 depicts the placement of data and redundant blocks in a system of 6 nodes. There are two redundant data blocks per parity group, denoted by $c_{i,0}$ and $c_{i,1}$ for parity group i. This enables the system to survive two node failures. Note that while the round-robin data placement order is shown in Figure 15.11, it is not a requirement. The exact ordering of the data blocks within a parity group is irrelevant to redundant data block generation. Thus, the RPDR algorithm is also compatible with redundant data generation.

After adding a new node to the system and reorganizing the data blocks, the new data block placements become the one shown in Figure 15.12. Again the order of data block placements is arbitrary and irrelevant to the process of redundant data generation. Under this new configuration the original redundant data $\{c_{i,j} \mid i, j \in 0, 1, \ldots\}$ will need to be updated.

In general, for a system of N nodes and h redundancies, we will need all $(N - h)$ data blocks in a parity group to compute the corresponding h redundant blocks. As data and redundant blocks of the same parity group are all stored individually in separate nodes, the system will need to send all data blocks to the redundant nodes to generate the new redundant blocks. This simple algorithm is clearly very inefficient.

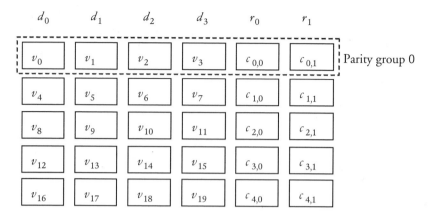

Figure 15.11   Data placement before addition of a new node

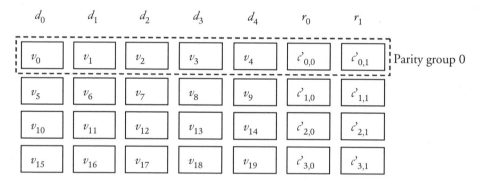

Figure 15.12   Data placement after adding one data node

On the other hand, if a central archive server storing a copy of all media data is available, then it can simply regenerate the new redundant data blocks locally and send them to the redundant nodes to replace the obsolete redundant data blocks. In this case, the number of block movements required will be equal to $(B/(N-h))$. Nevertheless maintaining this central archive server will incur additional costs and complicate management of the system. Depending on the particular application this approach may not be cost-effective.

15.4.2 Reuse of Original Redundant Data

To understand how we can reduce the redundant data update overhead, we first need to study the computation of the redundant blocks using the RSE code. Let B be the total number of fixed-size media data blocks in the system and denote the jth block of a media object by v_j. For simplicity we consider only one media object in the rest of the section. The results can be readily extended to multiple media objects.

Let $(N - h)$ and h be the number of data nodes and redundant nodes in the system respectively. Assuming the number of redundant nodes in the system is fixed, then we can apply the (N, h)-RSE code to compute the h redundant data blocks in each group of $(N - h)$ data blocks using

$$
F \cdot D = \begin{bmatrix}
f_{1,1} & f_{1,2} & f_{1,3} & \cdots & f_{1,N-h} \\
f_{2,1} & f_{2,2} & f_{2,3} & \cdots & f_{2,N-h} \\
\vdots & \vdots & \vdots & & \vdots \\
f_{h,1} & f_{h,2} & f_{h,3} & \cdots & f_{h,N-h}
\end{bmatrix}
\begin{bmatrix}
d_{i,0} \\
d_{i,1} \\
\vdots \\
d_{i,N-h-1}
\end{bmatrix}
$$

$$
= \begin{bmatrix}
1 & 1 & 1 & \cdots & 1 \\
1 & 2 & 3 & \cdots & N-h \\
\vdots & \vdots & \vdots & & \vdots \\
1 & 2^{h-1} & 3^{h-1} & \cdots & (N-h)^{h-1}
\end{bmatrix}
\begin{bmatrix}
d_{i,0} \\
d_{i,1} \\
\vdots \\
d_{i,N-h-1}
\end{bmatrix} \tag{15.2}
$$

$$
= \begin{bmatrix}
c_{i,0} \\
c_{i,1} \\
\vdots \\
c_{i,h-1}
\end{bmatrix} = C
$$

where the F, D, and C are the Vandermonde matrix [4], the media data vector, and the redundant data vector respectively; and $d_{i,j}$, $c_{i,k}$ represent data block j ($j = 0, 1, \ldots, N - h - 1$) and redundant block k ($k = 0, 1, \ldots, h - 1$) of parity group i. Elements in F are constants computed from $f_{i,j} = j^{i-1}$. Note that the matrix multiplication in equation (15.2) is computed over Galois Fields of 2^w where $N < 2^w$. For example, by setting $w = 16$, then the code can support up to 65,535 nodes.

Reconsider the generation of a redundant data block before and after addition of one data node in Figure 15.11 and Figure 15.12 respectively. We can observe that in many cases, the reorganized parity group still comprises some data blocks from the original parity group before reorganization. For example, in expanding a system from N nodes to $N + 1$ nodes, the first parity group will be reorganized from the composition of $\{v_0, v_1, \ldots, v_{N-h-1}\}$ to $\{v_0, v_1, \ldots, v_{N-h-1}, v_{N-h}\}$, which differs by only one data block v_{N-h}. This means that the resultant matrix computations in equation (15.2) will differ by only one variable in the data vector. This raises the question of whether we can reuse the original redundant data in computing the new redundant data, discussed next.

Consider the original configuration in Figure 15.11. The first two redundant data in redundant node r_1, denoted by $c_{0,1}$ and $c_{1,1}$, are computed from

$$
c_{0,1} = \sum_{j=0}^{3} f_{2,j+1} v_j \tag{15.3}
$$

and

$$
c_{1,1} = \sum_{j=4}^{7} f_{2,j-4+1} v_j \tag{15.4}
$$

respectively according to equation (15.2).

After a new node is added, the system will be reorganized to that in Figure 15.12. Now the two new redundant data block, denoted by $c'_{0,1}$ and $c'_{1,1}$, are computed from

$$c'_{0,1} = \sum_{j=0}^{4} f_{2,j+1} v_j \tag{15.5}$$

and

$$c'_{1,1} = \sum_{j=5}^{9} f_{2,j-5+1} v_j \tag{15.6}$$

Comparing equation (15.5) with equation (15.3) we can observe that they share four common terms in $v_j - v_0$, v_1, v_2, v_3. Hence we can rewrite equation (15.5) as follows:

$$c'_{0,1} = \sum_{j=0}^{3} f_{2,j+1} v_j + f_{2,5} v_4$$

$$= c_{0,1} + f_{2,5} v_4 \tag{15.7}$$

In other words, we can compute $c'_{0,1}$ using the original redundant data $c_{0,1}$ plus data block v_4. Thus, instead of sending all five data blocks to redundant node r_1 to perform redundant data update, we now only need to send one data block, i.e., v_4, thereby dramatically reducing the overheads in updating the redundant data $c'_{0,1}$.

15.4.3 Parity Group Reshuffling

In some cases, the previous straightforward reuse technique may not be applicable due to differences in the coefficients $f_{i,j}$. For example, $c'_{1,1}$ is computed from v_5 to v_9 and share common terms in v_5, v_6, and v_7 with $c_{1,1}$. Thus, it appears that we can reuse the common terms and send only v_8 and v_9 to r_1 to compute $c'_{1,1}$. However, comparing the equations for $c'_{1,1}$:

$$c'_{1,1} = \sum_{j=5}^{9} f_{2,j-5+1} v_j$$

$$= (f_{2,1} v_5 + f_{2,2} v_6 + f_{2,3} v_7) + f_{2,4} v_8 + f_{2,5} v_9 \tag{15.8}$$

and for $c_{1,1}$:

$$c_{1,1} = \sum_{j=4}^{7} f_{2,j-4+1} v_j$$

$$= f_{2,1} v_4 + (f_{2,2} v_5 + f_{2,3} v_6 + f_{2,4} v_7) \tag{15.9}$$

we found that the common terms v_5, v_6, and v_7 now have different coefficients $f_{i,j}$ (e.g., $f_{2,1} v_5$ versus $f_{2,2} v_5$). As a result, we cannot reuse $c_{1,1}$ in computing $c'_{1,1}$.

To tackle this problem, we can reshuffle the order of the elements in the data vector for computing $c'_{1,1}$ to obtain

$$c'_{1,1} = f_{2,1}v_8 + (f_{2,2}v_5 + f_{2,3}v_6 + f_{2,4}v_7) + f_{2,5}v_9 \qquad (15.10)$$

which then allows us to reuse $c_{1,1}$ in the computation:

$$c'_{1,1} = f_{2,1}v_8 + \left(\sum_{j=4}^{7} f_{2,j-4+1}v_j - f_{2,1}v_4 \right) + f_{2,5}v_9$$

$$= f_{2,1}v_8 + (c_{1,1} - f_{2,1}v_4) + f_{2,5}v_9 \qquad (15.11)$$

This reduces the number of data block transmissions from 5 to 3. Note that the client will also need to use the reshuffled order when decoding the parity group for playback. This parity group order information can either be generated dynamically, or sent along the video data blocks as meta-data.

Interestingly, there may be more than one way to reuse redundant block in updating the redundant data, and possibly with different redundant update overhead. For example, consider the computation of $c_{2,1}$:

$$c_{2,1} = \sum_{j=8}^{11} f_{2,j-8+1}v_j$$

$$= (f_{2,1}v_8 + f_{2,2}v_9) + f_{2,3}v_{10} + f_{2,4}v_{11} \qquad (15.12)$$

If we reshuffle the order of computations for $c'_{1,1}$ to

$$c'_{1,1} = (f_{2,1}v_8 + f_{2,2}v_9) + f_{2,3}v_5 + f_{2,4}v_6 + f_{2,5}v_7 \qquad (15.13)$$

then we can reuse $c_{2,1}$ in the computation:

$$c'_{1,1} = \left(\sum_{j=8}^{11} f_{2,j-8+1}v_j - f_{2,3}v_{10} - f_{2,4}v_{11} \right)$$

$$+ f_{2,3}v_5 + f_{2,4}v_6 + f_{2,5}v_7 \qquad (15.14)$$

$$= (c_{2,1} - f_{2,3}v_{10} - f_{2,4}v_{11})$$

$$+ f_{2,3}v_5 + f_{2,4}v_6 + f_{2,5}v_7$$

However, in this case the number of data block transmissions is five, two blocks more than that of reusing $c_{1,1}$. Thus, in the SRDU algorithm, the system will first compute the redundant data update overhead for all reusable redundant blocks and select the one with the lowest overhead for reuse.

15.4.4 Reuse of Cached Data Blocks

In addition to reusing old redundant data, we can also reduce data block movements by caching and reusing data blocks already received by the redundant node. Reconsider the previous example of computing $c'_{1,1}$ (cf. equation (15.11)), the data blocks needed are v_4, v_8, and v_9. However, v_4 has already been sent to the redundant node when computing $c'_{0,1}$ (cf. equation (15.7)) and thus can simply be reused if it is cached. As a redundant block is computed from a parity group of $(N - h)$ data blocks, we only need to cache the $(N - h)$ most recently received data blocks.

15.4.5 Redundant Data Update Overhead

In this section we evaluate performance of the SRDU algorithm using simulation. Beginning with a small system, we add new nodes to the system and then apply the SRDU algorithm to update the redundant data blocks. Performance is measured by the number of data blocks that need to be sent to the redundant nodes for redundant data update – or simply called redundant data update overhead. The total number of data blocks is 40,000 and is fixed throughout the simulation. For simplicity the redundant data update overhead for updating one redundant node is presented. We will return to the issue of updating multiple redundant data nodes in Section 15.5.

In the first experiment, we investigate the redundant data update overhead in continuous system growth. We begin with a system of five data nodes and one redundant node. Then we add a new node to the system one by one, each time the redundant data blocks are completely updated using the SRDU algorithm. This continues until the system grows to 400 data nodes. Figure 15.13 plots the redundant data update overhead versus data node size from 6 to 400. As expected, the simple algorithm (sending all data blocks to the redundant node) performs the worst. On the other hand, regenerating redundant data using a centralized archive server incurs the least overhead. These two curves serve respectively as the upper bound and lower bound for redundant data update overhead.

Surprisingly, direct reuse of the original redundant data does not result in significant savings. This is because the algorithm maintains the same data order within the parity group in computing the redundant data, and thus significantly reduces the opportunity for redundant data reuse. Once this restriction is relaxed by reshuffling the parity group, the redundant data update overhead is reduced by half to around 20,000 blocks. Reuse of cached data blocks further reduces the overhead by half to around 10,000 blocks. With all three techniques applied, the SRDU algorithm can reduce the redundant data update overhead by as much as 70%.

In the previous experiment, we completely update all redundant data blocks before adding another new node. Clearly this is inefficient if new nodes are added frequently or added to the system in a batch. To address this issue, we conduct a second experiment where multiple nodes, say W, are added to the system before the redundant data are updated. Figure 15.14 plots the redundant data update overhead versus the batch size W for initial system size of 80 nodes. The results show that we can reduce the per-node redundant data update overhead significantly by batching redundant data update for multiple new nodes.

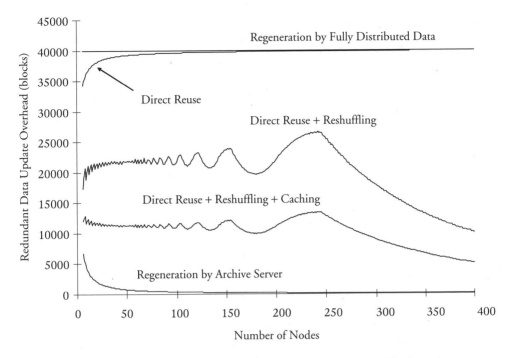

Figure 15.13   Redundant data update overhead versus data node size

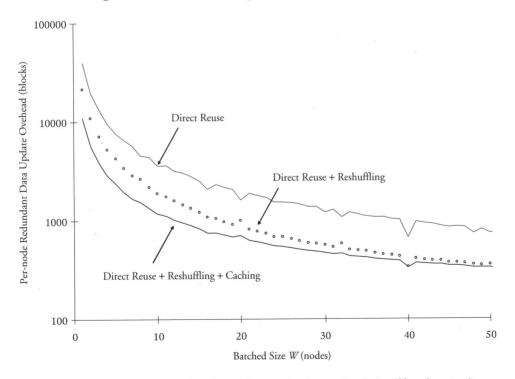

Figure 15.14   Per-node redundant data update overhead versus batch size (80-node system)

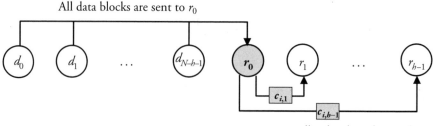

All data blocks are sent to r_0

r_0 computes all redundant data and
distributes them to all redundant nodes.

Figure 15.15   Update of multiple redundant nodes in the trivial redundant data generation algorithm

15.5 Multiple Redundant Nodes Update

In larger systems more than one redundant node will be needed to achieve good system reliability. We can update the individual redundant data blocks in each parity group independently.
However, if we jointly update the multiple redundant data blocks, then it is possible to achieve
further savings. In the following we assume there are N nodes, of which h of them are redundant
nodes storing redundant data blocks.

15.5.1 Redundant Data Regeneration

We first consider the trivial redundant data regeneration algorithm (cf. Section 15.4.1) depicted
in Figure 15.15. Instead of repeatedly sending all data blocks to the redundant nodes $r_0, r_1, \ldots,$
we can designate one of the redundant nodes, say r_0, to compute the redundant data for the other
redundant nodes $\{c_{i,j} \mid i = 0, 1, \ldots; \forall j \neq 0\}$ in addition to computing its own redundant data
$\{c_{i,0} \mid i = 0, 1, \ldots\}$. This is possible as all the elements in the RSE code computation equation
(15.2) are already available. Afterwards it simply sends the computed redundant data blocks
to the appropriate redundant nodes, thereby saving a significant number of data movements as
shown in Figure 15.15.

For a media object of B data blocks, this algorithm will need to send B data bocks to
the redundant node r_0, which then sends $(B/(N - h))$ computed redundant blocks to each
additional redundant node. Therefore, the total redundant data update overhead is equal to
$B + (h - 1)(B/(N - h))$. Note that it is not necessary to designate a single redundant node to
compute all the redundant blocks. Instead we can divide the computations equally among all
the redundant nodes to balance the load and overhead across them. The resultant redundant
data update overhead will be the same.

On the other hand, if there is a central archive server storing a copy of all the media
data blocks, then it can regenerate locally all the updated redundant data blocks $\{c_{i,j} \mid i =
0, 1, \ldots; j = 0, 1, \ldots, h - 1\}$ and then transmit them to the redundant nodes (Figure 15.16).
In this case the redundant data update overhead is equal to $h(B/(N - h))$. Obviously this is
also the lower bound.

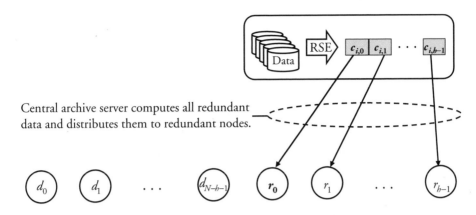

Figure 15.16   Update of multiple redundant nodes directly by a central archive server

15.5.2 Sequential Redundant Data Update

When updating multiple redundant nodes in SRDU, we observe that the data blocks required are in fact the same for all the redundant nodes. For example, consider the computation of the redundant block $c'_{0,j}$ in the redundant node r_j. The equation for computing $c'_{0,j}$ is

$$c'_{0,j} = c_{0,j} + f_{j+1,5}v_4 + f_{j+1,6}v_5 \tag{15.15}$$

where v_4 and v_5 are the required data blocks and are the same for all j's.

Therefore, if the data blocks v_4 and v_5 have already been sent to redundant node r_k, then it can compute partial results for the other redundant nodes:

$$p_j = f_{j+1,5}v_4 + f_{j+1,6}v_5, \forall j \neq k \tag{15.16}$$

and then send them to the other redundant nodes $\{r_j \mid \forall j \neq k\}$ to complete the redundant data update computation:

$$c'_{0,j} = c_{0,j} + p_j, \forall j \neq k \tag{15.17}$$

as shown in Figure 15.17. In this case, the overhead is reduced from sending two data blocks v_4 and v_5 to sending just the partial result p_j. In general, we always need to send only the partial results no matter how many data blocks are required. Therefore, the total redundant data update overhead is equal to the SRDU overhead in updating one redundant node plus the overhead in transmitting the partial results: $(h - 1)(B/(N - h))$.

Table 15.1 summarizes the total redundant data update overhead of the algorithms discussed earlier. An important observation is that the overhead is dominated by the overhead in updating the first redundant node due to the large number of data blocks required to generate the new redundant data blocks. Once these are cached in the redundant node, new redundant blocks of other redundant nodes can be computed with much lower overhead.

Consider an example of adding one data node to a system with 199 data nodes. Figure 15.18 plots the total redundant data update overhead versus number of redundant nodes in the system.

Table 15.1   Total overhead in studied algorithms

Algorithms	Total Overhead
Redundant Data Regeneration	$B + (h - 1)(B/(N - h))$
Regeneration by archive server	$B/(N - h) + (h - 1)(B/(N - h))$
Sequential Redundant Data Update (SRDU)	Block movement under SRDU+
	$(h - 1)(B/(N - h))$

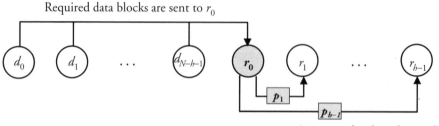

Required data blocks are sent to r_0

r_0 computes its own redundant data and
partial results p_j for other redundant nodes.

Figure 15.17   Update of multiple redundant nodes in SRDU

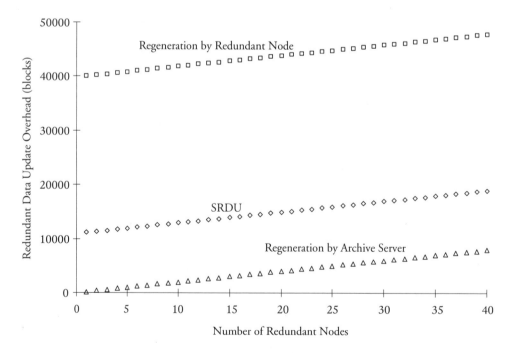

Figure 15.18   Redundant data update overhead versus redundant node size

The overhead of SRDU with one redundant node is 11,123 blocks while the total overhead for 10 redundant nodes is only 12,923 blocks. Therefore, the overhead for updating additional redundant nodes can be substantially reduced by computing and sending partial results.

15.6 Summary

In this chapter, we investigated the system expansion problem in systems employing striped storage, such as disk arrays, RAID, parallel servers, and even peer-to-peer systems. We presented efficient algorithms to carry out the two essential operations in expanding the system to include more nodes, namely a Row-Permutated Data Reorganization (RPDR) algorithm for reorganizing and redistributing media data blocks to the new nodes so that streaming load balance can be restored; and a Sequential Redundant Data Update (SRDU) algorithm for the efficient update of redundant data to support the new data organization after one or more new nodes are added.

Note that this chapter did not address issues in the transmission of the data blocks, redundant blocks, and partial results during the redundant data update process. As the system is online serving active media streams, some of the nodes from time to time may not have sufficient bandwidth to receive or transmit the data blocks. Thus proper scheduling must be done to avoid congesting a receiver node while maximizing utilization of the idle bandwidth in the system to shorten the system reorganization time.

References

[1] D.A. Patterson, G. Gibson, and R.H. Katz, A Case for Redundant Arrays of Inexpensive Disks (RAID), *Proc. of 1988 ACM SIGMOD Conference on Management of Data*, Chicago, June 1988.

[2] S. Ghandeharizadeh and D. Kim, On-line Reorganization of Data in Scalable Continuous Media Servers, *Proceedings 7th International Conference on Database and Expert Systems Applications*, Sept. 1996.

[3] A. Goel, C. Shahabi, S.-Y. Yao, and R. Zimmerman, SCADDAR: An Efficient Randomized Technique to Reorganize Continuous Media Blocks, *Proc. International Conference on Data Engineering*, 2002.

[4] J.S. Plank, A Tutorial on Reed-Solomon Coding for Fault-Tolerance in RAID-like Systems, *Software Practice and Experience*, vol. 27, no. 9, Sept. 1997, pp. 995–1012.

[5] L. Rizzo, Effective Erasure Codes for Reliable Computer Communication Protocols, *ACM Computer Communication Review*, vol. 27, no. 2, Apr. 1997, pp. 24–36.

[6] A.J. McAuley, Reliable Broadband Communication Using a Burst Erasure Correcting Code, *Proceedings of the ACM Symposium on Communications Architectures and Protocols*, Sept. 1990, pp. 297–306.

Part Three

Multicast Streaming Architectures

16

Overview of Multicast Streaming

The parallel server architectures solved the scalability problem of the media streaming server by combining multiple low-cost media servers into one high-capacity streaming system. In very-large-scale systems such as providing video-on-demand services to millions of users, the resultant network traffic generated will be very substantial. For example, to stream a media of 4Mbps (e.g., high-quality video encoded in MPEG2 or MPEG4) to just 10,000 users, we will need an aggregate bandwidth of 40Gbps which only optical fiber can carry in today's networks.

Interestingly, as many studies have shown, most users will probably request a small portion of popular media contents, such as popular songs, movies, or documentaries. Thus, within the huge volume of network traffic there are in fact a great deal of duplicated data being transferred from the servers to the clients. Now if the data are the same, can we send a single copy instead of sending many duplicated copies to the clients?

This is the question being addressed in Part III of the book. In particular, we investigate the use of network multicast to deliver media data from the servers to the clients. Unlike the unicast model used in today's Internet, network multicast allows a sender to multicast a single packet to be received by multiple receivers. This has the potential to significantly reduce the network traffic in sending duplicated media data to a large number of users. In this chapter, we give an overview of network multicast in general, and IP multicast in particular, and discuss how media streaming can be carried over network multicast. In the subsequent chapters we will cover in more detail three classes of multicast streaming algorithms, namely, closed-loop algorithms, open-loop algorithms, and hybrid algorithms.

16.1 Introduction

Today's Internet employs a form of data delivery known as unicast. In unicast a sender sends a data packet through the network which is eventually received by one and only one receiver, thus unicast is also known as point-to-point communications. This unicast data delivery model works well for a wide variety of applications, including the World-Wide-Web, email, file transfer, and even streaming continuous media such as audio and video.

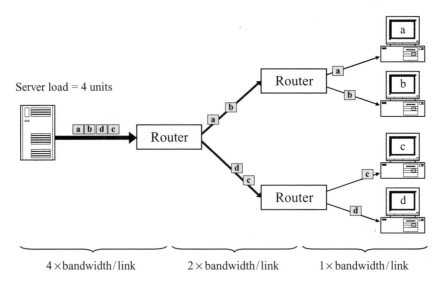

Server load = 4 units

4 × bandwidth / link   2 × bandwidth / link   1 × bandwidth / link

Figure 16.1   Sending the same content to four receivers using unicast data delivery

Nevertheless, the point-to-point delivery mechanism in unicast may impose challenges in some applications. Take TV broadcasting as an example. Traditional TV broadcasting is done over the air using the wireless transmission medium, which is inherently a broadcasting medium, i.e., the signal can be received by any receivers within range. This is an important characteristic not only from a technical point of view, but also from an economic point of view.

Specifically, in TV broadcasting the set-up cost and the operating cost are relatively constant regardless of the number of receivers (assuming sufficient coverage) tuning to the broadcasting channel. This allows the operator to benefit from the economy of scale such that the per-user cost can be reduced to a sufficiently low level to make the operation economically feasible (and profitable).

Now imagine provisioning similar video broadcasting services in the Internet. As the Internet does not support broadcast, the service provider will need to set up media servers to stream the contents to the individual receivers as illustrated in Figure 16.1. Suppose there are four receivers linked up to the media server as shown in Figure 16.1, then the media server will need to send four separate streams of data, identified by "a", "b", "c", and "d" in Figure 16.1 to each of the four receivers. This is radically different from the TV broadcasting model as the media server in this case consumes four times the bandwidth of the data rate of the content. This means that the media server capacity, as well as the intermediate network links, will all need multiple times the bit-rate of the media content to carry the contents to the receivers.

Obviously, this is far less efficient than broadcasting from an engineering point of view. Moreover, the cost to the server provider will likely increase near linearly with the number of receivers, thus lacking the economy of scale inherent in broadcasting. The alternative is to use native network multicast [1–6].

Figure 16.2 illustrates the delivery of content to four receivers using multicast data delivery. Compared to the unicast model in Figure 16.1, we can see that the media server and all the intermediate network links now only need to carry one copy of the media data to the multiple receivers. When a multicast packet reaches a router, the router will replicate and send over each

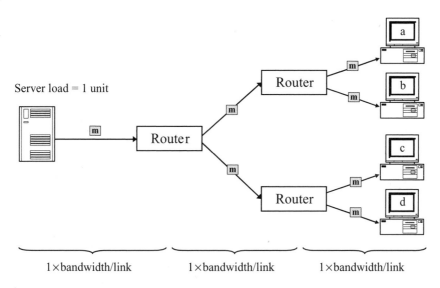

Figure 16.2   Sending the same content to four receivers using multicast data delivery

outgoing link a copy of the packet, without incurring any additional bandwidth overhead at the upstream links/routers or the media server. In other words, the cost to the service provider will not scale up linearly with the number of receivers in the system. In fact, the media server cost is fixed, irrespective of the size of the user population. Depending on the pricing model, the network cost may still increase when there are more receivers because more bandwidth will be consumed in the downstream networks but nonetheless the total network bandwidth consumed will still be substantially lower than using the unicast delivery model.

The current Internet, however, does not yet support native multicast due to a number of practical limitations. First, as Figure 16.2 illustrates routers/switches in the network must all have multicast capabilities to support end-to-end multicast data delivery. The early routers deployed in the Internet do not support native multicast, and the replacement of them by newer multicast-capable routers (which are only widely available in recent years) will likely take a long time.

Second, in addition to having multicast-capable routers, we also need to configure the routers to enable multicast, such as setting up appropriate multicast routing protocols, addressing allocation schemes, installing multicast-capable network monitoring systems, etc. This is a very difficult issue to resolve in the Internet as the Internet is not owned by a single party and so co-operation from many Internet service providers will be needed. Therefore, this becomes an administrative and business issue rather than a technical issue. It is worth noting that the Internet2 [7–8] does have native multicast support although it is not yet available to the general public. Nevertheless it shows the feasibility of multicast data delivery in the global Internet once the administrative and business issues are resolved.

Finally, to make use of network multicast, the application software (e.g., media server software and media client software) will need to be modified accordingly. This is less of an issue compared to enabling multicast in the network as many programming libraries (e.g., sockets) already have support for multicast data delivery.

16.2 Operational Issues

To facilitate discussion of the multicast streaming algorithms to be covered in the subsequent chapters, we will first review the interactions between the media server/client and the network. In the following we will use IP multicast [1–5] as an example to illustrate the operations of multicast application.

First, the network is likely to comprise a number of routers, linking up the media server to the media clients as illustrated in Figure 16.2 for a small-scale network. The network routers will run two types of protocols, one or more control protocols to manage the routes for forwarding multicast data; and the IP protocol to transport the multicast data to the intended recipients.

Over the years a number of multicast routing protocols have been developed, such as DVMRP [9], MOSPF [10], PIM-SM [11], CBT [12], and so on. It is beyond the scope of this book to cover these routing protocols and interested readers are referred to the literature for more details. We will simply assume that the network has been properly configured to run one or more of the routing protocols, and thus be capable of forwarding multicast data packets from the sender to all the receivers.

Second, similar to the concept of a channel in terrestrial TV broadcasting, a network multicast group address serves similar purposes. In IP multicast the address range from 224.0.2.0 to 238.255.255.255 is reserved for multicast data delivery (see Figure 16.3). Unlike an ordinary IP address, which identifies a unique network interface of a host in the Internet, an IP multicast group address is not bound to a specific host or network interface. Instead, similar to a channel in terrestrial TV broadcasting, any multicast-capable receivers can 'tune' into the channel and start receiving the multicast data by *joining* the multicast group. This join-group action is performed by the software application (e.g., by calling an appropriate API in the socket programming library) and then executed by the operating system by sending out an IGMP [13, 14] join-group request. The router, upon receiving the request, will set up the multicast

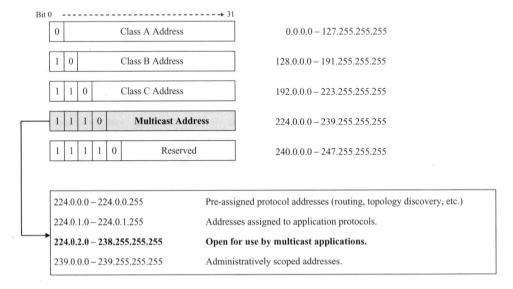

Figure 16.3   Multicast address range in IP Multicast

data flow with its upstream counterparts and begin forwarding to the receiver a copy of the data currently being multicast to the requested group address. A host can also join more than one multicast group at the same time to receive data from multiple groups simultaneously.

Note that in IP multicast the data delivery model is many-to-many, i.e., any host can send/receive data to/from the multicast group. Any data sent to the multicast group (i.e., with the multicast group address as the destination address in the IP packet header) will be forwarded to all hosts that have joined the multicast group, including the sender itself. To leave a multicast group, the application again can call an appropriate API function provided by the operating system, which will stop forwarding data from the multicast group to the application. Note that depending on the version of IGMP used [13, 14], the router may keep forwarding multicast data to the host (which are then discarded by the operating system) for a period of time until it discovers that the host no longer wants to receive data from the multicast group.

From the above discussions we can see that IP multicast involves processing at the application, the operating system, and the routers. The implication is that the processing will take some time to complete, depending on the implementation. For example, after the media client calls an API function to join a multicast group, it will take some time for the operating system and the network routers to set up the multicast dataflow and then forward the multicast packets to the receiving host. Thus, when implementing a multicast streaming application, we will need to take such processing delay into account in order to provide smooth and glitch-free playback to the end user.

16.3 Multicast Media Streaming

Generally speaking, we can classify multicast media streaming into three categories: broadcasting, on-demand multicast streaming, and interactive multicast streaming. Media broadcasting is similar to the terrestrial counterpart in that the objective is to deliver the media contents to anyone who wants to receive the data subject to conditions set forth by the service provider (e.g., subscription, geographical boundary, etc.).

In this broadcasting model the viewer can choose among the list of available media broadcasting channels which channel to view and when to view. However, the user has no control over the contents being broadcast in a channel and can only receive whatever is being broadcast in the channel. This is illustrated in Figure 16.4 where two clients join the same multicast channel at different times. Client j having joined the multicast channel later will have missed the content segment "A" and can only play back segment "B" and onwards. Moreover, like the TV, the user cannot alter the playback schedule, such as jumping to a future playback point. However, limited controls such as pause/resume or even rewinding to a previous playback point (or time-shifted playback) can be made possible by storing in local storage (e.g., hard disk) the media data that have already been played back.

In on-demand multicast streaming the user is allowed to choose when to begin playback of the selected contents similar to unicast-based video-on-demand services. While this is common in unicast-based streaming systems (also known as true-video-on-demand or TVoD), allowing the users to choose their own playback schedule inherently conflicts with the nature of network multicast (see Figure 16.5), where every user of the same multicast group receives the same set of media data. Researchers have conducted many studies in recent years to address this challenge, and some of the techniques will be presented in subsequent chapters.

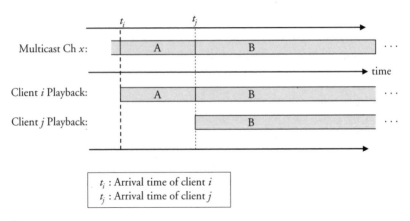

Figure 16.4   In the broadcasting model users can only receive and playback whatever is being multicast in the selected multicast channel

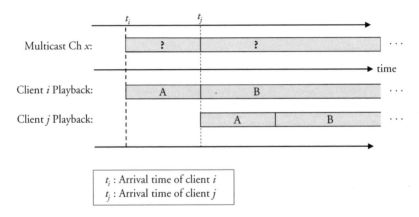

Figure 16.5   A single multicast stream cannot support different viewing schedules

The general principle is the same – by tradeoffs in one or more dimensions so that different users can share the same multicast data transmissions. The tradeoffs can be in time, e.g., by delaying the playback of some users to synchronize their playback schedule; in space, e.g., by caching multicast data locally for later playback; in quality, e.g., by changing the media playback rate so that multiple streams can eventually be merged into a single stream; and so on.

Finally, interactive multicast streaming presents the biggest challenge to system designers. When a user performs an interactive playback control, such as jumping to a different playback point in the media stream, the system cannot change the transmission schedule of the multicast data stream because it is likely being shared by many other users (who have not performed the same interactive playback control). This means that the break-away user will no longer be able to continue playback at the new playback point unless the required data can be received from another source as illustrated in Figure 16.6. Obviously the system can send the required

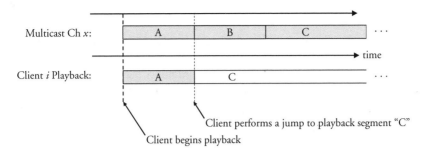

Figure 16.6 A client changing the playback point can no longer receive data from the same multicast channel

data to this user using a separate data stream but then the new data stream is not shared with other users, thus defeating the efficiency gains of using multicast in the first place. This is still an open problem that warrants more research.

16.4 Techniques for On-Demand Multicast Streaming

To support on-demand multicast streaming, it is necessary to trade off other dimensions such as time, space, and quality to permit multiple clients to share the same multicast data transmissions. The principles are to schedule the transmission, reception, and playback of media data to enhance sharing of multicast data among multiple clients. These techniques can be broadly classified into three categories: closed-loop, open-loop, and hybrid algorithms.

In closed-loop algorithms, the multicast media streams are dynamically scheduled according to the user arrival pattern to merge them into as few shared multicast data streams as possible. An important characteristic of closed-loop algorithm is that the transmission schedule may be modified as new clients arrive and as existing clients depart to exploit any opportunities to merge more clients into fewer multicast channels.

By contrast, in open-loop algorithms (also known as periodic broadcasting), all multicast transmissions are prescheduled in a fixed pattern irrespective of the user arrival pattern. Thus, the bandwidth requirement at the media server is fixed regardless of the number of users in the system. In other words, open-loop algorithms in principle can serve as many concurrent users as the network will allow. Unlike closed-loop algorithms where transmission schedules are dynamically determined at the media server (or a separate scheduler), in open-loop algorithms the complexity is shifted to the clients, who will need to determine which media data and when to receive media data from one or more multicast channels to sustain continuous media playback.

Comparing closed-loop and open-loop architectures, the performance (e.g., start-up latency) of closed-loop architectures depends on the system load (i.e., user arrival rate), and generally the performance deteriorates with higher system load. By contrast, open-loop architectures have invariant performance irrespective of the system load. Consequently, at light system load closed-loop architectures can achieve better performance while open-loop architectures perform better at high system load.

16.5 Summary

This chapter has presented an overview of network multicast in general and multicast streaming in particular. The focus of Part II will be on techniques to support on-demand multicast streaming, which are classified into closed-loop algorithms (Chapter 17), open-loop algorithms (Chapter 18), and hybrid algorithms (Chapter 19). In Chapter 20 we address the I/O issues in implementing hybrid media servers supporting both periodic and aperiodic data retrievals.

References

[1] S. Deering, *Host Extensions for IP Multicasting*, RFC 1112, August 1989.

[2] B. Quinn and K. Almeroth, *IP Multicast Applications: Challenges and Solutions*, RFC 3170, September 2001.

[3] K. Hastings and N. Nechita, Challenges and Opportunities of Delivering IP-based Residential Television Services, *IEEE Communications Magazine*, vol. 38, no. 11. Nov. 2000, pp. 86–92.

[4] B. Williamson, *Developing IP Multicast Networks*, 2000, Cisco Press.

[5] M. Goncalves and K. Niles, *IP Multicasting: Concepts and Applications*, 1998, McGraw-Hill.

[6] R.T. Bagwell, J.R. McDearman, and D.T. Marlow, A Comparison of Native ATM-Multicast to IP-multicast with Emphasis on Mapping Between the Two, *Proceedings of the Twenty-Seventh Southeastern Symposium on System Theory*, 12–14 March 1995, pp. 189–193.

[7] K.C. Almeroth, The Evolution of Multicast: From the MBone to Interdomain Multicast to Internet2 Deployment, *IEEE Network*, vol. 14, no. 1. Jan.–Feb. 2000, pp. 10–20.

[8] Internet2 Multicast Working Group, URL: http://multicast.internet2.edu/.

[9] D. Waitzman, C. Partridge, and S. Deering, *Distance Vector Multicast Routing Protocol*, RFC 1075, November 1988.

[10] J. Moy, *Multicast Extensions to OSPF*, RFC 1584, March 1994.

[11] D. Estrin, D. Farinacci, A. Helmy, D. Thaler, S. Deering, M. Handley, V. Jacobson, L. Liu, P. Sharma, and L. Wei, *Protocol Independent Multicast-Sparse Mode (PIM-SM): Protocol Specification*, RFC 2362, June 1998.

[12] A. Ballardie, *Core Based Trees (CBT version 2) Multicast Routing*, RFC 2189, September 1997.

[13] W. Fenner, *Internet Group Management Protocol, Version 2*, RFC 2236, November 1997.

[14] B. Cain, S. Deering, I. Kouvelas, B. Fenner, and A. Thyagarajan, *Internet Group Management Protocol, Version 3*, RFC 3376, October 2002.

17

Closed-Loop Algorithms

This chapter presents closed-loop algorithms to support on-demand multicast streaming. In closed-loop algorithms, the transmission schedule, reception schedule, and/or playback schedule are dynamically adjusted according to the current workload, i.e., number of active media streams, their playback points, etc. The objective is to merge as many clients onto as few multicast data streams as possible to reduce resource utilization. This chapter illustrates this closed-loop approach by describing the techniques of batching, patching, prefix caching, and piggybacking. These techniques not only can be applied individually, but also can be combined to achieve even more performance gains.

17.1 Introduction

In this chapter we present four techniques to improve the efficiency of using multicast to support on-demand media streaming. Compared to media broadcasting, on-demand media streaming allows the users to begin viewing a selected content at their chosen time. This model is applicable to many applications, such as video-on-demand, online education, digital library, and so on. Supporting arbitrary start times for different users, however, conflicts with the constraint of network multicast, where all users of the same multicast group will receive the same set of media data.

To resolve this conflict, we will need to accept trade-offs in other dimensions. The following sections describe four techniques to improve multicast efficiency by trade-offs in different dimensions. These four techniques are chosen to illustrate the possibilities of different trade-offs and the resultant performance gains, and as such are not meant to be exhaustive. In fact, there are many other sophisticated closed-loop algorithms in the literature which are not covered in this chapter. We will briefly discuss some of them in Section 17.6 and interested readers are referred to the literature for more details.

17.2 Batching

Batching is a general technique that has been applied to many engineering problems. The principle is to group together similar tasks so that they can be processed in a more efficient

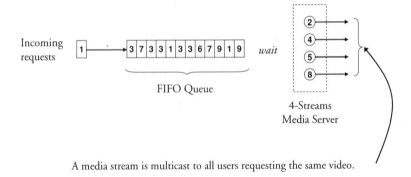

A media stream is multicast to all users requesting the same video.

Figure 17.1   First-come-first-serve batching

way. In multicast media streaming, the batching technique can also be applied to improve the bandwidth efficiency of multicast, i.e., to increase the sharing of multicast data by more users [1–4].

Figure 17.1 illustrates the batching principle using a media server with a capacity of four concurrent multicast channels – a channel represents the retrieval and transmission capacity to support one multicast media stream (we assume all multicast channels are of the same data rate). In Figure 17.1 all four multicast channels are busy serving on-going streaming sessions. As users arrive to find the server busy, they will queue up in a first-in-first-out (FIFO) queue. The numbers in the box in Figure 17.1 represents the identity of the media object being requested. Thus, the four on-going channels are streaming data from media object "2", "4", "5", and "8" respectively.

Now suppose after some time the media channel streaming media object "8" is completed and thus becomes available again as shown in Figure 17.2. Then in a unicast-based media streaming system the server will simply pick the head-of-line user, i.e., the one requesting media object "9", and serve it using the free channel. However, if we take a closer look at the users waiting in the queue, then we can find that there is another user in the third position of the queue who is also requesting media object "9". Thus instead of dedicating the free channel to serve the head-of-line user, we can serve both users simultaneously using multicast streaming. The two users can then simply join the same multicast group and the network will forward the media data to both of them.

In this case the system served two users using just one multicast media stream instead of two unicast media streams. Conceivably, we can achieve better efficiency if there are more users in the queue and/or there are fewer choices of media objects. Note that to the end user the only trade-off is the added complexity of supporting multicast in the client application (ignoring interactive playback control).

Obviously, in practice there may not be many users waiting in the queue when a channel becomes available. If there is only one queueing user requesting a media object, then multicast will not offer any advantage over unicast. One way to further improve the batching efficiency is to artificially delay users to wait for more users to join the batch, even if a free channel is available.

On the other hand, the system is also not limited to serve the waiting users in a FIFO manner. Reconsidering the example in Figure 17.1, we can observe that while there are two queueing

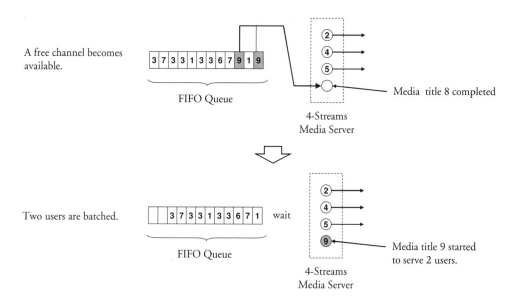

Figure 17.2   FCFS batching is not optimized for batching efficiency

users requesting media object "9", there are five other queueing users requesting media object "3". In other words, if the system allocates the available multicast channel to serve media object "3" instead of "9", even more bandwidth resources will be saved (four instead of one). More generally, instead of using a single queue, we can use separate queues assigned to individual media objects as shown in Figure 17.3. When a free channel becomes available, the system will simply serve the media object with the most number of users waiting to maximize batching efficiency. This technique is called Maximum Queue Length (MQL) [1].

The MQL queueing discipline, however, introduces another issue. Suppose that one of the media objects is very unpopular and rarely get requested by any user. If a user happens to request this unpopular media object during a period of heavy system load, i.e., there are many waiting users, then it may experience a very long delay before it can receive service. Thus, under the MQL queueing discipline the waiting time will be shorter for popular media objects and longer for unpopular media objects, creating a fairness issue.

This fairness issue and other related issues have been addressed in other studies. Interested readers are referred to the work by Dan *et al.* [1–2], Aggarwal *et al.* [3], Shachnai and Yu [4], and Liao and Li [5] for more details.

17.3 Patching

Unlike batching, patching (or stream merging) [5–17] is a technique to enable a client to share media data from an existing on-going multicast stream. Figure 17.4 illustrates the patching technique for a client arriving at time t_a. Instead of starting a new media stream to serve this client, we can apply patching to enable the client to share data from an existing multicast stream started earlier at time t_0. Referring to Figure 17.4, the new client upon arrival immediately

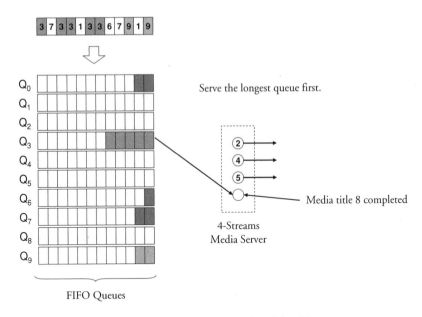

Figure 17.3   Maximal queue length batching

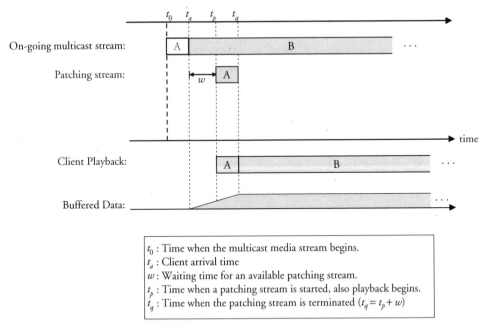

t_0 : Time when the multicast media stream begins.
t_a : Client arrival time
w : Waiting time for an available patching stream.
t_p : Time when a patching stream is started, also playback begins.
t_q : Time when the patching stream is terminated ($t_q = t_p + w$)

Figure 17.4   Operation of the patching technique

begins caching media data from the on-going multicast stream, and waits for the system to allocate a patching stream. The patching stream, once available, will transmit those media data that the client has missed from the on-going multicast stream, i.e., the "A" segment. Using the media data received from the patching stream the client then begins playback.

Eventually, the patching stream will complete the transmission of the 'A' segment, which includes media data from the beginning to playback time $(t_a - t_0)$. After that, the client can continue playback using the cached data and data received from the on-going multicast stream. Thus the client is effectively merged to the on-going multicast stream. Compared to starting a new media stream, the cost is reduced from a full stream of L seconds – length of the media content, to the patching stream of $(t_a - t_0)$ seconds.

Note that patching is orthogonal to batching and thus they can be used together to achieve even better efficiency. For example, if more clients arrive during the waiting time for an available patching stream (i.e., within $[t_a, t_p]$ in Figure 17.4), then they can be served using the same patching stream as shown in Figure 17.5. Two clients arrive at time t_a and t_b respectively, missing the 'A' and 'AB' portions of the on-going multicast stream respectively. The clients are batched together to use the same patching stream to begin playback. Note that in this case the patching stream will need to be extended to transmit segment 'B' required by client 2 in addition to segment 'A' required by both clients. In general, the duration of the patching stream will be determined by the last client arriving before the patching stream begins.

Patching has two trade-offs. First, during the patching duration the client will need to receive two media streams simultaneously. This obviously doubles the client's bandwidth requirement. Second, the client will need additional buffers to cache the data received from the on-going

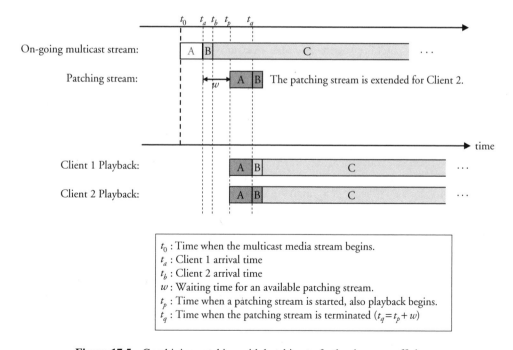

t_0 : Time when the multicast media stream begins.
t_a : Client 1 arrival time
t_b : Client 2 arrival time
w : Waiting time for an available patching stream.
t_p : Time when a patching stream is started, also playback begins.
t_q : Time when the patching stream is terminated ($t_q = t_p + w$)

Figure 17.5   Combining patching with batching to further improve efficiency

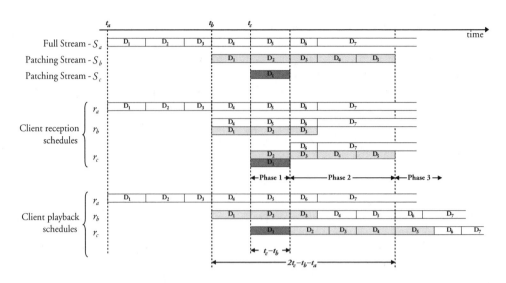

Figure 17.6　Transition patching – patching from both full and patching streams

multicast stream. Assuming the client discards media data after playback, then the extra buffer needed will be equal to $(t_a - t_0)R$ bytes, where R is the media bit-rate.

17.3.1 Transition Patching

In addition to sharing data from a full-length on-going multicast stream, it is possible to share data from another patching stream as well – *transition patching* [11]. Figure 17.6 illustrates the transition patching technique. There are three clients, denoted by r_a, r_b and r_c, that arrive at the system at time instants t_a, t_b, and t_c respectively requesting the same video. We assume that the length of the media object is L seconds, encoded at a constant bit-rate of R bps. To facilitate discussion, we divide the media object into 7 logical segments (D_1 to D_7) and denote the group of segments from the rth segment to the sth segment by $[D_r, D_s]$.

Assuming the system is idle when client r_a arrives, then the system will assign a *regular stream* (R-stream), denoted by S_a, to stream the whole media object from the beginning to the end (i.e., $[D_1, D_7]$) to client r_a. The cost of serving client r_a is thus equal to the bandwidth-duration product LR. For client r_b arriving at time t_b, it clearly cannot begin playback by receiving data from stream S_a as it has missed the first $(t_b - t_a)$ seconds of the media object, i.e., $[D_1, D_3]$. Instead of starting another R-stream for client r_b, the system then assigns a *patching stream* (P-stream) S_b to transmit only the first $(t_b - t_a)$ seconds (i.e., $[D_1, D_3]$) of missed media data to enable client r_b to begin playback. This is the patching technique described previously.

Now consider client r_c arriving at time t_c. We note that for client r_c, it has already missed media segments $[D_1, D_4]$ multicast from the R-stream S_a. To patch these missed media segments, we can apply transition patching as shown in Figure 17.6. We divide the process into three phases. In Phase 1, a P-stream S_c is allocated to stream the initial media segment D_1 to client r_c to begin playback. At the same time, the client caches media segment D_2 being multicast by the P-stream S_b. In Phase 2, the P-stream S_c is released and the client begins caching media segments [D_6, D_7] from the R-stream S_a. Note that the client also continues to receive media segments [D_3, D_5] from the P-stream S_b. Finally, in Phase 3 the remaining P-stream S_b is released and the client simply continues playback using cached data and data received from the R-stream S_a.

This transition patching technique differs from simple patching in two aspects. First, the P-stream S_c allocated for client r_c is occupied for a duration of ($t_c - t_b$) seconds, which is shorter than the duration when simple patching is used, i.e., ($t_c - t_a$) seconds. Second, the P-stream S_b is extended from ($t_b - t_a$) seconds to ($2t_c - t_a - t_b$) seconds to support client r_c. This stream is called a *transition stream* (T-stream) [11]. Thus, the net gain in resource reduction is equal to $(((t_c - t_a) - (t_c - t_b)) - ((2t_c - t_a - t_b) - (t_b - t_a))) = 3t_b - 2t_c - t_a$.

For example, suppose L, t_a, t_b and t_c equal to 7200, 0, 200 and 250 seconds respectively. Then the costs of supporting r_a, r_b and r_c are $7200R$, $200R$ and $150R$ respectively, representing resource savings of 97.22% and 97.92% for clients r_b and r_c.

17.3.2 Recursive Patching

In transition patching we allow a client to share data from an existing patching stream in addition to a full stream. If there are multiple on-going patching streams it is possible to further reduce resource consumption by allowing the client to share data from more than one patching streams – *recursive patching*. Figure 17.7 illustrates the recursive patching technique using a fourth client r_d which arrives at the system at time t_d in addition to the three clients r_a, r_b, and r_c considered in Figure 17.6. To facilitate discussion, we divide the whole media stream into 6 segments denoted by D_1 to D_6.

As client r_d has already missed the initial ($t_d - t_a$) seconds of the media stream, the cost of serving this client using simple patching will be equal to ($t_d - t_a$)R bytes. If we apply transition patching by sharing data from the patching stream S_b then the cost will become $(3t_d - 2t_c - t_b)R$ bytes.

Now consider the use of recursive patching, which in this case is divided into four phases as shown in Figure 17.7. In Phase 1, the client caches media segment D_2 from S_c while playing

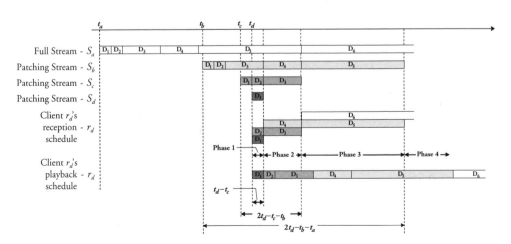

Figure 17.7 Operation of 4-phase recursive patching

back media segment D_1 using data received from S_d. In Phase 2, client r_d continues to receive media segment D_3 from S_c but releases S_d and replaces it with S_b to receive media segment D_4. In Phase 3, client r_d continues to receive media segment D_5 from S_b but releases S_c and replaces it with S_a to receive media segment D_6. Finally, in Phase 4 the client releases the remaining P-stream S_b and continues playback till the end of the media stream by receiving data from S_a.

Therefore, the additional cost (i.e., excluding the cost of the existing streams) to serve r_d is equal to $[(5t_d - 2t_c - 2t_b - t_a) - (3t_c - 2t_b - t_a)]R = 5(t_d - t_c)R$. Compared to the cost of $(3t_d - 2t_c - t_b)R$ bytes in transition patching, there is a gain of $(3t_c - 2t_d - t_b)R$ bytes. For example, if t_d equals 260 seconds, the cost of serving r_d is $50R$, resulting in 99.31% resource saving over serving with a new regular stream.

Note that for the example in Figure 17.7, the client caches up to two streams at any time so the client bandwidth requirement is the same as simple patching and transition patching. In the recursive patching process, the client caches media data through a total of three P-streams and one R-stream. In general, a client can cache media data through even more P-streams as long as there are eligible P-streams that will result in further resource savings.

More generally, if a client caches data from k streams (i.e., one R-stream plus $k - 1$ P-streams) to complete the recursive patching process, then we call it k-phase recursive patching (kP-RP). It is worth noting that simple patching and transition patching are equivalent to 2P-RP and 3P-RP respectively under this definition. We illustrate the performance differences in the next section.

17.3.3 Performance Gains

To illustrate the performance gains of various patching techniques, we developed a simulator to generate numerical results for comparisons. We assume Poisson client arrivals and all clients play back the media stream from the beginning to the end without performing interactive playback operations. For simplicity, we ignore network delays and processing delays in the simulator. Table 17.1 lists the system parameters used in generating the numerical results. With the system resources (i.e., number of multicast channels) fixed we use start-up latency – defined as the time from client arrival to the time playback can begin, as the performance metric for comparison.

Figure 17.8 plots the start-up latency versus arrival rate ranging from 0.1 to 1 client/second. There are three curves plotting the start-up delay for 3-phase, 4-phase, and 5-phase recursive patching respectively.

Compared to transition patching (i.e., with $k = 3$), 4-phase recursive patching can achieve significantly lower start-up latency under the same system load. For example, the latency is

Table 17.1 Parameters used in simulations

Parameter	Range of values
Request arrival rate (arrivals/sec)	0.1 – 1.0
Media stream length (seconds)	7,200
Number of server channels	20

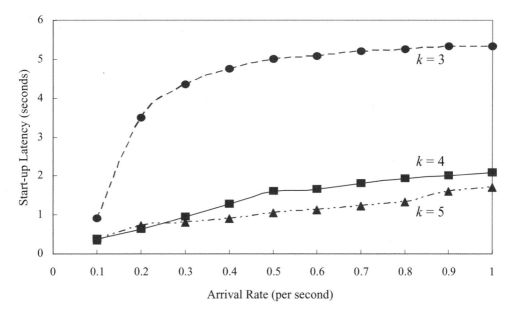

Figure 17.8 Performance of k-phase recursive patching

reduced by 78%, 67% and 62% at arrival rates of 0.3/s, 0.6/s and 0.9/s respectively. The improvement is particularly significant at higher arrival rates. This can be explained by the observation that at higher arrival rates, the streams are more closely spaced in time and thus enable more data sharing through recursive patching.

The latency is further reduced when 5-phase recursive patching is employed although the reduction is less significant. Compared to transition patching, 5P-RP can achieve latency reductions of 81%, 78% and 70% at arrival rates of 0.3/s, 0.6/s and 0.9/s respectively. Larger values of k will likely result in further resource reductions but the marginal improvements will decrease.

The patching techniques illustrated in this section are by no means exhaustive. In addition to applying patching during start-up of a new media stream, it is also possible to further reduce the resources consumption by continuously merging existing on-going streams. Interested readers are referred to the literature [5–17] for more details.

17.4 Caching

In actual deployment of media streaming services over a large geographical area it is common to structure the network in a hierarchical manner. In particular, the operator may place the media servers in the central office, which links up with the customers through a high-speed WAN connection. Once reaching the customer neighborhood, the regional distribution network will deliver the media data to the individual customer.

In this scenario, the media data will go through two types of networks, the long-distance WAN link and the short-distance residential network. Obviously, the bandwidth cost will be

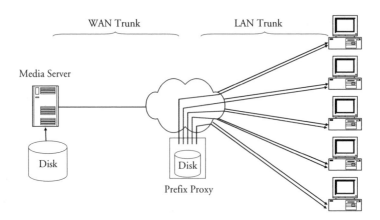

Figure 17.9   Using a prefix proxy to cache and serve the initial portion of the media stream

much higher for the WAN link than the residential network. This motivates the use of regional caches to reduce the amount of data transferred over the WAN link.

One immediate application of caching is to integrate it with patching. Observe that in patching the patching streams are primarily used to send the initial portion of the media stream – called the *prefix* of the media stream. Thus, if the prefix data are stored in the cache as shown in Figure 17.9, the patching streams can then be served entirely from the cache through the prefix proxy. In this way only full streams that are shared by many users will be transferred over the costly WAN link while the numerous patching streams are served by the prefix proxy. Because of the proximity of the prefix proxy to the clients, the latency in starting the patching streams can also be shortened.

More generally, we can extend caching beyond just the media prefix. Prefix caching, however, is particularly effective as the prefix duration is short and common across all clients requesting the same media stream, thus reducing the cache size requirement. If the prefix is sufficiently small, it is even possible to bypass the disk altogether and simply store the prefix in physical memory, thus eliminating another potential I/O bottleneck. Interested readers are referred to the literature for many sophisticated caching strategies [18–21].

17.5 Piggybacking

The previous techniques are all based on changing the transmission schedule or the reception schedule to merge clients onto a shared multicast data stream. In the fifth technique described in the following, called *piggybacking*, the principle is to change the playback schedule so that multiple clients will eventually converge to the same playback point, at which they can be merged and served using a single multicast data stream [22–25].

To change the playback schedule it means the client plays back the media stream either faster or slower than normal playback speed. Note that here playback speed does not equate to data rate. For example, in video the media playback speed is typically defined in frames per second (fps). Thus to increase playback speed the client can display the video frames in a frame rate higher than normal, which also consume media data at a higher data rate, or the media server can send a video stream with frames discarded periodically, thus achieving higher

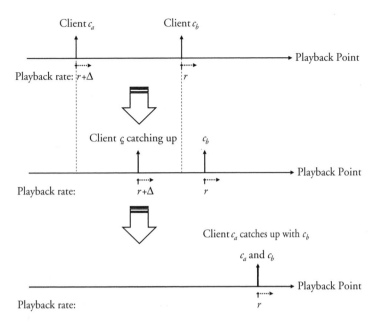

Figure 17.10   Illustration of the piggybacking technique

playback speed (relative to the actual event recorded by the video) without consuming media data at higher data rate. On the other hand, to decrease the playback speed, the client can simply reduce the playback frame rate, which also results in lower media data consumption rate.

Figure 17.10 illustrates the piggybacking process. There are two clients c_a and c_b in the system initially in different playback points of the media stream. To merge them into a single multicast stream we can increase the playback rate of client c_a from the normal speed r to $r + \Delta$ where $\Delta \geq 0$. Thus the difference in playback point between the two clients will decrease and eventually client c_a will catch up with client c_b at the same playback point. At this time the system can then use a single multicast stream to serve both clients, which now continue playback at the *same* playback speed.

In addition to speeding up the playback speed of the latecomer, we can also slow down the playback speed of the early-starter (e.g., client c_b in Figure 17.10), or apply both speed-up and slowdown simultaneously. Clearly, the larger the rate increase/decrease (i.e., Δ in Figure 17.10), the faster the clients can be merged. However, too much playback rate variation will be noticeable or even objectionable to the end users. Previous studies have suggested that playback rate variation up to 5% can be applied without objectionable effect on the end users, and this can be further extended using advanced signal processing techniques such as time-scale modification for audio [26–29].

17.6 Summary

In this chapter we have reviewed four orthogonal closed-loop algorithms for multicast streaming, namely batching, patching, caching, and piggybacking. These four approaches are complementary and hence can be combined to form even more sophisticated architectures. For example, Liao and Li [5], Hua *et al.* [8], and Cai *et al.* [9] have investigated integrating batching with patching to avoid the long start-up delay due to batching. Gao and Towsley [30] proposed a *controlled multicast* technique that integrated patching with dynamically scheduled multicast streaming. This is further refined by Gao *et al.* [31] in their *catching* and *selective catching* schemes. In another study, Ramesh *et al.* [32] proposed the multicast with cache (*Mcache*) approach that integrated batching, patching, and prefix caching. They proposed placing regional cache servers close to the users to serve the initial portion (prefix) of the videos. In this way, a client can start video playback immediately by receiving prefix data streamed from a regional cache server. The server will then dynamically schedule a patching channel for the client to continue the patching process beyond the prefix, and also identify an existing multicast channel for the client to cache and eventually merge into. This architecture has been shown to outperform prefix-cached versions of dynamic skyscraper, GDB, and selective catching. We refer interested readers to the literature for more details of these advanced closed-loop and hybrid algorithms.

References

[1] A. Dan, D. Sitaram, and P. Shahabuddin, Scheduling Policies for an On-Demand Video Server with Batching, *Proceedings of 2nd ACM International Conference on Multimedia*, 1994, pp. 15–23.

[2] A. Dan, D. Sitaram, and P. Shahabuddin, Dynamic Batching Policies for an On-demand Video Server, *ACM Multimedia Systems*, no. 4, 1996, pp. 112–121.

[3] C.C. Aggarwal, J.L. Wolf, and P.S. Yu, On Optimal Batching Policies for Video-on-Demand Storage Servers, *Proc. International Conference on Multimedia Systems*, June 1996.

[4] H. Shachnai and P.S. Yu, Exploring Wait Tolerance in Effective Batching for Video-on-Demand Scheduling, *Proc. 8th Israeli Conference on Computer Systems and Software Engineering*, June 1997, pp. 67–76.

[5] W. Liao and V.O.K. Li, The Split and Merge Protocol for Interactive Video-on-Demand, *IEEE Multimedia*, vol. 4, no. 4, 1997, pp. 51–62.

[6] S.W. Carter, D.D.E. Long, K. Makki, L.M. Ni, M. Singhal, and N. Pissinou, Improving Video-on-Demand Server Efficiency Through Stream Tapping, *Proc. 6th International Conference on Computer Communications and Networks*, Sep. 1997, pp. 200–207.

[7] H.K. Park and H.B. Ryou, Multicast Delivery for Interactive Video-on-Demand Dervice, *Proc. 12th International Conference on Information Networking*, Jan. 1998, pp. 46–50.

[8] K.A. Hua, Y. Cai, and S. Sheu, Patching: A Multicast Technique for True Video-on-Demand Services, *Proc. 6th International Conference on Multimedia*, Sep. 1998, pp. 191–200.

[9] Y. Cai, K.A. Hua, and K. Vu, Optimizing Patching Performance, *Proc. SPIE/ACM Conference on Multimedia Computing and Networking*, San Jose, CA, Jan. 1999, pp. 204-215.

[10] D.L. Eager, M.K. Vernon, and J. Zahorjan, Optimal and Efficient Merging Schedules for Video-on-Demand Servers, *Proc. 7th ACM International Multimedia Conference (MULTIMEDIA '99)*, pp. 199–203.

[11] Y. Cai and K.A. Hua, An Efficient Bandwidth-Sharing Technique for True Video on Demand Systems, *Proc. 7th ACM International Multimedia Conference (ACM MULTIMEDIA '99)*, Orlando, FL, Nov. 1999, pp. 211–214.

[12] D.L. Eager, M.K. Vernon, and J. Zahorjan, Bandwidth Skimming: A Technique for Cost-Effective Video-on-Demand, *Proc. IS&T/SPIE Conference on Multimedia Computing and Networking 2000 (MMCN 2000)*, San Jose, CA, Jan. 2000, pp. 206–215.

[13] E.G. Coffman, P. Jelenkovic, and P. Momcilovic, Provably Efficient Stream Merging, *Proc. 6th International Workshop on Web Caching and Content Distribution*, 2001.

[14] B.N. Amotz and E.L. Richard, Competitive On-line Stream Merging Algorithms for Media-on-Demand, *Proc. 12th Annual ACM-SIAM Symposium on Discrete Algorithms*, Jan. 2001.

[15] Z. Shi and C.C.J. Kuo, Recursive Patching for Video-on-Demand (VOD) Systems with Limited Client Buffer Constraint, *Proc. IEEE International Symposium on Circuits and Systems*, vol. 1, 2002, pp. 373–376.

[16] V.O.K. Li and W. Liao, Interactive Video-on-Demand System, *U. S. Patent No. 6,543,053 B1*, issued 1 Apr., 2003.

[17] Y.W. Wong and Jack Y.B. Lee, Recursive Patching – An Efficient Technique for Multicast Video Streaming, *Proc. 5th International Conference on Enterprise Information Systems (ICEIS 2003)*, Angers, France, April 23–26, 2003.

[18] W. Ma and D.H.C. Du, Design a Progressive Video Caching Policy for Video Proxy Servers, *IEEE Transactions on Multimedia*, vol. 6, no. 4, Aug. 2004, pp. 599–610.

[19] W. Tavanapong, M. Tran, J. Zhou, and S. Krishnamohan, Video Caching Network for On-Demand Video Streaming, *IEEE Global Telecommunications Conference*, vol. 2, 17–21 Nov. 2002, pp. 1723–1727.

[20] J. Kangasharju, F. Hartanto, M. Reisslein, and K.W. Ross, Distributing Layered Encoded Video Through Caches, *IEEE Transactions on Computers*, vol. 51, no. 6, June 2002. pp. 622–636.

[21] Z. Miao and A. Ortega, Scalable Proxy Caching of Video Under Storage Constraints, *IEEE Journal on Selected Areas in Communications*, vol. 20, no. 7, Sep. 2002, pp. 1315–1327.

[22] L. Golubchik, J.C.S. Lui, and R. Muntz, Reducing I/O Demand in Video-On-Demand Storage Servers, *Proc. 1995 ACM SIGMETRICS Joint International Conference on Measurement and Modeling of Computer Systems*, Ottawa, Canada, May 1995, pp. 25–36.

[23] L. Golubchik, J.C.S. Lui, and R.R. Muntz, Adaptive Piggybacking: A Novel Technique for Data Sharing in Video-on-Demand Storage Servers, *ACM Multimedia Systems*, vol. 4, no. 30, 1996, pp. 14–55.

[24] S.W. Lau, J.C.S. Lui, and L. Golubchik, Merging Video Streams in a Multimedia Storage Server: Complexity and Heuristics, *ACM Multimedia Systems*, vol. 6, no. 1, 1998, pp. 29–42.

[25] C.C. Aggarwal, J.L. Wolf, and P.S. Yu, On Optimal Piggyback Merging Policies for Video-on-Demand Systems, *Proc. International Conference on Multimedia Systems*, June 1996, pp. 253–258.

[26] D. Dorran and R. Lawlor, Time-scale Modification of Music Using a Subband Approach Based on the Bark Scale, *Proc. 2003 IEEE Workshop on Applications of Signal Processing to Audio and Acoustics*, 19–22 Oct. 2003, pp. 173–176.

[27] M. Covell, M. Slaney, and A. Rothstein, FastMPEG: Time-scale Modification of Bit-Compressed Audio Information, *Proc. 2001 IEEE International Conference on Acoustics, Speech, and Signal Processing*, vol. 5, 7–11 May 2001, pp. 3261–3264.

[28] P.H.W. Wong and O.C. Au, Fast SOLA-based Time Scale Modification Using Modified Envelope Matching, *Proc. IEEE International Conference on Acoustics, Speech, and Signal Processing*, vol. 3, 13–17 May 2002, pp. 3188–3191.

[29] M. Kalman, E. Steinbach, and B. Girod, Adaptive Media Playout for Low-Delay Video Streaming over Error-Prone Channels, *IEEE Transactions on Circuits and Systems for Video Technology*, vol. 14, no. 6, June 2004, pp. 841–851.

[30] L. Gao and D. Towsley, Supplying Instantaneous Video-on-Demand Services Using Controlled Multicast, *Proc. IEEE International Conference on Multimedia Computing and Systems*, Florence, Italy, vol. 2, June 1999, pp. 117–121.

[31] L. Gao, Z.L. Zhang, and D. Towsley, Catching and Selective Catching: Efficient Latency Reduction Techniques for Delivering Continuous Multimedia Streams, *Proc. 7th ACM International Multimedia Conference*, Orlando, USA, November 1999, pp. 203–206.

[32] S. Ramesh, I. Rhee, and K. Guo, Multicast with Cache (Mcache): An Adaptive Zero-delay Video-on-demand Service, *IEEE Transactions on Circuits and Systems for Video Technology*, vol. 11, no. 3, March 2001, pp. 440–456.

18

Open-Loop Algorithms

This chapter introduces open-loop multicast streaming algorithms that all have fixed media transmission schedules irrespective of the workload of the system. This implies that the complexity will be shifted to the clients, where the reception schedules and the playback schedules are to be optimized to selectively receive media data from one or more of the multicast channels to sustain continuous media playback. We first define a taxonomy and then briefly review some of the existing open-loop algorithms. Next we present a Consonant Broadcasting algorithm in detail to illustrate the design constraints and performance trade-offs in open-loop algorithms. Finally, we present numerical and experimental results to illustrate the characteristics, performance, and practical issues of open-loop multicast streaming algorithms.

18.1 Introduction

Open-loop architectures, also known as periodic broadcasting, have fixed schedules for all media streaming channels irrespective of the user arrival pattern. A new user will receive media data from one or more of the pre-scheduled multicast channels to sustain continuous playback. This implies that the server load is constant regardless of the number of concurrent users in the system. Therefore as long as the network can support it, in principle, there is no limit to how many users an open-loop multicast streaming algorithm can support.

In the past decade researchers have developed many innovative open-loop multicast streaming algorithms. Some examples are the Pyramid Broadcasting scheme proposed by Viswanathan and Imielinski [1], the Skyscraper Broadcasting scheme proposed by Hua and Sheu [2], the Greedy Disk-Conserving Broadcasting scheme proposed by Gao *et al.* [3], the Staircase Data Broadcasting scheme proposed by Juhn and Tseng [4], the Harmonic Broadcasting scheme proposed by Juhn and Tseng [5], the Poly-harmonic Broadcasting scheme proposed by Paris *et al.* [6], the Pagoda Broadcasting scheme proposed by Paris *et al.* [7], and many others [8–11]. Interested readers are referred to the study by Hu [12] for a comprehensive study of the existing periodic broadcasting schemes.

Scalable Continuous Media Streaming Systems   Jack Y. B. Lee
© 2005 John Wiley & Sons, Ltd.

In the next section, we first describe a taxonomy to classify the design dimensions of open-loop multicast streaming algorithms, and in Section 18.3 we present a known performance bound of open-loop algorithms in general. To further illustrate the design choices and performance trade-offs, we present in Section 18.4 the design of the Consonant Broadcasting algorithm. In addition, we also address some practical issues in deploying open-loop algorithms in a multicast-enabled network (e.g., using IP multicast) and present some experimental results obtained from a system implementation.

18.2 A Taxonomy

Fundamental to all periodic broadcasting schemes are four design dimensions. First, a media stream is divided into a number of smaller segments according to a *data partition scheme*. Second, the system (i.e., server and network) bandwidth is divided into a number of logical channels according to a *bandwidth partition scheme*. Third, a predetermined and fixed *broadcasting schedule* defines when the server should broadcast (or multicast, we will use these two terms interchangeably in the rest of the chapter) a media segment over which logical channels. Fourth, a client *reception schedule* defines when a client should receive media data from which logical channels.

Different designs of the four design dimensions result in different trade-offs between the three system resources, namely, system bandwidth, client access bandwidth, and client buffer requirement. Clever designs of the four design dimensions can result in significant resource savings compared to current unicast-based video streaming systems. More importantly, the resource requirements and performances of these periodic broadcasting systems are independent of the system scale. In other words, the same system can potentially serve an unlimited number of concurrent users, as long as the network infrastructure can accommodate them. This property is instrumental to deploying metropolitan-scale media streaming services as it reduces the per-user system cost when more users are added, thereby allowing the service provider to achieve the crucial economy of scale.

In this section, we review some of the existing periodic broadcasting schemes [2–4, 6, 7], and present some known performance bounds. Due to space limitations, interested readers are referred to the study by Hu [12] for a more comprehensive study and comparison of the existing periodic broadcasting schemes. We summarize in Table 18.1 the notations used throughout this chapter. Note also that we use *broadcast* and *multicast* interchangeably in this chapter.

18.2.1 Fixed-Segment Fixed-Bandwidth Schemes

In fixed-segment fixed-bandwidth schemes, a media stream is divided into fixed-size media segments. These segments are then broadcast over a group of fixed-bandwidth channels according to its broadcasting schedule. A notable example is the Pagoda Broadcasting scheme [7] proposed by Paris *et al.* in 1999. A media stream is divided into N fixed-sized media segments, based on the number of channels K, obtained from solving the equation $N = 4 \cdot (5^{\frac{K}{2}-1}) - 1$ if K is even or $N = 2 \cdot (5^{\frac{K-1}{2}}) - 1$ if K is odd. Each media segment is then broadcast over a fixed-bandwidth channel according to its broadcasting schedule at a defined broadcasting frequency.

Table 18.1 Summary of notations

Symbol	Definition
L	The length of the video (sec)
b	The playback rate of the video (Mbps)
K	The total number of logical channels
L_i	The size of the ith video segment (sec)
N	The total number of video segments
B	The total network bandwidth (Mbps)
B_i	The network bandwidth for the video segment L_i (Mbps)
C	The client access bandwidth constraint (Mbps)
T	The maximum start-up latency (sec)
H	The maximum client buffer requirement (Mb)

The client receives from the beginning of a media segment as soon as it encounters it in any of the broadcasting channels. In the worst case, the client has to receive data from all channels simultaneously. The maximum start-up latency T is equal to the broadcast duration of the first media segment L_0.

18.2.2 Variable-Segment Fixed-Bandwidth Schemes

Variable-segment fixed-bandwidth schemes (e.g., [1–3]) divide a media stream into variable-size media segments for broadcast over fixed-bandwidth network channels (e.g., b Mbit/sec). A notable example is the Skyscraper Broadcasting scheme [2] proposed by Hua *et al.* in 1997 as an improvement on the Pyramid Broadcasting scheme proposed by Viswanathan and Imielinski [1]. Unlike the Pyramid Broadcasting scheme, where the media segment sizes increase according to a geometric series, the Skyscraper Broadcasting scheme divides a media stream into N video segments according to a predefined data partition function. They also limited the maximum media segment size to a given length W to reduce the client buffer requirement. The network bandwidth B is then divided equally into N channels (i.e., same as the number of media segments), each with a bandwidth equal to the media playback bit-rate b. Video segment L_i ($i = 0, 1, \ldots, N - 1$) is then repeatedly broadcast over channel i.

The client always caches media data from the beginning of a media segment (instead of from anywhere in between). The client begins by caching data from the next broadcast of the media segment L_0. Then it caches the subsequent media segments L_i in the order of $i = 1, 2, \ldots, N - 1$ at the earliest time after it started playing back the media segment L_i. The client receives data from up to two channels simultaneously and the maximum start-up latency T is equal to the broadcast duration of the first media segment L_0. The client buffer requirement is equal to $L_0 b(W - 1)$ [2].

Another notable example is the Greedy Disk-Conserving Broadcasting scheme [3] proposed by Gao *et al.* in 1998. It is a greedy algorithm that minimizes the number of server channels needed to guarantee a given maximum start-up latency T and client I/O bandwidth requirement. Unlike the Skyscraper Broadcasting scheme, GDB allows the client to receive media segments from $n-1$ channels simultaneously, where n is defined as the order of this scheme (denoted as GDBn). Again the maximum start-up latency T is equal to the broadcast duration of the first media segment L_0 and the client buffer requirement is equal to $L_0 \cdot b \cdot (f^C_{GDB(n)}(N) - 1)$ [3], where $f^C_{GDB(n)}(N)$ is the largest media segment size.

18.2.3 Fixed-Segment Variable-Bandwidth Schemes

Alternatively, we can broadcast fixed-size media segments over variable-bandwidth channels. Notable examples include the Harmonic Broadcasting scheme proposed by Juhn and Tseng [5] in 1997 and the Poly-harmonic Broadcasting scheme [6] proposed by Paris *et al.* in 1998.

In the Poly-harmonic Broadcasting scheme, the media stream is partitioned into N equal-size media segments. Given the desired start-up latency T and a control parameter m, one can choose N by solving the equation $T = (m \cdot L)/N$. The network bandwidth B is then divided into N channels (i.e., same as the number of media segments), with the bandwidth for channel i equal to $B_i = \frac{b}{m+i}, i = 0, 1, \ldots, N - 1$. Media segment L_i is then repeatedly broadcast over channel i. The client, on the other hand, is required to cache media segments from *all* channels simultaneously once it enters the system.

The Poly-harmonic Broadcasting scheme can achieve near-optimal performance when m is large. There are, however, also a few practical issues. First, as the client must receive all channels simultaneously, the client's access network bandwidth requirement is very large (same as the server bandwidth requirement). This may not be practical in all wired systems as in some cases the access bandwidth is substantially more limited than server bandwidth (e.g., ADSL, cable modem). Second, using a large value of m, while it improves performance, will generate a huge number of media segments, each requiring its own network channel for transmission. For some types of network (e.g., IP multicast), this may become a bottleneck as the number of network channels is limited (e.g., IP multicast addresses). We address these issues in the Consonant Broadcasting scheme described later in this chapter.

18.2.4 Variable-Segment Variable-Bandwidth Schemes

The final type of broadcasting scheme is to have both variable segment size and variable channel bandwidth. Juhn and Tseng proposed the first variable-segment variable-bandwidth scheme called Staircase Data Broadcasting [4] scheme in 1997. In Staircase Data Broadcasting, a media stream is first partitioned into N equal-size media segments, based on the number of channels K, derived from the equation $N = \sum_{j=0}^{K-1} 2^j = 2^K - 1$. The network bandwidth B is then divided equally into K channels, with the same bandwidth b for the ith logical channel. For each media segment L_i, it is further divided into 2^i continuous media sub-segments for $i = 0, 1, \ldots, K - 1$. Similarly, each logical channel i is further sub-divided into 2^i sub-channels, each with a bandwidth of $b/2^i$. Finally, each sub-segment is then broadcast repeatedly over a separate sub-channel.

The client begins by receiving data from the first occurrence of the beginning of media segment L_0 at time t_0. The 2^i continuous media sub-segments $L_{i,j}, j = 0, 1, \ldots, 2^i - 1$, within channel $i(i = 0, 1, \ldots, N - 1)$ are then cached at time $t_0 + (L \cdot j)/N$. The client access bandwidth requirement is equal to $2b$, the maximum start-up latency T is equal to the broadcast duration of the first media segment, and the client buffer requirement is bounded by 25% of the size of the media object.

18.3 Performance Bounds

Common to all periodic broadcasting schemes, the key system parameters are start-up latency, network bandwidth, client access bandwidth, and client buffer requirement. Different schemes can be considered as achieving different trade-offs among these four parameters, and thus the natural question is whether bounds on the system's performance exist.

This question has been investigated independently by Hu [12], and Birk and Mondri [13], and others. Although the approaches and the derivations are different, all studies arrive at the same result. Specifically, given a start-up latency of T, it can be shown that the minimum network bandwidth needed for any periodic broadcasting scheme, is given by

$$B = b \cdot \ln(\frac{L+T}{T}) \tag{18.1}$$

assuming there is no constraint on the client access bandwidth.

Additionally, for any optimal periodic broadcasting scheme achieving the performance bound in equation (18.1), it can be shown that the client buffer requirement is equal to

$$H(t') = \begin{cases} t' \cdot b \cdot \displaystyle\int_0^L \frac{dt}{t+T} = t' \cdot b \cdot \ln(\frac{L+T}{T}), & 0 \le t' < T \\[4mm] t' \cdot b \cdot \displaystyle\int_{t'-T}^L \frac{dt}{t+T} = t' \cdot b \cdot \ln(\frac{L+T}{t'}), & T \le t' \le T+L \end{cases} \tag{18.2}$$

where t' is the elapsed time after the client has entered the media streaming system and t is the time relative to the start of media stream. The upper bound of this client buffer requirement is 37% of the size of the media stream. Note that this is only a sufficient condition so it is still possible for a periodic broadcasting scheme to achieve lower client buffer requirement at the expense of increased latency or bandwidth.

18.4 A Generalized Consonant Broadcasting Algorithm

Starting from this section, we use an open-loop algorithm called Consonant Broadcasting (CB) to illustrate the design and trade-offs of open-loop multicast streaming algorithms. An important feature of CB is that it can be used in networks with limited client access bandwidth, which is the norm in typical metropolitan broadband networks. Figure 18.1 shows CB's broadcasting schedule and reception schedule. We divide a media stream into N equal-size segments and repeatedly broadcast them in separate variable-bandwidth multicast channels, i.e., media segment L_i is multicast in the ith logical channel, for $i = 0, 1, \ldots, N-1$. Thus CB belongs to the category of fixed-segment variable-bandwidth schemes. We assume the media stream is constant-bit-rate encoded and thus the playback duration for each media segment is the same, denoted by U seconds.

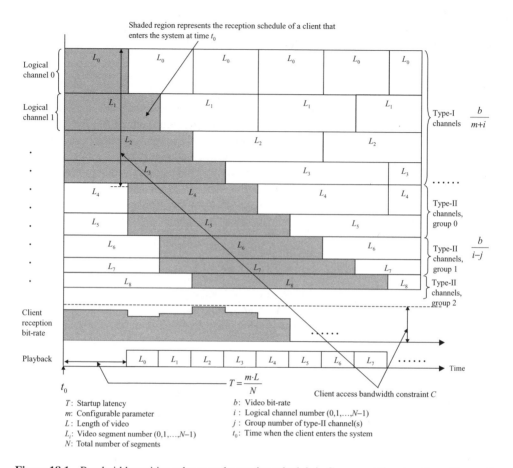

Figure 18.1 Bandwidth partition scheme and reception schedule in Consonant Broadcasting with $m = 2$

To determine the bandwidth for the logical channels, we need to first set a target latency T in multiples of media segment duration U and the number of segments N in the following equation:

$$T = \frac{m \cdot L}{N} \tag{18.3}$$

where m is a configurable parameter to trade off between performance and system complexity. Given the same target latency T, increasing m will result in larger value of N (i.e., dividing the media stream into more segments of shorter duration) and this in turn will reduce the bandwidth requirement, and vice versa. While larger m is desirable from the bandwidth point of view, some network technologies (e.g., IP multicast) have limited number of logical multicast channels (e.g., multicast IP addresses) and thus m cannot be too large. We will return to this issue in Section 18.6.

Next, each media segment is multicast over a separate logical transmission channel (e.g., an IP multicast group address) in the network. There are two types of logical channels, namely

Type-I, and Type-II channels. We define their respective bandwidth partition schemes and reception schedules in the following chapters.

18.4.1 Type-I Channels

The set of Type-I channels begins with the first channel, with a bandwidth allocation of

$$B_0 = \frac{b}{m} \tag{18.4}$$

Subsequent channels are allocated with progressively less bandwidth as given by

$$B_i = \frac{b}{m+i}, \quad i = 0, 1, \ldots, n_1 - 1 \tag{18.5}$$

for the ith channel, where n_1 is the total number of Type-I channels. We can solve for n_1 so that the following two constraints are both satisfied:

$$\sum_{i=0}^{n_1-1} B_i \leq C \quad \text{and} \quad \sum_{i=0}^{n_1} B_i > C \tag{18.6}$$

The first constraint represents the requirement that the aggregate bandwidth must be smaller than the client access bandwidth. This allows the client to receive all Type-I channels simultaneously. The second constraint represents the requirement that we should allocate as many channels as the client access bandwidth will allow maximizing utilization of the client access bandwidth.

It is worth noting that if we remove the client access bandwidth constraint C, the number of Type-I channels n_1 will simply equal to N, i.e., all channels are of Type-I. In this special case, the bandwidth partition scheme in equation (18.5) will be identical to the Poly-harmonic Broadcasting scheme [6]. Therefore, Poly-harmonic Broadcasting can be considered as a special case of Consonant Broadcasting when there is no client access bandwidth constraint.

Figure 18.1 illustrates the operation of Type-I channels (channels 0 to 3). When a client enters the system to start a new media stream, it will immediately start caching data from all Type-I channels simultaneously. The client can start playback after a latency of T seconds as the first media segment L_0 will be completely received by then.

In general, let t_0 be the time the client enters the system, and let c_i be the playback time for media segment L_i, which can be computed from

$$c_i = t_0 + (m+i) \cdot U, \quad i = 0, 1,\ldots, n_1 - 1 \tag{18.7}$$

As the client caches all Type-I channels immediately at time t_0, it will have completely received media segment L_i by the time s_i given by

$$s_i = t_0 + \frac{(L \cdot b)/N}{b/(m+i)}, \quad i = 0, 1, \ldots, n_1 - 1 \tag{18.8}$$

$$= t_0 + (m+i) \cdot U, \quad \because \frac{L}{N} = U$$

which precisely meets the playback schedule c_i's and thus playback continuity is guaranteed.

18.4.2 Type-II Channels

Type-II channels are divided into groups of consecutive channels as shown in Figure 18.1. When a client completes receiving a media segment, the corresponding channel will be released. With the increased available client access bandwidth, the client can then begin to receive a group of Type-II channels. Channels within the same group have their bandwidth allocated according to equation (18.9) and subject to the same client access bandwidth constraint. For the example in Figure 18.1, at time $(t_0 + (m + 1)U)$, the client completes receiving media segment L_1 (releases channel 1) and then begins receiving data from channels 6 and 7. These two channels form the group 1 of Type-II channels.

It may appear that it is simpler to reallocate all the available bandwidth to a single channel instead of a group of channels. However, doing so will unnecessarily increase the bandwidth requirement because there is more than enough time to transmit the new media segment. To see why, consider video segment L_4 being broadcast in channel 4 in Figure 18.1. Channel 0 is released at time $t_0 + 2U$ and media segment L_4 will be playback at time $t_0 + 6U$, thus we have $4U$ seconds to transmit the media segment. However, since the bandwidth released by channel 0 is equal to $b/2$, media segment L_4 will be transmitted completely in just $2U$ seconds if all the available bandwidth is allocated for this logical channel. The extra $2U$ seconds available are then wasted and the network bandwidth is unnecessarily increased.

We tackle this deficiency by transmitting a media segment in a *just-in-time* manner. For the previous example, we can transmit video segment L_4 using the lowest possible bit-rate, i.e., $b/4$, to meet the playback schedule. Then we allocate the remaining bandwidth to the next media segment using the same *just-in-time* scheduling procedure until no more media segment transmissions can be added. These channels then form a group of Type-II channels.

Let $n_{2,j}$ be the number of channels in group j, where $j = 0, 1, \ldots$, etc. Then the bandwidth allocation for channels in group j is given by

$$B_i = \frac{b}{i - j}, \quad \text{for } i \geq n_1 \tag{18.9}$$

and the number of channels in group j can be determined from solving for $n_{2,j}$ in

$$\sum_{i=j+1}^{n_1+n_{2,0}\ldots+n_{2,j}-1} B_i \leq C \quad \text{and} \quad \sum_{i=j+1}^{n_1+n_{2,0}\ldots+n_{2,j}} B_i > C \tag{18.10}$$

which represents the client bandwidth constraints.

To prove playback continuity for media segments broadcast in Type-II channels, we consider an arbitrary Type-II channel i in group j. As the client begins receiving all channels in group j at the same time and it takes $(U \cdot b)/B_i$ seconds to completely receive media segment L_i, we can then compute the time s_i at which media segment L_i is ready for playback from

$$t_0 + (m + j) \cdot U \tag{18.11}$$

$$s_i = t_0 + (m + j) \cdot U + \frac{U \cdot b}{B_i} \tag{18.12}$$

Substituting B_i from equation (18.9) into equation (18.12) we obtain

$$s_i = t_0 + (m + j) \cdot U + (i - j) \cdot U \qquad (18.13)$$

$$= t_0 + (m + i) \cdot U \equiv c_i$$

which is equal to the playback schedule and thus playback continuity for media segments broadcast in Type-II channels is also guaranteed.

18.4.3 Client Buffer

As Figure 18.1 illustrates, the amount of media data accumulated in the client buffer can vary during the media streaming session. Assume a client arrives at the system at time t_0. Let t_i's be the time instants at which a change in the reception schedule occurs, e.g., when the client releases an existing channel (i.e., media segment completely received) and begins to receive data from a new group of Type-II channels. As media segments are of the same size U and channel bit-rates are integral fractions of the media bit-rate b, we can compute t_i ($i = 1, 2, \ldots$) from

$$t_i = T + (i - 1) \cdot U \qquad (18.14)$$

In particular, at time t_i, the client begins playback of media segment L_{i-1} and begins to receive group $i-1$ of Type-II channels (see Figure 18.1).

Let H_i be the amount of media data accumulated but not yet played back at time t_i. Then $H_0 = 0$, and we can compute H_1 from

$$H_1 = \sum_{i=0}^{n_1-1} m \cdot U \cdot B_i \qquad (18.15)$$

where n_1 is the total number of Type-I channels received and the B_i's are their respective bit-rates. Similarly, we can compute H_2 from

$$H_2 = H_1 - U \cdot b + \sum_{k=1}^{n_1+n_{2,0}-1} U \cdot B_k \qquad (18.16)$$

where the first term is the buffer occupancy at time t_1, the second term is the amount of media data consumed, and the last term is the amount of media data received from time t_1 to t_2 (i.e., U seconds).

In general, we can compute H_i ($i \geq 2$) recursively from

$$H_i = H_{i-1} - U \cdot b + \sum_{k=i-1}^{n_1+n_{2,0}+\ldots+n_{2,i-2}-1} U \cdot B_k \qquad (18.17)$$

As both media data consumption rate and total reception rate are constant within a given time interval from t_i to $t_i + 1$, the maximum client buffer requirement must occur at one of the time instants given by the t_i's. Hence we can determine the maximum client buffer requirement H simply by finding the maximum H_i:

$$H = \max\{H_i \mid \forall i = 0, 1, \ldots\} \qquad (18.18)$$

18.5 Performance Comparisons

To illustrate the performance trade-offs in various open-loop algorithms, we present in this
section performance results of Consonant Broadcasting together with Skyscraper Broadcasting (SB), Greedy Disk-Conserving Broadcasting (GDB), Staircase Data Broadcasting (SDB),
Poly-harmonic Broadcasting (PHB), and Pagoda Broadcasting (PB). In computing the numerical results, we use a media stream of length $L = 72,00$ seconds (2 hours) and assume the client
access bandwidth is equal to twice the media bit-rate, i.e., $2b$. For example, if the media bit-
rate is 3Mbps, then the client access bandwidth is 6Mbps, within the limit of current 10Mbps
Ethernet. All open-loop algorithms are optimized using procedure proposed by the original
studies [2–4, 6, 7] to configure their operating parameters. The following sections compare
these algorithms in terms of start-up latency and client buffer requirement, with respect to the
network bandwidth required.

18.5.1 Start-up Latency versus Network Bandwidth

Start-up latency is defined as the maximum time from a client entering the system to the time
media playback starts. With a client access bandwidth of $2b$, we plot in Figure 18.2 the start-up
latency versus the network bandwidth ranging from $2b$ to $10b$.

The results in Figure 18.2 show that PHB achieves the lowest start-up latency, close to the
theoretical lower bound when configured with large value of m (e.g., 16). Similarly, PB also
achieves very good performance, comparable to CB with $m = 1$. However, unlike the other
schemes, we did not apply the client access bandwidth constraint in computing results for PHB

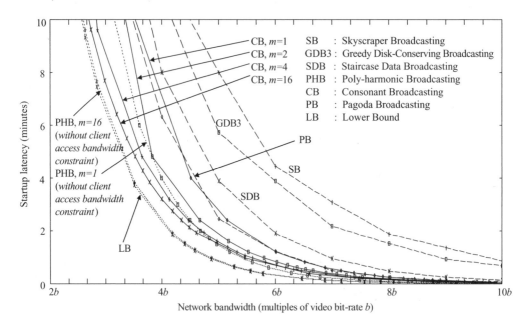

Figure 18.2   Start-up latency versus network bandwidth at large latency range

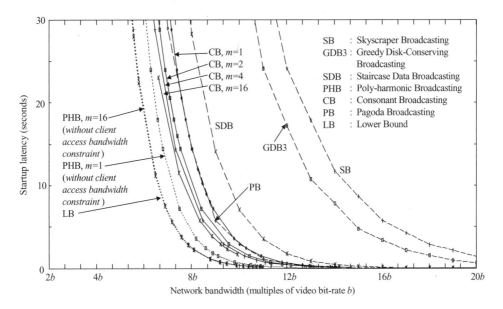

Figure 18.3   Start-up latency versus network bandwidth at small latency range

and PB and thus the results are not directly comparable. Nevertheless, this result shows the performance loss due to limited client access bandwidth.

Except for PHB and PB, CB achieves the lowest start-up latency. This is true even for $m = 1$, which generates the least number of media segments (and hence system complexity) given the same system parameters. Increasing m can further reduce the start-up latency but at the expense of higher system complexity. For a network bandwidth of $5b$, CB with $m = 4$ achieves start-up latency 81%, 74%, and 60% lower than SB, GCB, and SDB respectively.

Figure 18.3 compares the start-up latency of the broadcasting schemes for larger network bandwidth ranging from $2b$ to $20b$. At this range the start-up latency is reduced to seconds, well within the response time required in an on-demand media streaming service. Again the observation is consistent with the results in Figure 18.2, showing that CB achieving the lowest start-up latency. For example, with a network bandwidth of $10b$, CB with $m = 4$ can achieve a start-up latency of only 2 seconds, which is 96%, 95%, and 72% lower than SB, GCB, and SDB respectively.

18.5.2 Start-up Latency versus Client Access Bandwidth

Figure 18.4 plots the start-up latency versus the client access bandwidth ranging from $2b$ to $6b$, where b is the media bit-rate. The network bandwidth is equal to $6b$. There are three observations.

First, CB clearly outperforms the other schemes, especially when the client access bandwidth is low. This is a significant property as the client access network in practice will likely have substantially lower bandwidth than backbone networks. Second, the performances, of PHB and PB degrade significantly when the client access bandwidth is reduced. This is because both

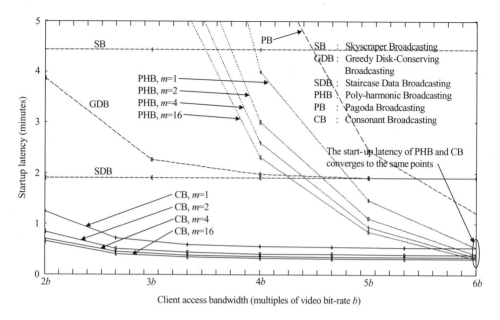

Figure 18.4   Start-up latency versus client access bandwidth (network bandwidth = $6b$)

broadcasting schemes require a client access bandwidth to be equal to the network bandwidth. Therefore, if the client access bandwidth is the bottleneck, the network bandwidth in fact cannot be fully utilized, leading to the performance degradation.

Finally, we note that the performance of PHB and CB converge when the client access bandwidth is increased to $6b$, i.e., same as the network bandwidth, as CB reduces to PHB when the client access bandwidth constraint is removed.

18.5.3   Client Buffer Requirement

Figure 18.5 plots the maximum client buffer requirement versus the network bandwidth, ranging from $2b$ to $10b$. The client buffer requirement is normalized and expressed as the ratio of the size of the media stream. For example, a ratio of 0.3 means that the client buffer must be large enough to store up to 30% of the whole media stream.

We can observe from Figure 18.5 that the maximum client buffer requirement for all the schemes are comparable, and varies within a range from 0.2 to 0.5. For example, at a network bandwidth of $5b$, the maximum client buffer requirements are 27%, 43%, 24%, and 32% for SB, GDB, SDB, and CB (with $m = 4$) respectively. The only broadcasting scheme that consistently achieves lower client buffer requirement is SDB. Therefore, the client buffer requirements of these broadcasting schemes are comparable.

18.6   Grouped Consonant Broadcasting

Results in the previous section show that the performance of CB continues to improve for larger values of the system parameter m in equation (18.3). The trade-off, however, is increased

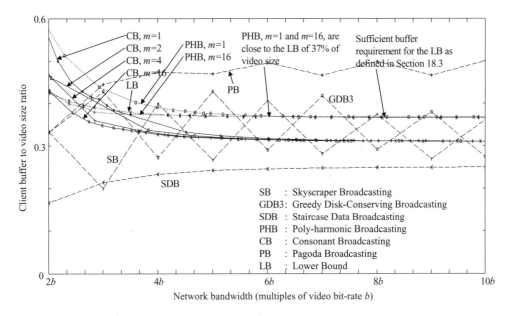

Figure 18.5 Client buffer to video size ratio versus network bandwidth

system complexity in terms of the number of channels required for broadcasting the media segments.

For example, given a client access bandwidth constraint of $2b$ and network bandwidth of $8.98b$, CB with $m = 4$ can achieve a start-up latency of only 5.76 seconds but this requires 5,000 network multicast channels. In networks with limited number of multicast channels (e.g., group addresses in IP multicast), this requirement can become a significant bottleneck. To tackle this problem, we present in the following chapter a Grouped Consonant Broadcasting (GCB) scheme to dramatically reduce the number of network channels required, with a small trade-off in performance.

18.6.1 Bandwidth Partitioning and Reception Schedule

Type-I channels in GCB are the same as the original CB as defined in equations (18.5) and (18.6). The difference is in the design of the Type-II channels. In CB, reception of Type-II channels in the same group begins at the same time but ends at different times due to the just-in-time scheduling principle. While this technique can reduce the bandwidth requirement, it also requires the use of a separate network transmission channel (e.g., an IP multicast address) for each of the Type-II channels.

To reduce the number of channels needed, we modify CB such that reception of Type-II channels in the same group all begins and ends at the same time as shown in Figure 18.6. Consequently, individual Type-II channels in the same group no longer need to be multicast over a separate network channel, but can be transmitted over a single shared channel.

Let n_1 be the total number of Type-I channels and $n_{2,j}$ be the number of Type-II channels in group j ($j = 0, 1, \ldots$) respectively. Then the bandwidth allocation of each channel in group j,

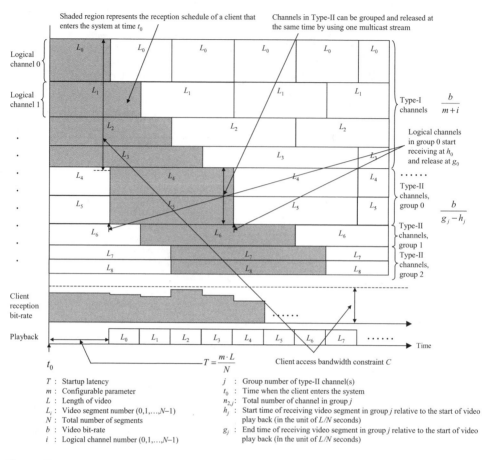

Figure 18.6 Bandwidth partition scheme and reception schedule in Grouped Consonant Broadcasting with $m = 2$

denoted by W_j, is given by

$$W_j = \frac{b}{g_j - h_j} \tag{18.19}$$

where g_j and h_j represent respectively the completion time and the start time for receiving the video segments in the group relative to the time video playback begins, in unit of U seconds, and are given by

$$g_j = \begin{cases} n_1, & \text{for } j = 0 \\ n_1 + n_{2,0} + \ldots + n_{2,j-1}, & \text{otherwise} \end{cases} \tag{18.20}$$

$$h_j = \begin{cases} j, & \text{for } j \leq n_1 \\ n_1 + n_{2,0} + \ldots + n_{2,j-n_1-1}, & \text{for } j > n_1 \end{cases} \tag{18.21}$$

We can determine the number of channels in group j from solving for $n_{2,j}$ in

$$\sum_{i=h_{j+1}}^{n_1+n_{2,0}+\ldots+n_{2,j}-1} B_i \leq C \quad \text{and} \quad \sum_{i=h_{j+1}}^{n_1+n_{2,0}+\ldots+n_{2,j}} B_i > C \tag{18.22}$$

where $B_i = W_j$ for all i's in the range $g_j \leq i < g_{j+1}$.

To prove playback continuity for media segments broadcast in Type-II channels, we consider an arbitrary Type-II group, say group j, comprising media segments $\{L_i \mid g_j \leq i < g_{j+1}\}$. As the client begins receiving all channels in group j at time

$$t_0 + (m + h_j) \cdot U \tag{18.23}$$

and it takes $(U \cdot b)/W_j$ seconds to completely receive the media segments, the time s_j at which all media segments in the group is ready for playback can be computed from

$$s_j = t_0 + (m + h_j) \cdot U + \frac{U \cdot b}{W_j} \tag{18.24}$$

Substituting W_j from equation (18.19) into equation (18.24) we obtain

$$s_j = t_0 + (m + h_j) \cdot U + (g_j - h_j) \cdot U$$
$$= t_0 + (m + g_j) \cdot U$$
$$\leq t_0 + (m + i) \cdot U, \quad \text{for } g_j \leq i < g_{j+1} \tag{18.25}$$

which is equal to or earlier than the playback schedule and thus guaranteeing playback continuity.

18.6.2 Client Buffer Requirement

Compared to CB, GCB generally requires more client buffer because all but the first media segments in a Type-II group are not transmitted in a just-in-time manner. Instead, they are transmitted at a higher rate so that reception can be completed at the same time as the first media segment. Consequently, these media segments are received completely before the playback time and thus occupy more client buffer.

Specifically, the client will play back media segment L_i at time $t_0 + (m + i) \cdot U$, where t_0 is the time the client entered the system. We define H_i as the amount of media data received but not yet played back at time t_i as defined in (18.14). As channel switching occurs only at the time instants $t_0 + (m + h_j) \cdot U$ for $j = 0, 1, \ldots$, we only need to consider the buffer occupancy at these instants. We can compute H_{h_j} recursively from

$$H_{h_j+1} = \begin{cases} \sum\limits_{k=0}^{n_1-1} m \cdot U \cdot B_k, & \text{for } j = 0 \\ H_{h_{j-1}+1} - (h_j - h_{j-1}) \cdot U \cdot b + (h_j - h_{j-1}) \cdot \sum\limits_{k=h_j}^{n_1+n_{2,0}+\ldots+n_{2,j-1}-1} U \cdot B_k, & \text{for } j > 0 \end{cases} \tag{18.26}$$

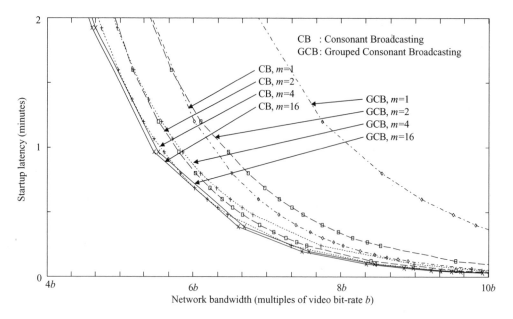

Figure 18.7   Start-up latency versus network bandwidth

where in the second case the first term is the buffer occupancy at time $t_{h_{j-1}+1}$ (i.e., when the last channel is released), the second term and the last term are the amount of media data consumed and the amount of media data received from time $t_{h_{j-1}+1}$ to t_{h_j+1} (i.e., $(h_j - h_{j-1}) \cdot U$ seconds) respectively. The maximum client buffer requirement can then be computed from $H = \max\{H_{h_j} | \forall j\}$.

18.6.3 Performance Trade-offs

As most Type-II channels in GCB are transmitted at higher than necessary bit-rate, we can expect it to require more network bandwidth as well as client buffer to achieve the same latency in CB. We first consider the bandwidth trade-off in Figure 18.7. The results clearly show that GCB has larger latency than CB at the same network bandwidth setting. However, the differences decrease significantly when m is large. For example, the network bandwidth required to achieve the same latency of 15 seconds is $7.48b$ and $8.47b$ respectively for CB and GCB with $m = 2$, and $7.14b$ and $7.23b$ respectively for CB and GCB with $m = 16$.

We also observe similar trade-offs in client buffer requirement as shown in Figure 18.8. As expected, GCB always requires more client buffer than CB under the same setting. Nevertheless the differences are again significantly smaller when m is large. Another observation is that variation in the client buffer requirement with respect to network bandwidth is substantially larger in GCB. Thus, more careful planning is needed to strike a balance between client buffer requirement and network bandwidth requirement.

In contrast to the two trade-offs, GCB gains in terms of the number of network transmission channels required. Figure 18.9 plots the number of channels required to achieve a given latency for CB and GCB. The results clearly show the significant reduction achieved by GCB. For

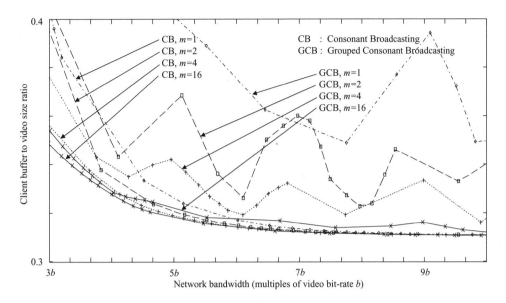

Figure 18.8   Client buffer to video size ratio versus network bandwidth

example, with a latency constraint of 15 seconds, CB and GCB requires 1,920 and 104 channels respectively with $m = 4$, and 7,680 and 434 channels respectively with $m = 16$. The trade-offs in network bandwidth and client buffer requirement in these two cases are 5.47% and 2.76% respectively for $m = 4$, but only 1.27% and 1.28% respectively when $m = 16$.

On the other hand, if the number of network channels available is the limiting factor, then GCB can achieve lower network bandwidth requirement than CB for a given start-up latency by observing the result in Figure 18.10. For example, given a maximum of 64 channels and start-up latency of 3.75 minutes, GCB requires $4b$ network bandwidth, while CB requires $4.12b$. Moreover, in Figure 18.11, we observe that, by further increasing the network bandwidth, CB cannot achieve lower start-up latency as the number of network channels limits the number of video segments N defined in equation (18.3), however, GCB is not subject to this limitation. Thus GCB will be particularly useful for networks having very limited supply of network transmission channels (e.g., IP multicast). Otherwise, CB can be employed to achieve better performance.

18.7 Implementation and Benchmarking

In this section we address some practical issues in the implementation and deployment of open-loop multicast streaming algorithms. Using CB/GCB as an example we explain the issues we encountered during implementation of the system and describe some experimental results obtained from benchmarking.

We developed the CB/GCB system implementation in C++, which runs in the Red Hat Linux 7.0 operating system. We use UDP over IP multicast as the network transmission protocol and set up a testbed with off-the-shelf PCs connected by an IP-multicast-ready FastEthernet

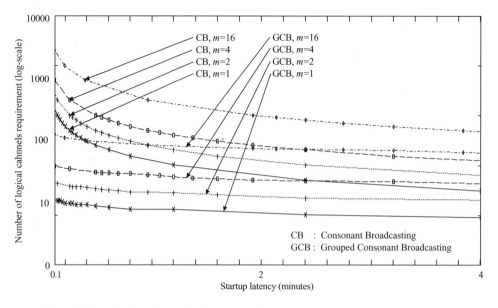

Figure 18.9   Number of logical channels requirement (log-scale) versus start-up latency

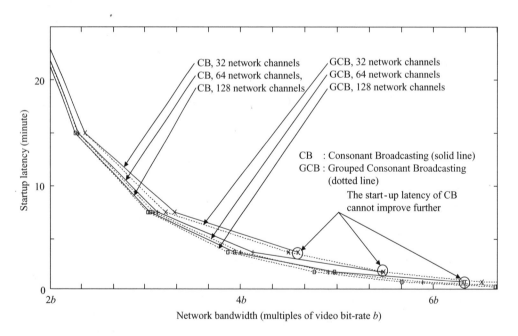

Figure 18.10   Start-up latency versus network bandwidth for fixed number of network channels

switch. The media server sends each of the channel or (group of channels in GCB) in CB/GCB using UDP over a separate IP multicast address. Each media packet carries 1,400 bytes of MPEG-1 compressed video data and an 8-byte header comprising sequence number. The packet size is chosen to match Ethernet's maximum frame size to prevent datagram fragmentation.

To begin a media streaming session, the media client software joins the corresponding multicast groups as defined by the CB/GCB algorithm by sending out an IGMP Join Group requests. The IGMP request will be handled by the network switch and thus has no effect on the video server. Upon receiving the IGMP request the network switch will begin forwarding to the client packets of the requested multicast group. The client then re-sequences the received packets based on the sequence number embedded in the packet header. Once a media segment is completely received, the client will send an IGMP Leave Group request to the network switch, which then stops the forwarding of data belonging to the requested multicast group.

18.7.1 Practical Issues

Implementing the system prototype reveals a number of practical issues not found in the theoretical model. First, the time in joining and leaving an IP multicast group is not precise but subject to delay variations in packet transmission and request processing. In joining a multicast group, the client may experience a small delay before data packets of the media segment are received. In a small local area network such as our experimental test-bed, the channel switching latency is in the order of 10^{-3} seconds. Given that the client has buffered media segment L_0 comprising multiple seconds of media data before commencing playback, this channel switching delay can readily be absorbed. For a larger network that involves multicast routing, the channel switching delay will be larger and thus extra measures (e.g., increasing the amount of prefetch data before playback commence) may be needed to prevent playback starvation during channel switching.

Similarly, the client may not be able to leave a multicast group immediately after receiving a media segment, and thus additional duplicate data packets may continue to arrive at the receiver. The client can detect and discard these duplicate packets. Alternatively, the client can simply process them normally as there is no harm overwriting existing data with the same data. However, these duplicate data do incur additional bandwidth usage at the client access link.

Another aspect where the implementation deviates from the theoretical model is in data transmission. Specifically, we have thus far modeled the transmission of media data as a continuous bit-stream, i.e., using a fluid-flow like model. In practice, the server must packetize media data into discrete UDP datagrams for transmission. In our implementation, we use a datagram size of 1,408 bytes (1,400-byte video data plus 8-byte header) excluding UDP and IP headers. Thus, with a configuration of $N = 1,000$ in our experiments, the inter-packet transmission time can vary from the order of 10^{-2} seconds (e.g., 0.02 seconds) to a few seconds (e.g., 5.7 seconds) depending on the broadcast duration.

Our experiments show that the large deviations in the inter-packet transmission time can result in substantial variations ($\sim$20% bit-rate variation averaged over a 1 second interval) in the bit-rate of the aggregate network traffic. We tackle this problem in GCB by combining all the media segments within the same group into a single data block, and then perform packetization for the combined data block instead of individually packetizing each media segment for transmission. Our experiments show that this can reduce the bandwidth variation to negligible levels (order of 10^{-1} percentage bit-rate variation averaged over 1 second interval) without any impact on other parts of the system.

18.7.2 Experimental Results

We conducted extensive benchmarking experiments to collect three performance results, namely start-up latency, aggregate bit-rate of all channels, and peak aggregate reception bit-rate to compare with the theoretical calculations. In all experiments, we use the system parameters of $L = 4,401, C = 2.84, m = 2$ and $b = 1.42$. The media stream is a MPEG-1 encoded system stream multiplexing one video stream with one audio stream. We conducted benchmarks for a total of 7 GCB system configurations, with the number of media segments N ranging from 50 to 1,000. For each configuration, we obtain the performance data by averaging data collected from 20 benchmark runs. The results are summarized in Table 18.2.

We first consider start-up latency that is measured from within the client software. The results show that the experimental results agree closely with the theoretical calculations. The minor differences are likely due to network delay and software processing delay. Next, we measured the aggregate network bit-rate of all channels using a hardware protocol analyzer connected to the Ethernet switch's mirroring port, which forwards all packets passing through the switch. The measured results exhibit a consistent 5% increase in bandwidth usage compared to the theoretical calculations. This increase is due to the header overheads in the application-layer protocol (8 bytes), UDP (8 bytes), IP (24 bytes), and Ethernet (18 bytes). With a UDP datagram payload of 1,400 bytes, the combined header overhead is equal to $(8 + 8 + 24 + 18)/1458 = 4\%$, which closely matches the measured results.

Finally, we measure the aggregate reception bandwidth usage in the client access link, again using a hardware protocol analyzer. Unlike the aggregate network bit-rate, the reception bit-rate is not constant and does vary depending on which media segments are being received. Nevertheless, we are more interested in the peak bandwidth usage and thus we measure the maximum bandwidth usage averaged over a 10-second window. The results show similar header overhead-induced bit-rate increases ($\sim$5%) for configurations with N up to 200. For larger values of N, the differences widen further up to 9.51%. Our study of the log data shows that two factors lead to the bit-rate increase.

Table 18.2 Comparison of theoretical and experimental results (with $m = 2$)

Config	Latency		Aggregate Bit-rate of all channels			Peak aggregate reception bit-rate		
N/N_G*	Theory	Measured	Theory	Measured	Difference (%)	Theory	Measured	Difference (%)
50/19	176.04	176.10	5.8208	6.11	+4.97	2.84	2.98	+4.93
80/21	110.03	110.05	6.8302	7.17	+4.97	2.84	2.98	+4.93
100/22	88.02	88.04	7.2779	7.64	+4.98	2.84	2.98	+4.93
200/28	44.01	44.05	8.6611	9.10	+5.07	2.84	3.00	+5.63
500/33	17.60	18.00	10.6277	11.16	+5.00	2.84	3.05	+7.39
800/37	11.00	11.20	11.5828	12.16	+4.98	2.84	3.11	+9.51
1000/38	8.80	9.00	12.1096	12.71	+4.96	2.84	3.09	+8.80

Note.*N and N_G are the number of video segments and number of channels respectively.

First, larger values of N result in more frequent channel switching, and as discussed earlier in Section 18.7.1, there is some delay from the time the client leaves a multicast group to the time the network switch stops forwarding the multicast data. This results in some duplicated data being transmitted to the client, only to be discarded by the client's operating system. The second reason is due to the specific network switch we used in the experiment. Our results show that there seems to be bugs in the switch's hardware, resulting in some random multicast data transmitted to the client *after* the switch has pruned the multicast tree. This specific problem is easy to miss because the random multicast data will be discarded by the client's operating system (as the client has left the multicast group already) and thus will not cause any data transmission or application error. We expect this problem to be resolved in future revisions of the switch hardware.

18.8 Summary

In this chapter we have reviewed some open-loop multicast streaming algorithms in the context of a taxonomy, which classifies the algorithms according to the media segmentation and transmission bandwidth schemes adopted. To further illustrate the design and trade-offs in developing an open-loop algorithm, we described in detail as well as analyzed the performance of the Consonant Broadcasting algorithm. We also addressed some practical issues in the implementation and deployment of the Consonant Broadcasting algorithm, which are likely to be applicable to other open-loop algorithms as well.

Unlike the closed-loop algorithms, the resources consumed by open-loop algorithms are fixed irrespective of the system load, i.e., number of concurrent users. The upside is that open-loop algorithms will be very cost-effective in serving popular media streams (e.g., popular movies) of a large user population. The downside, however, is that for unpopular media streams the resources requirement could exceed those of closed-loop algorithms, which are more efficient when the system load is lighter. In the next chapter, we illustrate a hybrid approach to multicast streaming, combining elements of both open-loop and closed-loop algorithms in the same architecture.

References

[1] S. Viswanathan and T. Imielinski, Metropolitan Area Video-on-Demand Service Using Pyramid Broadcasting, *IEEE Multimedia Systems*, vol. 4, 1996, pp. 197–208.

[2] K.A. Hua and S. Sheu, Skyscraper Broadcasting: A New Broadcasting Scheme for Metropolitan Video-on-Demand Systems, *Proceedings of the ACM SIG-COMM '97*, Cannes, France, Sept. 1997, pp. 89–100.

[3] L. Gao, J. Kurose, and D. Towsley, Efficient Schemes for Broadcasting Popular Videos, *Proceedings of the 8th International Workshop on Network and Operating Systems Support for Digital Audio and Video*, Cambridge, UK, July 1998.

[4] L.S. Juhn and L.M. Tseng, Staircase Data Broadcasting and Receiving Scheme for Hot Video Service, *IEEE Transactions on Consumer Electronics*, vol. 43, no. 4, Nov. 1997, pp. 1110–1117.

[5] L.S. Juhn and L.M. Tseng, Harmonic Broadcasting for Video-on-Demand Service, *IEEE Transactions on Broadcasting*, vol. 43, no. 3, Sept. 1997, pp. 268–271.

[6] J.F. Paris, S.W. Carter, and D.D.E. Long, A Low Bandwidth Broadcasting Protocol for Video on Demand, *Proceedings of the 7th International Conference on Computer Communications and Networks*, Lafayette, LA, USA, Oct. 1998, pp. 690–697.

[7] J.F. Paris, S.W. Carter, and D.D.E. Long, A Hybrid Broadcasting Protocol for Video on Demand, *Proceedings of the 1999 Multimedia Computing and Networking Conference*, San Jose, CA, Jan. 1999, pp. 317–326.

[8] C.C. Aggarwal, J.L. Wolf, and P.S. Yu, A Permutation-Based Pyramid Broadcasting Scheme for Video-on-Demand Systems, *Proceedings of the International Conference on Multimedia Computing and Systems*, Hiroshima, Japan, June 1996, pp. 118–126.

[9] J.F. Paris, S.W. Carter, and D.D.E. Long, Efficient Broadcasting Protocols for Video on Demand, *Proceedings of the International Symposium on Modeling, Analysis and Simulation of Computer and Telecommunication Systems*, Montreal, Canada, July 1998, pp. 127–132.

[10] L.S. Juhn and L.M. Tseng, Fast Data Broadcasting and Receiving Scheme for Popular Video Service, *IEEE Transactions on Broadcasting*, vol. 4, no. 4, Mar. 1998, pp. 100–105.

[11] J.F. Paris, S.W. Carter, and P.E. Mantey, Zero-Delay Broadcasting Protocols for Video-on-Demand, *Proceedings of the 1999 ACM Multimedia Conference*, Orlando, FL, Nov. 1999, pp. 189–197.

[12] A.Hu, Video-on-Demand Broadcasting Protocols: A Comprehensive Study, *Proceedings of the IEEE Infocom 2001*, Anchorage, AK, Apr. 2001, pp. 508–517.

[13] Y. Birk and R. Mondri, Tailored Transmissions for Efficient Near-Video-on-Demand Service, *Proceedings of the IEEE International Conference on Multimedia Computing and Systems*, Florence, Italy, June 1999, pp. 226–231.

19

A Hybrid Architecture

This chapter illustrates a hybrid approach to developing multicast media streaming systems. In particular, we describe the design, analysis, and implementation of a Super-Scalar Video-on-Demand (SS-VoD) system that integrates the techniques of batching, patching, and periodic broadcasting. Instead of striving for maximum multicast efficiency, we focus on a number of practical issues in designing the SS-VoD architecture, such as implementation and deployment complexity, buffer requirement, multicast channel switching frequency, support of interactive playback controls, etc. As well as describing the design choices, we also devise a technique to model this relatively complex system so that approximate performance results can be obtained without requiring lengthy simulations. Finally, we briefly describe the implementation of the system and present some experimental results.

19.1 A Super-Scalar Architecture

Figure 19.1 depicts the overall architecture of the Super-Scalar Video-on-Demand (SS-VoD) system. The system comprises a number of service nodes connected via a multicast-capable network to the clients. The clients form clusters according to their geographical proximity. An admission controller in each cluster performs authentication and schedules requests for forwarding to the service nodes.

Each service node operates independently of the rest, having its own disk storage, memory, CPU, and network interface. Hence a service node is effectively a mini video server, albeit serving a small number of video titles to the *entire* user population. This modular architecture can simplify the deployment and management of the system. For example, since the configuration of each service node is decoupled from the scale of the system and each service node carries just a few movies, a service provider simply deploys the right number of service nodes according to the desired video selections. Additional service nodes can be added when more movie selections are needed, with the existing nodes remain unchanged.

SS-VoD achieves scalability and bandwidth efficiency with two techniques. The first technique is through the use of multicast to serve multiple clients using a single multicast channel. However, simple multicast such as those used in a near-video-on-demand (NVoD) system,

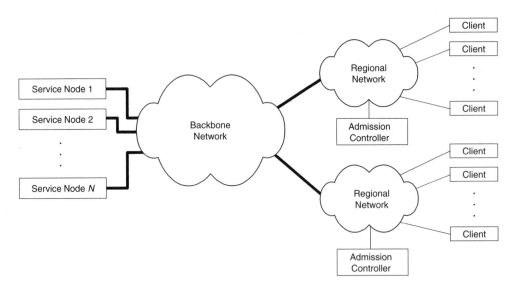

Figure 19.1 The super-scalar video-on-demand architecture

where the same video is repeatedly multicast over several channels in a time-staggered manner, limits the time during which a client may start a new video session. Depending on the
number of multicast channels allocated to a video title, this start-up delay can range from
a few minutes to tens of minutes. To tackle this initial delay problem, we employ batching
and patching to enable a client to start video playback at any time using a dynamic multicast
channel until it can be merged back into an existing multicast channel. The following sections
present these techniques in detail.

19.1.1 Transmission Scheduling

Each service node in the system streams video data into multiple multicast channels. Let M
be the number of video titles served by each service node and let N be the total number of
multicast channels available to a service node. For simplicity, we assume N is divisible by M
and hence each video title is served by the same number of multicast channels, denoted by
$N_M = N/M$. These multicast channels are then divided into two groups of N_S static multicast
channels and $N_D = N_M - N_S$ dynamic multicast channels.

The video title is multicast repeatedly over all N_S static multicast channels in a time-staggered
manner as shown in Figure 19.2. Specifically, adjacent channels are offset by

$$T_R = \frac{L}{N_S} \tag{19.1}$$

seconds, where L is the length of the video title in seconds. Transmissions are continuously
repeated, i.e., restart from the beginning of a video title after transmission completes, regardless
of the load of the server or how many users are active. These static multicast channels are used
as the main channel for delivering video data to the clients. A client may start out with a
dynamic multicast channel but it will shortly be merged back into one of these static multicast
channels.

Figure 19.2   Transmission schedules for static and dynamic multicast channels

We note that while using more sophisticated open-loop multicast algorithms (cf. Chapter 18) can achieve better resource savings, they often require more client-side bandwidth and client-side buffer. More importantly, these multicast schedules require the client to switch between multiple multicast channels during a video session to achieve the resource savings. For large-scale systems comprising millions of users, the channel switching overhead can present a significant burden to the network.

Let us use IP multicast as an example. A client wishing to switch from one multicast channel to another will need to send an IGMP message to the edge router to stop it from forwarding data in the current multicast group. Another IGMP message will then be sent to request the edge router to start forwarding data from the new multicast group. Unlike processing data packets, these control messages and group management processing are performed in software running on the router CPU. Hence, the more channel switching it requires, the more chance that a router could become overloaded. This could lead to missed schedule and/or data loss, resulting in client playback hiccup and/or visual quality degradation.

Another advantage of using the simple staggered multicast schedule in SS-VoD is in the support of interactive playback control. In particular, interactive controls such as pause-resume, slow motion, and seeking can be supported in SS-VoD without incurring any additional resources or processing at the video server, nor any additional buffer at the client (cf. Section 19.2).

The next section presents the admission procedure for starting a new video session and we explain in Section 19.1.3 how the client is merged back into one of the static multicast channels.

19.1.2 Admission Control

To reduce the response time while still leveraging the bandwidth efficiency of multicast, SS-VoD allocates a portion of the multicast channels and schedules them dynamically according to the request arrival pattern.

Specifically, a new request always goes to the admission controller. Knowing the complete transmission schedule for the static multicast channels, the admission controller then determines if the new user should wait for the next upcoming multicast transmission from a static

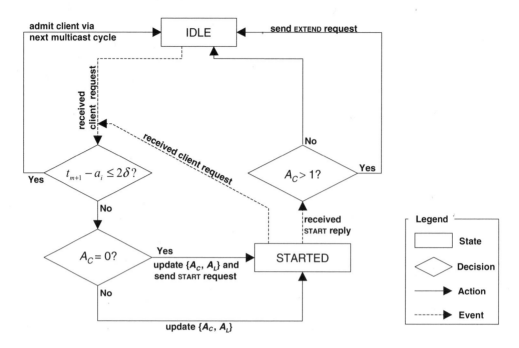

Figure 19.3   State-transition diagram for the admission controller

multicast channel, or should start playback with a dynamic multicast channel. In the former case, the client just waits for the next multicast cycle to begin, without incurring any additional load on the backend service nodes. In the latter case, the admission controller performs additional processing to determine if a new request needs to be sent to the appropriate service node to start a new dynamic multicast stream.

Figure 19.3 depicts the state-transition diagram for the admission procedure. Beginning from the IDLE state, suppose that a new request arrives at time a_i, which is between the start time of the previous multicast cycle, denoted by t_m, and the start time of the next multicast cycle, denoted by t_{m+1}. Now a predefined admission threshold, denoted by δ, determines the first admission decision made by the admission controller: the new request will be assigned to wait for the next multicast cycle to start playback if the waiting time, denoted by w_i, is equal to or smaller than 2δ, i.e.

$$w_i = t_{m+1} - a_i \leq 2\delta \tag{19.2}$$

We call these requests *statically-admitted* and the admission controller returns to the IDLE state afterwards. This admission threshold is introduced to reduce the amount of load going to the dynamic multicast channels. Optimization procedures for this admission threshold will be presented in Section 19.3.3.

If equation (19.2) does not hold, then the admission controller will proceed to determine if a request needs to be sent to the appropriate service node to start a new dynamic multicast stream – *dynamically-admitted*. The service nodes and admission controllers each keep a counter-length tuple: $\{A_C, A_L\}$, where $A_C = \{0, 1\}$ is the counter, and $A_L, 0 \leq A_L \leq (T_R - 2\delta)$, is the length of service for each video title being served. Therefore, each service node will have M such admission tuples and each admission controller will have MK such admission tuples, where K is the total number of service nodes in the system. Both the counter and the length fields are initially set to zero.

Now with the admission tuples, the admission procedure proceeds as follows. For requests that cannot be statically-admitted, the admission controller will first check the counter in the admission tuple for the requested video title. If the counter A_C is zero, then the counter is incremented by 1, and the length field is set according to

$$A_L = a_i - t_m \tag{19.3}$$

which is the length of time passed since the beginning of the last multicast. In other words, this particular client will occupy the dynamic channel for a duration of A_L seconds for patching purpose. At the same time a START request carrying the requested video title and the length field A_L will be sent to a service node and the admission controller enters the STARTED state.

If another request for the same video title arrives during the STARTED state, say, at time a_{i+1}, the admission controller will not send another request to the service node, but will just update the local length field according to

$$A_L = a_{i+1} - t_m \tag{19.4}$$

This process repeats for all subsequent requests arrived during the STARTED state. As a result, only one START request will be sent to the service node regardless of how many requests are received during the STARTED state, thereby significantly reducing the processing overhead at the service node.

At the service node side, upon receiving a START request from the admission controller, the service node will wait for a free channel from the N_D dynamic multicast channels to start transmitting the video title for a duration of A_L seconds as shown in Figure 19.4. Once a channel becomes available, a START reply will be sent back to all admission controllers to announce the commencement of the new transmission.

The admission controllers, upon receiving the START reply, will do one of two things. If the local counter value is 1, then both the counter and the length fields are zeroed and the admission process is completed. If the counter is greater than 1, then the admission controller will send an EXTEND request to the service node to extend the transmission duration according to the value of the local length field A_L. Note that in this case, the length field at the admission controller will be larger than the length field at the service node because only the length field at the admission controller is updated for subsequent requests for the same video title. The length field at the service node is always the one for the first request. After receiving EXTEND requests from the admission controllers, the service node will update the transmission duration to the largest one among all EXTEND requests. Transmission will stop after the specified transmission duration expires. Note that the service node does not need to wait for any EXTEND request to begin streaming. Streaming will begin as soon as a free dynamic channel becomes available. The purpose of the EXTEND request is to increase the transmission time of the dynamic channel to cater for subsequent requests in the same batch that require a longer patching duration.

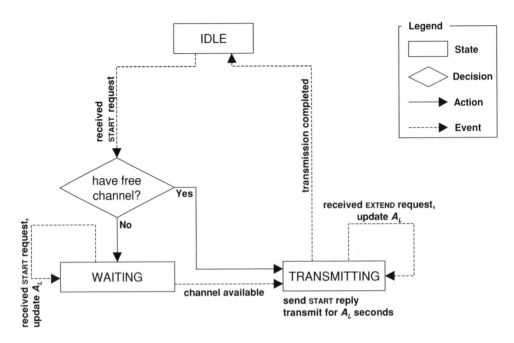

Figure 19.4 State-transition diagram for a service node

It may appear that the previous admission procedure is unnecessarily complex and the clients would be better off sending requests directly to the service nodes. However, this direct approach suffers from poor scalability. In particular, recall that each service node serves a few video titles to the entire user population. Therefore, as the user population grows, the volume of requests directed at a service node will increase linearly and eventually exceed the service node's processing capability.

By contrast, an admission controller generates at most two requests, one START request and one EXTEND request, for each dynamically-started multicast transmission, irrespective of the actual number of client requests arriving in an admission cycle (i.e., from receiving the first request in a batch to sending the EXTEND request). Given that the number of admission controllers is orders of magnitude smaller than the user population, the processing requirement at the service nodes is substantially reduced. For extremely large user populations where even requests from admission controllers can become overwhelming, one can extend this request-consolidation strategy into a hierarchical structure by introducing additional layers of admission controllers to further consolidate requests until the volume becomes manageable by the service nodes.

19.1.3 Channel Merging

According to the previous admission control policy, a statically-admitted client starts receiving streaming video data from a static multicast channel for playback as depicted in Figure 19.5. For dynamically-admitted clients, video playback starts with video data received

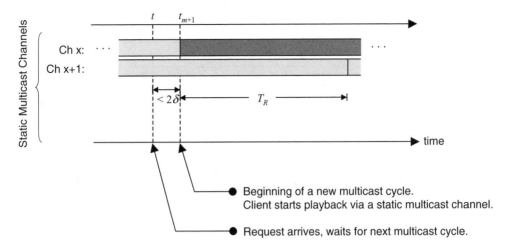

Figure 19.5 Timing diagram for a statically-admitted client

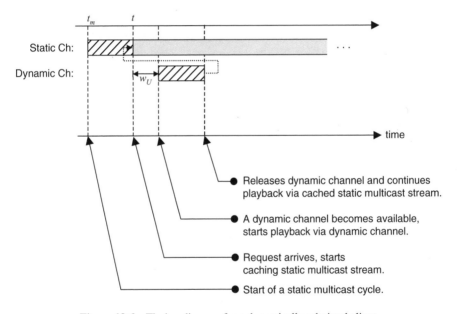

Figure 19.6 Timing diagram for a dynamically-admitted client

from a dynamically-allocated multicast channel. To merge the client back into an existing static multicast channel, the client concurrently receives and caches video data from a nearby (in time) static multicast channel as illustrated in the timing diagram in Figure 19.6. Eventually, playback will reach the point where the cached data began and the client can then release the dynamic multicast channel. Playback then continues using data received from the static multicast channel.

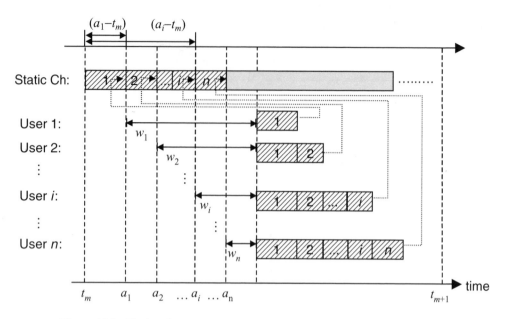

Figure 19.7 Timing diagram for admitting a group of dynamically-admitted users

As an illustration, consider a dynamic multicast channel serving n dynamically-admitted clients as shown in Figure 19.7. Let a_i be the time client i arrives in the system and the nearest multicast cycle starts at t_m and t_{m+1} respectively, where $t_m < a_1 < a_2 \ldots < a_n < (t_{m+1} - 2\delta)$. Each client upon arrival will begin caching data from a static multicast channel while waiting for an available dynamic channel to begin playback. Note that the later a client arrives in the batch, the longer it must receive data from the dynamic multicast channel to make up for the missed data transmitted by the static multicast channel. Eventually all clients in the batch will reach their cached data position and the dynamic multicast channel is released. Therefore, the channel holding time of the dynamic multicast channel is equal to $(a_n - t_m)$, i.e., dominated by the last client joining the batch.

Compared to TVoD systems, a SS-VoD client must have the capacity to receive two multicast channels concurrently and have a local buffer large enough to hold up to T_R seconds of video data. Given a video bit-rate of 3Mbps (e.g., high-quality MPEG4 video), a total of 6Mbps downstream bandwidth will be needed for the initial portion of the video session. For a two-hour movie served using 25 static multicast channels, the buffer requirement will become 108MB. This can easily be accommodated today using a small hard disk in the client, and in the near future simply by using memory as technology improves.

19.2 Interactive Controls

To provide a complete VoD service, interactive playback controls such as pause-resume, slow motion, seeking, etc. will also need to be supported. Among these, pause–resume is likely to be the control most frequently performed in typical movie-on-demand applications. Intuitively, performing an interactive control in SS-VoD essentially breaks the client away from the current static multicast channel and then restarts it at another point within the video stream. Hence a simple method to support interactive control is to treat them just like a new request. Clearly, this approach will increase loads at the dynamic multicast channels and result in increased waiting time for both new session and interactive control requests. As there is no generally accepted user-activity model, we do not attempt to quantify the performance impact of this approach in this study.

In the following sections, we present algorithms that take advantage of the staggered static multicast schedule to support pause–resume, slow motion, and seeking in SS-VoD in a resource-free way. In other words, no additional server resource or client buffer is needed to support these interactive controls in SS-VoD.

19.2.1 Pause–Resume

We use a simple channel-hopping algorithm to implement pause–resume in SS-VoD. Specifically, since a client has a buffer large enough to cache T_R seconds of video, it can just continue buffering incoming video data after the user has paused playback. If the user resumes playback before the buffer is full, then no further action is required. By contrast, if the buffer becomes full, then the client simply stops receiving data and enters an idle state.

When the user resumes playback, the client can resume playback immediately and at the same time determine the nearest multicast channel that is currently multicasting the video. Since a movie is repeated every T_R seconds and the client buffer already contains T_R seconds' worth of video data, we can guarantee that the client can locate and merge back into an existing static multicast channel.

19.2.2 Slow Motion

Slow motion is playback at a rate lower than the normal playback rate. As video data are always being transmitted and received at the normal video bit-rate, it is easy to see that once slow motion is started, data will begin to accumulate in the client buffer. Now if the user resume normal speed playback before the buffer is full, then no additional action needs to be undertaken.

However, if playback continues in slow motion state for a sufficiently long time, the client buffer will eventually be completely filled with video data. Note that at the instant when the buffer becomes full, the buffer will contain T_R seconds' worth of video data. This is equivalent to the buffer full state in performing a pause operation. The only difference is that in performing a pause, the client will stop receiving data until the user resumes playback, at which time a nearby (in time) multicast channel will be located to merge back into. For slow motion, however, playback continues at that instant and hence it is necessary to immediately locate a nearby multicast channel other than the current one to merge back into. As any play point is at most T_R seconds away due to the staggered static multicast schedule, the T_R seconds' worth of data in the buffer guarantees that the client can locate and merge back into a static multicast channel. If slow motion continues after merging, then data will begin to accumulate in the buffer again and the cycle repeats until normal playback speed is resumed.

Using this algorithm, slow motion at any rate slower than the normal playback rate can be supported without the need for any additional resource from the server. Client buffer requirement also remains the same.

19.2.3 Seeking

Seeking is the change from one playback point to another. Typically, the user initiates seeking either by giving a new destination time offset or by means of using a graphical user interface such as a slider or a scroll bar. SS-VoD can support different types of seeking depending on the seek direction, seek distance, and the state of the client buffer and static multicast channels. Specifically, due to patching, the client buffer typically has some advance data cached. Moreover, some past video data will also remain in the client buffer until being overwritten with new data. Hence, if the new seek position is within the range of video data in the client buffer, seeking can be implemented simply by changing the playback point internally.

Now if the seek position, denoted by t_s, lies outside the client buffer, then the client may need to switch multicast channels to accomplish the seek. Let t_i, $i = 0, 1, \ldots, N_S - 1$ be the current playback points of the N_S static multicast channels and assume the client is currently on channel x. Then the client will choose the nearest channel to restart playback by finding the channel j so that the seek error $\varepsilon = \min\{|t_j - t_s|, |t_s - t_j|\}$ is minimized. Note that the current channel may happen to be the nearest channel and in this case, the client simply seeks the oldest data in the buffer if t_s is earlier than the current playback point, or seeks the newest data in the buffer otherwise.

Clearly in the previous case the seek operation may not end up in the precise location specified by the user and the seek error can be up to $T_R/2$ seconds. In return, this seeking algorithm can be supported without incurring server overhead or additional client buffer. If more precise seeking is needed, then one will need to make use of a dynamic multicast channel to merge the client back into an existing static multicast channel. Further research will be needed to develop efficient yet precise seeking algorithms.

19.3 Performance Modeling

In this section we present an approximate performance model for the SS-VoD architecture. While an exact analytical solution does not appear to be tractable, we were able to derive an approximate model that can be solved numerically. The purpose of this performance model is to assist system designers to quickly evaluate various design options and to perform preliminary system dimensioning. Once the approximate system parameters are known, one can turn to a more detailed simulation to obtain more accurate performance results.

The primary performance metric we use in this study is start-up latency, defined as the time from when a client submits a request to the admission controller to the time when the beginning of the requested video starts streaming. For simplicity, we assume there is a single video title stored in a service node and ignore network delay, transmission loss, and processing time at the admission controller.

In the following sections, we will first derive the average waiting time for statically-admitted clients and dynamically-admitted clients, and then investigate the configuration of the admission threshold and the channel partitioning policy. We will compare results computed using this approximate performance model with the simulation results in Section 19.4.1.

19.3.1 Waiting Time for Statically-Admitted Clients

As described in Section 19.1.2, there are two ways to admit a client into the system. The first way is admission through a static multicast channel as shown in Figure 19.5. Given that any clients arriving within the time window of 2δ seconds will be admitted this way, it is easy to see that the average waiting time for statically-admitted clients, denoted by $W_S(\delta)$, is equal to half of the admission threshold:

$$W_S(\delta) = \delta \qquad (19.5)$$

assuming it is equally probable for a request to arrive at any time within the time window.

19.3.2 Waiting Time for Dynamically-Admitted Clients

The second way to admit a new client is through a dynamic multicast channel as shown in Figure 19.6. Unlike static multicast channels, dynamic multicast channels are allocated in an on-demand basis according to the admission procedure described in Section 19.1.2. Specifically, if there are one or more free channels available at the time a request arrives, a free channel will be allocated to start transmitting video data to the client immediately and the resultant waiting time will be zero.

On the other hand, if there is no channel available at the time a request arrives, then the resultant waiting time will depend on when a request arrives and when a free dynamic multicast channel becomes available. Specifically, requests arriving at the admission controller will be consolidated using the procedure described in Section 19.1.2 where the admission controller will send a consolidated START request to a service node to initiate video transmission.

Figure 19.8 illustrates this admission process. This example assumes that there is no request waiting and all dynamic multicast channels are occupied before client request 1 arrives. After receiving request 1, the admission controller sends a START request to a service node to initiate a new multicast transmission for this request. However, as all channels are occupied, the transmission will not start until a later time t_1 when a free channel becomes available. During

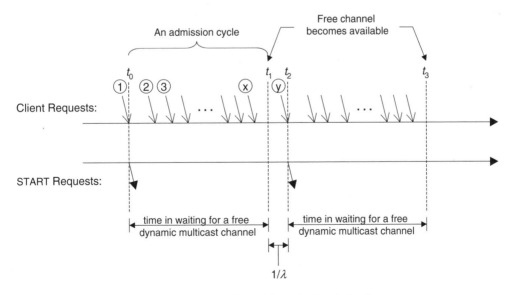

Figure 19.8   Classification of dynamically-admitted users

this waiting time, additional client requests such as requests 2, 3, and so on, arrive but the admission controller will not send additional START requests to the service node. This process repeats when a new request arrives at time t_2.

Based on this model, we first derive the average waiting time experienced by a START request at the service node. For the arrival process, we assume that user requests form a Poisson arrival process with rate λ. The proportion of client requests falling within the admission threshold is given by

$$P_S = \frac{2\delta}{T_R} \tag{19.6}$$

and these clients will be statically-admitted.

Correspondingly, the proportion of dynamically-admitted clients is equal to $(1 - P_S)$. We assume that the resultant arrival process at the admission controller is also Poisson, with a rate equal to

$$\lambda_D = (1 - P_S)\lambda \tag{19.7}$$

Referring to Figure 19.8, we observe that the time between two adjacent START requests is composed of two parts. The first part is the waiting time for a free dynamic multicast channel, and the second part is the time until a new dynamically-admitted client request arrives. For the first part, we let $W_C(\delta)$ be the average waiting time for a free dynamic multicast channel given δ. To derive the second part, we first note that the mean inter-arrival time between the two requests (request x and y in Figure 19.8) immediately before and after a free dynamic channel becomes available, called event E, is equal to $2/\lambda_D$, or *twice* the normal mean inter-arrival time. This counter-intuitive result is due to the fact that longer interval is more likely to be encountered by the event E. With an inter-arrival time that is exponentially distributed with mean $1/\lambda_D$, the length-biased mean inter-arrival time as observed by the event E will become $2/\lambda_D$ [1]. Next we observe that the event E is equally likely to occur within the interval between the two requests, thus the mean time until the next arrival is simply half the length of the interval, or $1/\lambda_D$.

Therefore, the inter-arrival time for START requests is given by

$$\frac{1}{\lambda_S} = W_C(\delta) + \frac{1}{\lambda_D} \tag{19.8}$$

where λ_S is the arrival rate for START requests. For simplicity, we assume that the arrival process formed from START requests is also a Poisson process.

For the service time of START request, it depends on the last user joining the batch (Figure 19.7). In particular, the service time of the last user equals to the arrival time a_n minus the time t_{m-1} for the previous multicast of the requested video title. The service time, denoted by s, can range from 0 to $(T_R - 2\delta)$. We assume the service time s is uniformly distributed between

$$0 < s < T_R - 2\delta \tag{19.9}$$

Therefore, the dynamic multicast channels form a multi-server queueing system with Poisson arrival and uniformly distributed service time. As no close-form solution exists for such a queueing model, we turn to the approximation by Allen and Cunneen [2] for G/G/m queues to obtain the average waiting time for a dynamic multicast channel:

$$W_C(\delta) = \frac{E_C(N_D, u)}{N_D(1 - \rho)} \left(\frac{C_A^2 + C_S^2}{2} \right) T_S \tag{19.10}$$

where $C_A^2 = 1$ is the coefficient of variation for Poisson process,

$$C_S^2 = \frac{(T_R - 2\delta)^2}{12} \left(\frac{2}{T_R - 2\delta} \right)^2 = \frac{1}{3} \tag{19.11}$$

is the coefficient of variation for uniformly-distributed service time, and T_S is the average service time, given by

$$T_S = \frac{T_R - 2\delta}{2} \tag{19.12}$$

Additionally, $u = \lambda_S T_S$ is the traffic intensity, $\rho = u/N_D$ is the server utilization, and $E_C(N_D, u)$ is the Erlang-C function:

$$E_C(N_D, u) = \frac{u^{N_D}/N_D!}{u^{N_D}/N_D! + (1 - \rho) \sum_{k=0}^{N_D-1} \frac{u^k}{k!}} \tag{19.13}$$

Since the traffic intensity depends on the average waiting time, and the traffic intensity is needed to compute the average waiting time, equation (19.10) is in fact recursively defined. Due to equation (19.13), equation (19.10) does not appear to be analytically solvable. Therefore, we apply numerical methods to solve for $W_C(\delta)$ in computing the numerical results presented in Section 19.4.

Now that we have obtained the waiting time for a START request, we can proceed to compute the average waiting time for dynamically-admitted client requests. Specifically, we assume the waiting time for a START request is exponentially distributed with mean $W_C(\delta)$. We classify client requests into two types. A Type-1 request is the first request that arrives at the beginning of the admission cycle. Type-2 requests are the other requests that arrive after a Type-1 request. For example, request 1 in Figure 19.8 is a Type-1 request, and request 2 and 3 are Type-2 requests.

We first derive the average waiting time for Type-2 requests. Let $W_2(\delta)$ be the average waiting time for Type-2 requests which can be shown to be (please refer to the Appendix):

$$W_2(\delta) = W_C(\delta) \left(1 - \left(\frac{1 + (T_R - 2\delta)/2W_C(\delta)}{1 - e^{\frac{-(T_R - 2\delta)}{W_C(\delta)}}} \right) \frac{(T_R - 2\delta)}{W_C(\delta)} e^{\frac{-(T_R - 2\delta)}{W_C(\delta)}} \right) \tag{19.14}$$

Next for Type-1 requests, the average waiting time, denoted by $W_1(\delta)$, is simply equal to $W_C(\delta)$. Therefore, the overall average waiting time, denoted by $W_D(\delta)$, can be computed from a weighted average of Type-1 and Type-2 requests:

$$W_D(\delta) = \frac{W_1(\delta) + M_2(\delta)W_2(\delta)}{1 + M_2(\delta)} \tag{19.15}$$

where $M_2(\delta)$ is the expected number of Type-2 requests in an admission cycle and can be computed from

$$M_2(\delta) = W_C(\delta)\lambda_D \qquad (19.16)$$

19.3.3 Admission Threshold

In the previous derivations, we have assumed that the admission threshold value is given *a priori*. Consequently, the resultant average waiting time for statically-admitted and dynamically-admitted users may differ. To maintain a uniform average waiting time for both cases, we can adjust the admission threshold so that the average waiting time difference is within a small error ε:

$$\delta = \min\{x \,|\, (W_S(x) - W_D(x)) \le \varepsilon, T_R \ge x \ge 0\} \qquad (19.17)$$

As adjusting the admission threshold does not affect existing users, the adjustment can be done dynamically while the system is online. In particular, the system can maintain a moving average of previous users' waiting time as the reference for threshold adjustment. This enables the system to maintain a uniform average waiting time for both statically-admitted and dynamically-admitted users. The term *latency* in this chapter refers to this uniform average waiting time.

19.3.4 Channel Partitioning

An important configuration parameter in SS-VoD is the partitioning of available channels for use as dynamic and static multicast channels. Intuitively, having too many dynamic multicast channels will increase the traffic intensity at the dynamic multicast channels due to increases in the service time (cf. equations (19.1) and (19.12)). On the other hand, having too few dynamic multicast channels may also result in higher load at the dynamic multicast channels. We can find the optimal channel partitioning policy by enumerating all possibilities, which in this case is O(N). Unlike the related Unified Video-on-Demand (UVoD) architecture [3], where the optimal channel partition policy is arrival-rate dependent, we found that the optimal channel partitioning policy is relatively independent of the user arrival rate in SS-VoD. This will be studied in more detail in Section 19.4.2.

19.4 Performance Evaluation

In this section, we present simulation and numerical results to evaluate performance of the SS-VoD architecture. We first validate the analytical performance model using simulation results and then proceed to investigate the effect of the channel partitioning policy, to compare latency and channel requirement between TVoD, NVoD, UVoD [3], with SS-VoD, and finally investigate the performance of SS-VoD under extremely light loads. The focus of the comparisons is on the server and backbone network resource requirements, represented by the number of channels required to satisfy a given performance metric such as latency. Note that for simplicity, we do not distinguish between unicast and multicast channels and assume they have the same cost. In practice, a multicast channel will incur higher costs in the access network where the network routers will need to duplicate and forward the multicast video data to multiple recipients. Nevertheless, this additional cost is not present at the server (e.g., using IP multicast) and at the backbone network before fanning out to the access sub-networks and therefore will be ignored in this study.

19.4.1 Model Validation

To verify accuracy of the performance model derived in Section 19.3, we developed a simulation program using CNCL [4] to obtain simulation results for comparison. A set of simulations is run to obtain the latency over a range of arrival rates. Each run simulated a duration of 1,440 hours (60 days), with the first 24 hours of data skipped to reduce initial condition effects. There is one movie in the system, with a length of 120 minutes. We divide available multicast channels equally into static-multicast and dynamic-multicast channels. We do not simulate user interactions and assume all users play back the entire movie from start to finish.

Figure 19.9 shows the latency versus arrival rate ranging from 1×10^{-3} to 5.0 requests per second. We observe that the analytical results are reasonable approximations for the simulation results. At high arrival rates (e.g., over 1 request per second), the analytical results over-estimate the latency by up to 5%. As discussed in the beginning of Section 19.3, the analytical model is primarily used for preliminary system dimensioning. Detailed simulation, while lengthy (e.g., hours), is still required to obtain accurate performance results.

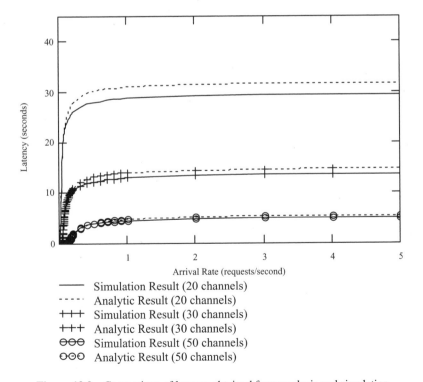

Figure 19.9 Comparison of latency obtained from analysis and simulation

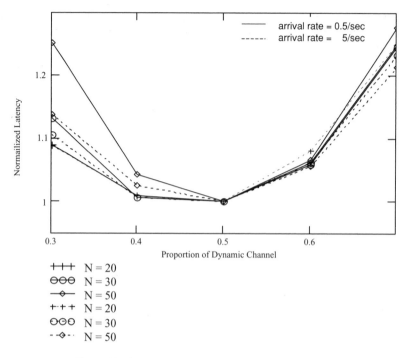

Figure 19.10 Effect of channel partitioning on latency

19.4.2 Channel Partitioning

To investigate the performance impact of different channel allocations, we conducted simulations with the proportion of dynamic multicast channels, denoted by r, ranging from 0.3 to 0.7. The results are plotted in Figure 19.10. Note that we use a normalized latency instead of actual latency for the y-axis to facilitate comparison. Normalized latency is defined as

$$\frac{w(r)}{\min\{w(r), \forall r\}} \tag{19.18}$$

where $w(r)$ is the latency with $r \times N$ dynamic multicast channels.

We simulated three sets of parameters with $N = 20$, 30, and 50 for two arrival rates, namely, heavy load at 5 requests/second and light load at 0.5 requests/second. Note that normalized latency obtained from two different values of N cannot be compared directly as the denominator in equation (19.18) is different.

Surprisingly, the results show that in all cases the latency is minimized by assigning half of the channels to dynamic multicast and the other half to static multicast. For comparison, UVoD exhibits a different behavior and requires more channels allocated for static multicast to minimize latency at high loads as shown in Figure 19.11 for a 50-channel configuration.

UVoD's behavior is explained by the observation that at higher arrival rates, the waiting time for a free unicast channel increases rapidly near full utilization. Therefore, it is desirable to allocate more multicast channels to reduce the traffic intensity (arrival rate $\times T_R$) routed to

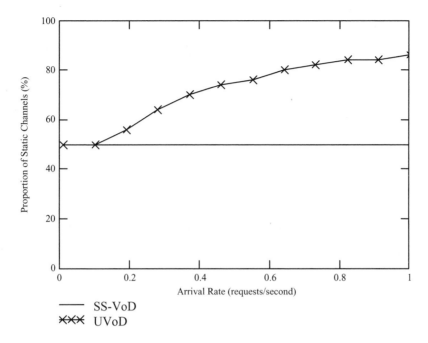

Figure 19.11   Comparison of optimal channel allocation in SS-VoD and UVoD

the unicast channel to prevent operating the unicast channels near full utilization. By contrast, the same situation does not occur in SS-VoD because a dynamic multicast channel can batch and serve multiple waiting requests. Moreover, the batching efficiency increases for longer waiting time, thus compensating for the increases in the arrival rate. This remarkable property of SS-VoD greatly simplifies system deployment as one will not need to reconfigure the system with a different channel partition policy if the user demand changes.

19.4.3 Latency Comparisons

Figure 19.12 plots the latency for SS-VoD, UVoD, TVoD, and NVoD for arrival rates up to 5 requests per second. The service node (or video server for TVoD/NVoD) has 50 channels and serves a single movie of length 120 minutes. The first observation is that except for NVoD, which has a constant latency of 72 seconds, the latency generally increases with higher arrival rates as expected. For TVoD, the server overloads for arrival rates larger than 1.16×10^{-4} requests per second. UVoD performs significantly better with the latency asymptotically approaches that of NVoD. SS-VoD performs even better than UVoD, and the latency levels off and approaches 5.6 seconds, or a 92% reduction compared to UVoD.

It is also worth noting that the performance gain of SS-VoD over UVoD does not incur any trade-off at the client side. Specifically, the buffer requirement and bandwidth requirement are the same for both SS-VoD and UVoD. The only differences are the replacement of the dynamic unicast channels in UVoD with dynamic multicast channels in SS-VoD, and the addition of the more complex admission procedure in the admission controller.

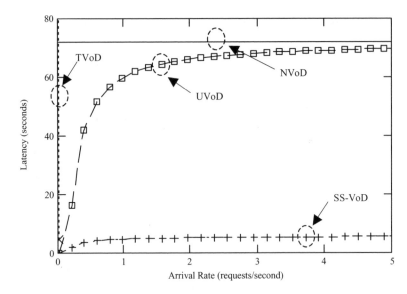

Figure 19.12   Comparison of latency for different arrival rates

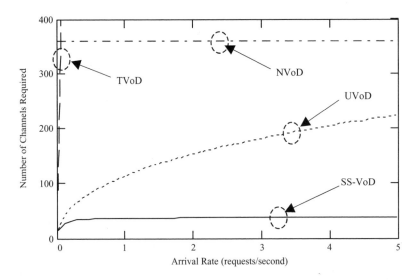

Figure 19.13   Comparison of channel requirement for different arrival rates

19.4.4 Channel Requirement

Channel requirement is defined as the minimum number of channels needed to satisfy a given latency constraint at a certain arrival rate. Figure 19.13 plots the channel requirements of SS-VoD, UVoD, TVoD, and NVoD versus arrival rates from 0.01 to 5 requests per second. The latency constraint is set to 10 seconds.

The number of channels required for NVoD is a constant value and equal to 360. The channel requirement of TVoD increases with the arrival rate and quickly exceeds that of NVoD. The channel requirements of SS-VoD and UVoD are significantly lower than both TVoD and NVoD. For higher arrival rates, SS-VoD outperforms UVoD by a wide margin. For example, the channel requirements at one request per second are 114 and 36 for UVoD and SS-VoD respectively; and the channel requirements at 5 requests per second are 225 and 38 for UVoD and SS-VoD respectively. This result demonstrates the performance gain achieved by replacing the dynamic unicast channels in UVoD with dynamic multicast channels in SS-VoD.

19.4.5 Performance at Light Loads

The previous results are computed using relatively high arrival rates. Intuitively, the performance gains will decrease at lower arrival rates as fewer requests will be served by a dynamic multicast channel. To investigate this issue, we define a percentage of channel reduction over TVoD, denoted by G, as

$$G = \frac{\min\{n | W_{TVoD}(n) \leq 1, \forall n = 0, 1, \ldots\} - \min\{n | W(n) \leq 1, \forall n = 0, 1, \ldots\}}{\min\{n | W_{TVoD}(n) \leq 1, \forall n = 0, 1, \ldots\}} \times 100\%$$

(19.19)

where $W_{TVoD}(n)$ and $W(n)$ are the latency, given there are n channels, for TVoD and SS-VoD/UVoD respectively.

Figure 19.14 plots the channel reduction for arrival rates from 1×10^{-4} to 0.01 for SS-VoD and UVoD. The results show that SS-VoD requires fewer channels than TVoD as long as arrival rates are over 1.8×10^{-4} requests per second. Note that at this low arrival rate, both TVoD and

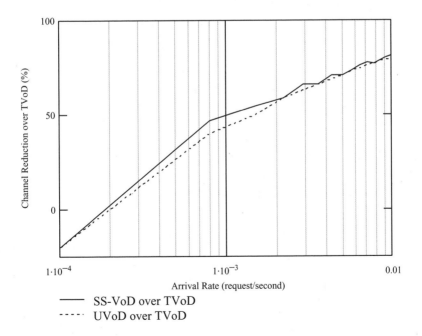

Figure 19.14 Channel reduction over TVoD at very low arrival rates

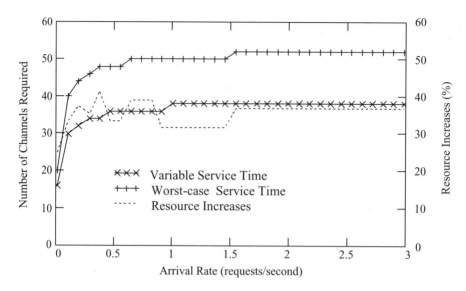

Figure 19.15   Performance trade-off for using worst-case service time for dynamic channels

SS-VoD require only six channels. This suggests that SS-VoD will likely outperform TVoD in practice.

19.4.6 Simplicity versus Performance Trade-off

The admission controller is among the more complex components in the SS-VoD architecture. One way to simplify the admission controller is to use a constant service time of $T_R - 2\delta$ seconds for the dynamic channels. As this is the worst-case service time, the admission controller no longer needs to maintain the counter-length tuple $\{A_C, A_L\}$ and also does not need to send an EXTEND update request to the service node. The trade-off for this simplification is increased channel requirement as the dynamic channel will be occupied for a time longer than necessary. Figure 19.15 compares the two cases, showing that using the worst-case service time of $T_R - 2\delta$ seconds results in resource increases of over 30%. This shows that the more complex admission procedure is still desirable unless system complexity must be minimized.

19.5 Implementation and Benchmarking

We implemented a SS-VoD prototype using off-the-shelf software and hardware. There are three components in the prototype: service node, admission controller, and video clients. Both the service node and the admission controller are implemented using the C++ programming language running on Red Hat Linux 6.2. Two client applications have been developed, one is implemented using the Java programming language and the Java Media Framework (JMF) 2.1, while the other is implemented using C++ on the Microsoft Windows platform. Both the service node and the admission controller are video format independent. The Java-version client supports MPEG1 streams, while the Windows-version client supports MPEG1, MPEG2, as well as basic MPEG4 streams. We also implemented the interactive playback controls presented in Section 19.2, namely pause–resume, slow motion, and seeking.

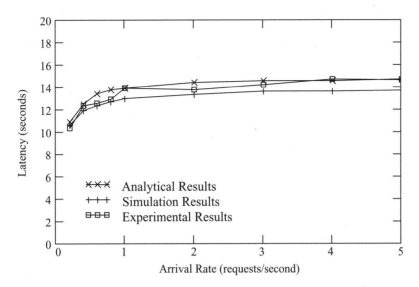

Figure 19.16   Comparison of latencies obtained from analysis, simulation and benchmarking

With the SS-VoD prototype, we conducted extensive experiments to obtain measured benchmark results to verify against the analytical and simulation results. We developed a traffic generator in order to simulate a large number of client requests. The service node runs on a Compaq Proliant DL360 serving one movie of length 120 minutes with 30 channels, each at 1.5Mbps. The clients are ordinary PCs and all machines are connected using a layer-3 IP switch with hardware IP multicast support. We measured the start-up latency for arrival rates ranging from 1 to 5 requests per second. Each benchmark test runs for a total of six hours. Benchmark data collected during the first hour is discarded to reduce initial condition effect.

Figure 19.16 compares the start-up latencies obtained from analysis, simulation, and benchmarking respectively. We observe that the benchmarking results agree very well with the analytical results and simulation results. Note that the latencies obtained from benchmarking are consistently larger than those obtained from simulation. We believe that this is due to the non-zero processing delay and network delay in the system, both of which have been ignored in the simulation model.

19.6 Summary

In this chapter, we investigated a Super-Scalar Video-on-Demand (SS-VoD) architecture that can achieve super-linear scalability by integrating techniques of batching, patching, and periodic broadcasting. In designing the SS-VoD architecture the focus is on its practicality and the implementation and deployment complexities. For example, instead of adopting more sophisticated open-loop algorithms to schedule the static multicast channels, we employed the simple staggered periodic multicast schedule that enables us to implement interactive playback control such as pause–resume, slow motion, and seeking in a simple yet efficient way. Moreover, the staggered schedule also requires significantly lower client buffer requirement and more importantly, eliminates the need to switch multicast channels during a video session.

On the other hand, hybrid multicast streaming architectures present unique challenges to media server design due to the use of both periodic and aperiodic media retrievals. In the next chapter we address the issues in designing efficient media streaming server that supports hybrid multicast streaming algorithms.

Appendix

In this Appendix, we derive the mean waiting time for Type-2 users, denoted by $W_2(\delta)$. The complication is due to length biasing as a Type-2 user is more likely to observe a longer Type-1 wait than a shorter Type-1 wait. First, we compute the waiting time distribution for Type-1 users, denoted by $f_C'(t)$, *as observed by a Type-2 user* using the results from Kleinrock [1]:

$$f_C'(t) = \frac{t f_C(t)}{W_C(\delta)} \tag{19.20}$$

where $f_C(t)$, and $W_C(\delta) = E[f_C(t)]$ is the actual waiting time distribution and mean waiting time of Type-1 users respectively. Let $W_C'(\delta)$ be the mean of $f_C'(t)$:

$$W_C'(\delta) = \int_{-\infty}^{\infty} t f_C'(t) dt \tag{19.21}$$

Substituting equation (19.20) into equation (19.21) we then have:

$$W_C'(\delta) = \int_{-\infty}^{\infty} \frac{t^2 f_C(t)}{W_C(\delta)} dt \tag{19.22}$$

We note that the waiting time can only range from zero to $(T_R - 2\delta)$, so we can rewrite equation (19.22) as:

$$W_C'(\delta) = \int_0^{T_R - 2\delta} \frac{t^2 f_C(t)}{W_C(\delta)} dt \tag{19.23}$$

Motivated by simulation results, we assume that $f_C(t)$ is truncated exponentially distributed:

$$f_C(t) = \left((1 - e^{\frac{-(T_R - 2\delta)}{W_C(\delta)}}) W_C(\delta) \right)^{-1} e^{\frac{-t}{W_C(\delta)}} \tag{19.24}$$

Substituting equation (19.24) into equation (19.23) we have

$$W_C'(\delta) = \int_0^{T_R - 2\delta} \frac{t^2 e^{\frac{-t}{W_C(\delta)}}}{(1 - e^{\frac{-(T_R - 2\delta)}{W_C(\delta)}}) W_C(\delta)^2} dt \tag{19.25}$$

Solving the integral and after a series of simplifications equation (19.25) becomes

$$W_C'(\delta) = 2 W_C(\delta) \left(1 - \left(\frac{1 + (T_R - 2\delta)/2W_C(\delta)}{1 - e^{\frac{-(T_R - 2\delta)}{W_C(\delta)}}} \right) \frac{(T_R - 2\delta)}{W_C(\delta)} e^{\frac{-(T_R - 2\delta)}{W_C(\delta)}} \right) \tag{19.26}$$

Finally, as a Type-2 user is equally likely to arrive any time during a Type-1 wait, the mean waiting time is simply equal to half of the Type-1 mean wait:

$$W_2(\delta) = \frac{W_C'(\delta)}{2} = W_C(\delta)\left(1 - \left(\frac{1 + (T_R - 2\delta)/2W_C(\delta)}{1 - e^{\frac{-(T_R - 2\delta)}{W_C(\delta)}}}\right)\frac{(T_R - 2\delta)}{W_C(\delta)}e^{\frac{-(T_R - 2\delta)}{W_C(\delta)}}\right) \quad (19.27)$$

References

[1] L. Kleinrock, *Queueing Systems*, vol. I: *Theory*, Wiley-Interscience, 1975, pp. 171.

[2] A.O. Allen, *Probability, Statistics, and Queueing Theory with Computer Science Applications*, 2nd edn. Academic Press, New York, 1990.

[3] J.Y.B. Lee, On a Unified Architecture for Video-on-Demand Services, *IEEE Transactions on Multimedia*, vol. 4. no. 1. March 2002, pp. 38–47.

[4] ComNets Class Library and Tools: http://www.comnets.rwth-aachen.de/doc/cncl.html

20

Efficient Server Design for Hybrid Multicast Streaming

This chapter investigates the issues in designing efficient media servers for hybrid multicast streaming algorithms that integrate both closed-loop and open-loop algorithms. Existing media server designs are either optimized for closed-loop algorithms, which generate aperiodic data retrievals or optimized for open-loop algorithms, which generate periodic data retrievals. However, hybrid architectures such as the Super-Scalar Video-on-Demand (SS-VoD) system in Chapter 19 require both periodic and aperiodic data retrievals which the existing server designs are sub-optimal. This chapter presents an efficient server design to address this problem, which can achieve up to 60% capacity gains compared to conventional server designs.

20.1 Introduction

In Chapter 19 we presented a Super-Scalar Video-on-Demand (SS-VoD) architecture combining the batching, patching, and periodic broadcasting for implementing scalable and efficient VoD services. In a SS-VoD system, multicast channels are divided into two types – static channels and dynamic channels. Each channel transmits video data at the video playback rate using network multicast. Static channels are organized in a time-staggered manner to stream the whole video repeatedly and periodically. Dynamic channels are scheduled with batching and patching to enable clients to begin playback quickly. By simultaneously caching data from a static channel, the client can eventually merge back to an existing static channel and release the dynamic channel for reuse by other clients.

In this chapter, we present an efficient disk-array-based server design for implementing the video server in a SS-VoD system. The video server in a SS-VoD system is unique in that there are both statically scheduled and dynamically scheduled video channels. Existing video servers in general [1–5], and disk schedulers in particular [6–9], are designed either for systems with statically scheduled video channels (i.e., open-loop algorithms in Chapter 18), or for systems with dynamically scheduled video channels (i.e., closed-loop algorithms in Chapter 17). The former simply cannot be applied to a SS-VoD system as the video placement policy and I/O scheduler typically do not allow random data retrievals. The latter, on the other hand, can still be applied to a SS-VoD system but the efficiency will be sub-optimal as the static channels' periodic retrieval patterns are not exploited to increase retrieval efficiency.

To support both static batching channels and dynamic patching channels, we develop a new video placement policy and retrieval algorithm that can support random data retrievals, while still be able to exploit the periodic retrieval pattern to increase disk efficiency. To tackle the disk-zoning problem, we develop a Weighted Group Segment Pairing Scheme (WSGP) based on SGP [8] for video placement. By pairing an outer zone with an inner zone and allocating video data according to the zone's storage capacity, we can achieve increased disk utilization without sacrificing disk storage capacity.

Compared to conventional server designs using round-based schedulers, this efficient server design can increase the system capacity by as much as 60% with the same buffer requirement. This chapter presents details of this new server design, derives a performance model, and analyzes it using numerical results.

20.2 Background

Video server design has been studied extensively in the literature. Gemmell *et al.* [2] provide an excellent overview of the area, explaining the key challenges and reviewing existing solutions. Most of the existing server designs for TVoD systems are centered around round-based algorithms, such as the CSCAN scheduler [7] and the Grouped Sweeping Scheme (GSS) scheduler [6]. Common among these round-based schedulers is the assumption that there is no correlation between the active video streams, i.e., the video streams play back video independently at arbitrary schedules. While this assumption is valid and necessary for TVoD systems, it is sub-optimal for a SS-VoD system where some of the multicast video streams are prescheduled and thus have a fixed temporal relation with one another.

At the other extreme of the spectrum is NVoD systems where all video streams are broadcast repeatedly and periodically in a fixed schedule. Armed with complete knowledge of the broadcasting schedules, one can then design an optimized video placement and disk retrieval scheme to increase disk efficiency. The principle is to take advantage of the fixed temporal relation between broadcast video streams and place video data in an interleaved manner so that the server can retrieve video data continuously with minimal disk seeking.

For example, Chen and Manu [8] proposed a video placement policy called Segment Group Pairing (SGP) to allocate video data in zone-bit-recording (ZBR) disk for NVoD servers. With SGP, data blocks that are to be retrieved in the same round are divided evenly into two groups. The first group is stored continuously in the outer zone while the second group is stored continuously in the inner zone. This continuous placement reduces seeking overhead in data retrieval. In a service round, the disk head will first retrieve the group of blocks located in the outer zone and then seek the inner zone to retrieve the remaining data blocks. This scheme enables the data rates of both zones to be averaged and thus results in a higher deterministic disk throughput.

Nevertheless, this algorithm did not account for capacity differences among different zones. In particular, inner zones usually have lower capacity compared to outer zones. As a result, SGP will likely fill up the inner zone before the outer zone is fully utilized and thus the remaining storage capacity in the outer zones becomes unused. To tackle this limitation, we extend the SGP policy to account for zone capacity differences. This new policy, called Weighted Segment Group Pairing (WSGP), allocates data blocks to the inner and outer zones in proportion to the zone capacities, thus eliminating the above-mentioned limitation.

Nevertheless, the WSGP policy still does not address the problem of supporting both random and periodic video streams. In our server design, we develop a new placement policy and I/O scheduler incorporating the virtues of CSCAN/GSS for random video streams, and data interleaving for periodic video streams. In the next section, we first devise a GSS-based server design and then introduce our new server design in Section 20.4. We then compare their performances using numerical results in Section 20.5 and give a summary in Section 20.6.

20.3 A GSS-based Server Design

In this section, we apply the well-known Grouped Sweeping Scheme (GSS) scheduler for use in a SS-VoD server. This design will serve as a baseline to compare the efficient server design to be presented in Section 20.4.

Let N be the number of disks in the system, assuming the disks are homogeneous. The disk's storage is divided into fixed-size blocks of Q bytes each, and a service group is defined to consist of all the data blocks at the same location from each of the N disks.

Video data are striped across the N disks as shown in Figure 20.1, effectively forming a RAID-4 [10] disk array without parity. Denote the jth data block of video i by $b_{i,j}$. Then, the first N blocks of video i, $[b_{i,1}, b_{i,2}, b_{i,3} \ldots b_{i,N}]$ are allocated to the first service group. This storage allocation scheme ensures load balance among all N disks.

Figure 20.2 depicts the Grouped Sweeping Scheme (GSS). In GSS, a macro-round is divided evenly into G micro-rounds, with each micro-round serving a separate group of video streams. Assuming all the videos are encoded using constant-bit-rate (CBR) encoding method with the same bit-rate R_V, then in each micro-round the server retrieves one data block from each disk for each channel, and this data block is then multicast over the next G micro-rounds (i.e., one

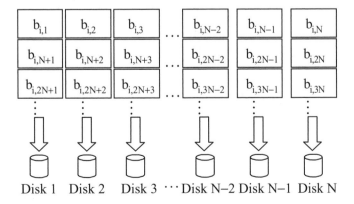

Figure 20.1 Allocation of video blocks among disks for video i

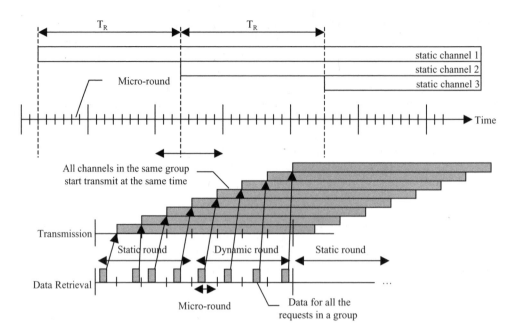

Figure 20.2   Transmission and retrieval schedule of GSS

macro-round). Reducing G we can pack more video streams in a group and this results in increased disk efficiency, albeit at the expense of increased buffer requirement and scheduling delay and vice versa. In the extreme case with $G = 1$, GSS reduces to SCAN; and in the other extreme case with $G = n$, where n is the maximum number of streams that the system can support, GSS reduces to first-come-first-serve.

In SS-VoD, static channel starts once every T_R seconds, where T_R is given by

$$T_R = \frac{L}{N_S} \tag{20.1}$$

and L is the length of the video title in seconds. However, transmission in GSS can only start at the beginning of a micro-round and hence does not necessary match the transmission schedule of the static channels. In particular, when T_R is not divisible by the micro-round time, some of the static channels will not be able to transmit precisely at the scheduled time. To avoid this problem, we can use additional buffers to perform read-ahead to absorb the time differences. However, this will increase the buffer requirement up to 50% and is thus not desirable. Alternatively, we can choose the value of G and Q so that T_R is an integer multiple of the duration of a micro-round to avoid the additional buffer requirement. For each video stream, N data blocks, one from each disk, are retrieved in a micro-round for transmission over the next G micro-rounds. Thus, the service round length T_r is given by

$$T_r = \frac{NQ}{R_V}, \tag{20.2}$$

and the total server buffer requirement [6] is given by

$$B_{Server} = CNQ(1 + \frac{1}{G}),\tag{20.3}$$

where C is the total number of multicast channels in the server.

As the static channels' offset T_R is integer multiples of the micro-round length, it is easy to see that the static channels will be equally distributed to all G groups. The remaining disk capacity is then used to support dynamic channels.

20.4 An Efficient Server Design

In SS-VoD, there are two types of multicast channels – static and dynamic multicast channel. Static multicast channels stream the whole video while dynamic channels serve clients with up to the first T_R seconds of the video only. In terms of data access pattern, static channels retrieve data in a fixed schedule. By contrast, the data access pattern of dynamic channels is random. In this section, we present an efficient design for the SS-VoD server. Specifically, this efficient server design has three distinctive features. First, the disk storage is organized using an improved Weighted Segment Group Pairing (WSGP) scheme to exploit disk zoning to increase disk throughput and storage utilization. Second, an interleaving data placement policy is used to store video data to be transmitted over the static channels to exploit the static channels' periodicity. Third, the first T_R seconds of the videos are replicated for placement in the outermost zones to increase disk throughput for serving the dynamic channels. We design a new scheduler to schedule the data retrievals for both static and dynamic channels and quantify its performance. These are presented in details in the following sections.

20.4.1 The Weighted Segment Group Pairing (WSGP) Scheme

Today's hard disks commonly employ zone-bit-recording (ZBR) technique to increase disk capacity. In zoning, outer tracks are equipped with more sectors than inner tracks to exploit the increased disk surface area available. With a constant rotation speed, the data transfer rate of the outer zone is also higher than the inner zones. Traditional deterministic performance analysis limits one to using the lowest data transfer rate in the innermost zone for system dimensioning and thus wastes the higher transfer rate available in the outer zones.

To improve disk throughput, we devise a Weighted Segment Group Pairing (WSGP) scheme based on the Segment Group Pairing (SGP) proposed by Chen and Manu [8]. In WSGP, the group of data blocks to be retrieved in the same round (i.e., a $G_{j,k}$) is divided into two sub-groups, denoted by $G_{j,k}^1$ and $G_{j,k}^2$. The first group is placed in an outer zone with higher transfer rate and the second group is placed in an inner zone with lower transfer rate. For a disk with Z zones, zone h and zone $(Z - h + 1)$ are paired together and the two sub-groups, $G_{j,k}^1$ and $G_{j,k}^2$, are allocated to these two zones respectively. In the original SGP [8] the groups are of equal size. This is undesirable as the zones often have different capacities and the extra capacity in the larger zone will be wasted. Thus, we extend SGP to WSGP by dividing the group into sub-groups of sizes proportional to the zone capacities. Results show that with WSGP, we can achieve 100% disk storage utilization while achieving the same throughput as SGP.

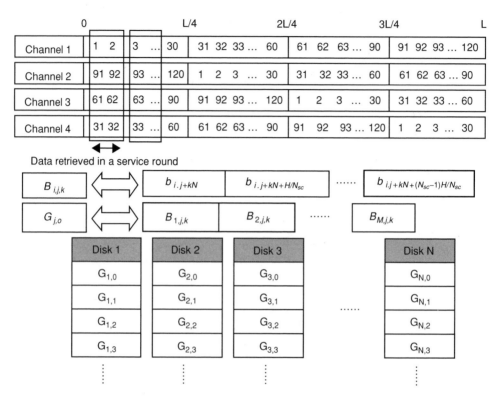

Figure 20.3 Data access pattern and placement of static channels

20.4.2 Interleaving of Data Blocks

Next we consider the interleaving data placement policy for serving the static channels. Consider a system with $N = 2$, $N_S = 4$ and $k = 120$ as shown in Figure 20.3, illustrating the data access pattern for the static channels. We observe that data blocks $b_{i,j}$, $b_{i,j+30}$, $b_{i,j+60}$, $b_{i,j+90}$ of video i stored in disk 1 and data blocks $b_{i,j+1}$, $b_{i,j+31}$, $b_{i,j+61}$, $b_{i,j+91}$ stored in disk 2 are always retrieved together in the same service round, with $j \in [1, 3, 5, \ldots 29]$. Thus, by placing these data blocks in a continuous portion of the disk surface, we can effectively eliminate the disk seeks required in conventional round-based schedulers. However, this placement policy only works for static channels where the transmission schedules are known and fixed. We address data retrievals for dynamic channels using replication in the next section.

20.4.3 First T_R Seconds Replication

Data retrievals for dynamic channels are more random in nature and hence the interleaving data placement policy offers no advantage. Moreover, with the WSGP policy in place, serving the dynamic channels using the interleaving data placement policy will result in the disk constantly seeking between outer tracks and inner tracks, further degrading disk throughput. To tackle this problem, we note that dynamic channels have one crucial property – it only serves up to

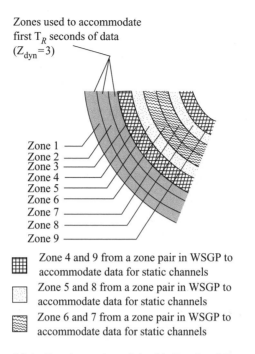

Zones used to accommodate
first T_R seconds of data
$(Z_{dyn}=3)$

Zone 1
Zone 2
Zone 3
Zone 4
Zone 5
Zone 6
Zone 7
Zone 8
Zone 9

Zone 4 and 9 from a zone pair in WSGP to
accommodate data for static channels

Zone 5 and 8 from a zone pair in WSGP to
accommodate data for static channels

Zone 6 and 7 from a zone pair in WSGP to
accommodate data for static channels

Figure 20.4   Data layout for a disk with $Z = 9$ and $Z_{dyn} = 3$

the first T_R seconds of a video. Therefore, we propose replicating the first T_R seconds of each
video in the outermost zones of the disk, thereby taking advantage of the higher transfer rate
of the outermost zones. Assume the first Z_{dyn} zones are used to store the replicated video data,
then the WSGP algorithm will begin pairing zone $(Z_{dyn} + 1)$ and Z. Figure 20.4 illustrates the
overall data layout.

20.4.4 An Integrated Scheduler

To support data retrievals for both static and dynamic channels, we devise a new integrated
scheduler based on the three design features previously discussed. The integrated scheduler
is still round-based but each round is divided into two parts – a *static round* and a *dynamic
round* as shown in Figure 20.5. In a static round, two continuous data retrievals are performed,
one for an outer-track block and one for an inner-track block. The retrieved data will then be
used for transmission over the static channels. The dynamic round is further sub-divided into
G_D dynamic micro-rounds and the dynamic channels are then assigned to these G_D dynamic
micro-rounds as in the GSS case. Retrievals within a dynamic micro-round will be executed
using SCAN and the retrieved data will be transmitted over the dynamic channels. The buffer
requirement of this scheduler is thus given by

$$NQ(2N_S + N_D) \tag{20.4}$$

as illustrated in Figure 20.6.

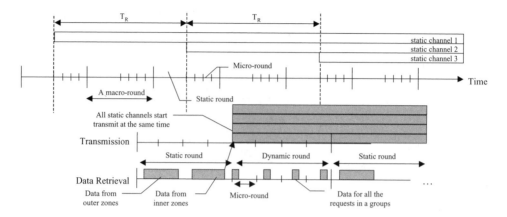

Figure 20.5 Scheduling disk retrieval and network transmission of WSGP

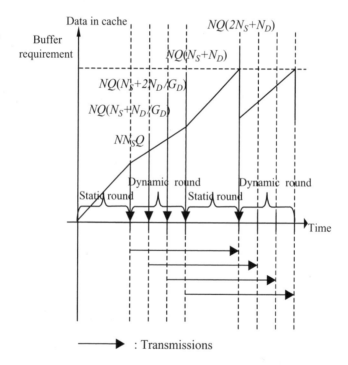

Figure 20.6 Data in cache at different temporal positions in a service round

This scheduler, however, has a subtle problem. We found that the server buffer requirement is dominated by the memory used to cache data for the static channels, which involved continuous retrievals for two large data blocks, and this increases the buffer requirement significantly. To reduce the buffer requirement, we sub-divide the static round into G_S micro-rounds of equal durations, where G_S equals to the number of videos. In each static micro-round, static channels belonging to the same video are scheduled and data are transmitted at the end of the static

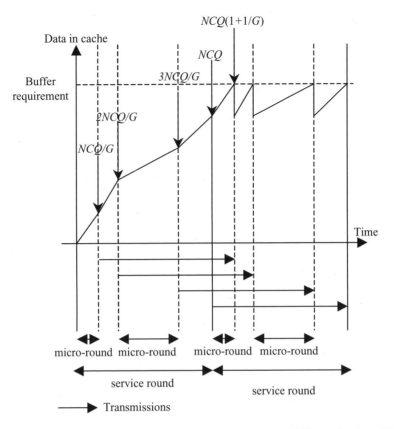

Figure 20.7   Server buffer requirement when micro-rounds are of different durations (G = 4)

micro-round. Since different groups retrieve data from different zones of the disk, the static micro-round will be of different durations.

As illustrated in Figures 20.7 and 20.8, and using derivations similar to GSS we can derive the new total buffer requirement, which is reduced to

$$NQ \left(N_S \left(1 + \frac{1}{G_S} \right) + N_D \left(1 + \frac{1}{G_D} \right) \right). \tag{20.5}$$

In practice, this modification can reduce the buffer requirement of static channels by as much as 40%. In the next section we evaluate and compare performance of the proposed server design with the GSS-based server design.

20.5 Performance Evaluation

In this section, we evaluate the performance of the presented efficient server design and compare it against the GSS-based server design. Table 20.1 lists the key system parameters used in the numerical calculations and Table 20.2 gives the specification of the disks used in performance evaluation.

Table 20.1 System parameters used in performance evaluation

System parameter	Symbol	Value
Video data rate	R_V	4Mb/s (MPEG2)
Number of disk	N	8
Length of video	L	7200s
Static channel per video	N_S	20
Dynamic channel per video	N_D	20

Table 20.2 Specification of different disks used in performance evaluation

Disk Parameter	Value			
Disk Model	Atlas 10K	Barracuda	Cheetah 9LP	IBM 18es
Disk rotation speed (rpm)	10,025	7,200	10.045	7200
Full strobe seek time (ms)	10.828	16.679	10.627	12.742
Track-to-track seek time (ms)	1.245	1.943	0.831	1.086
Head switching time (ms)	0.176	0.100	0.030	0.062
Number of data surfaces	6	5	12	5
Blocks per disk	17,938,986	4,110,000	17,783,240	17,916,240
Total number of tracks	10,022	5,172	6,962	11,474
Number of zones	24	11	11	55

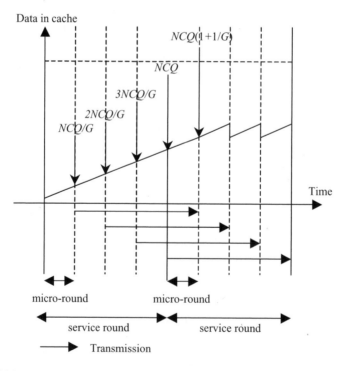

Figure 20.8 Server buffer requirement when micro-rounds are of same durations (G = 4)

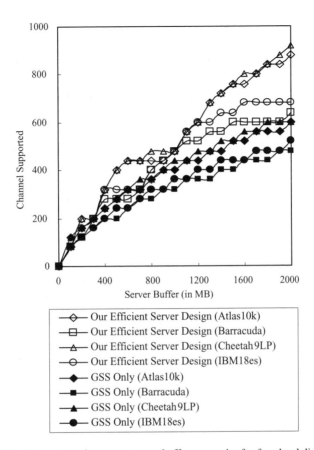

Figure 20.9   Server capacity versus server buffer constraint for four hard disk models

20.5.1 Server Capacity

Figure 20.9 plots the server capacity versus the server buffer size constraint for our efficient server design and the GSS-based server design for four different disks. Compared to GSS-based design, our design can increase the server capacity by up to 60%. Moreover, the performance gain increases to about 40% on average when the buffer available is more than 1GB as evident in Figure 20.10. This is because the interleaved data placement policy reduces disk seeks substantially and thus the gain in I/O efficiency due to larger block size becomes more significant.

20.5.2 Utilization of Disk Capacity

The major advantage of using WSGP is that while achieving full utilization of disk storage capacity, it provides similar performance improvement as SGP. Results show that with the same system configuration, SGP can only utilize 91% storage capacity of the disk while WSGP can achieve 100% utilization.

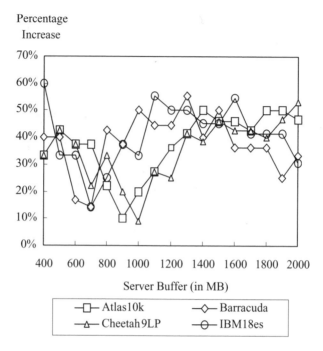

Figure 20.10   Percentage increase in number of channels supported compared with GSS

20.6 Summary

In this chapter, we presented an efficient disk-array-based server design for the Super-Scalar VoD system. We proposed a placement scheme to exploit disk zoning and the characteristics of static and dynamic channels. Coupled with an integrated scheduler, we were able to increase the server capacity by as much as 60% compared to the conventional GSS-based server design.

While the server design presented in this chapter is specifically targeted for use in a SS-VoD system, the design principles are general and thus can be applied to other multicast video streaming architectures with both periodic and aperiodic multicast streaming channels.

References

[1] K. Breidler, H. Kosch, and L. Böszörményi, A Comparative Study of Selected Parallel Video Servers, *Proceedings of IEEE 2000 11th International Workshop on Database and Expert Systems Applications*, September 6–8, 2000, Greenwich, London.

[2] D.J. Gemmell, H.M. Vin, D.D. Kandlur, P.V. Rangan, and L.A. Rowe, Multimedia Storage Servers: A Tutorial, *IEEE Computer*, vol. 28, no. 5, May 1995, pp. 40–49.

[3] W.J. Bolosky, J.S. Barrera III, R.P. Draves, R.P. Fitzgerald, G.A. Gibson, M.B. Jones, S.P. Levi, N.P. Myhrvold, and R.F. Rashid, The Tiger Video Fileserver, *Proceedings of Sixth International Workshop on Network and Operating System Support for Digital Audio and Video*, IEEE Computer Society Press, Los Alamitos, CA, 1996.

[4] Z.-R. Lin and M.-S. Chen, Design and Performance Study of Scalable Video Storage in a Disk-Array-Based Video Server, *Proc. of International Conference on Multimedia and Expo (ICME 2000)*, vol. 3, 2000, pp. 1341–1344.

[5] J.B. Kwon and H.Y. Yeom, Generalized Data Placement for Periodic Broadcast of Videos, *Proc. IEEE International Conference on Multimedia and Expo (ICME 2001)*, August, 2001, Tokyo, Japan.

[6] P.S. Yu, M.S. Chen and D.D. Kandlur, Grouped Sweeping Scheduling for DASD-Based Multimedia Storage Management, *ACM Multimedia Systems*, vol. 1, no. 3, 1993, pp. 99–109.

[7] N. Reddy and A.L. Wyllie, 'I/O Issues in a Multimedia System,' *IEEE Computer*, vol. 27, no. 3, March 1994, pp. 69–74.

[8] S. Chen and T. Manu, A Novel Video Layout Strategy for Near Video-on-Demand Servers, *Proceedings of IEEE International Conference on Multimedia Computing and Systems' 97*, Ottawa, Canada, June 1997, pp. 37–45.

[9] S.L. Tsao and Y.M. Huang, An Efficient Storage Server in Near Video-on-Demand Systems, *IEEE Transactions on Consumer Electronics*, vol. 44, no. 1, Feb. 1998, pp.

[10] G.A. Gibson, *Redundant Disk Arrays: Reliable, Parallel Secondary Storage*, MIT Press, 1992.

Index